EARTH FROM SPACE

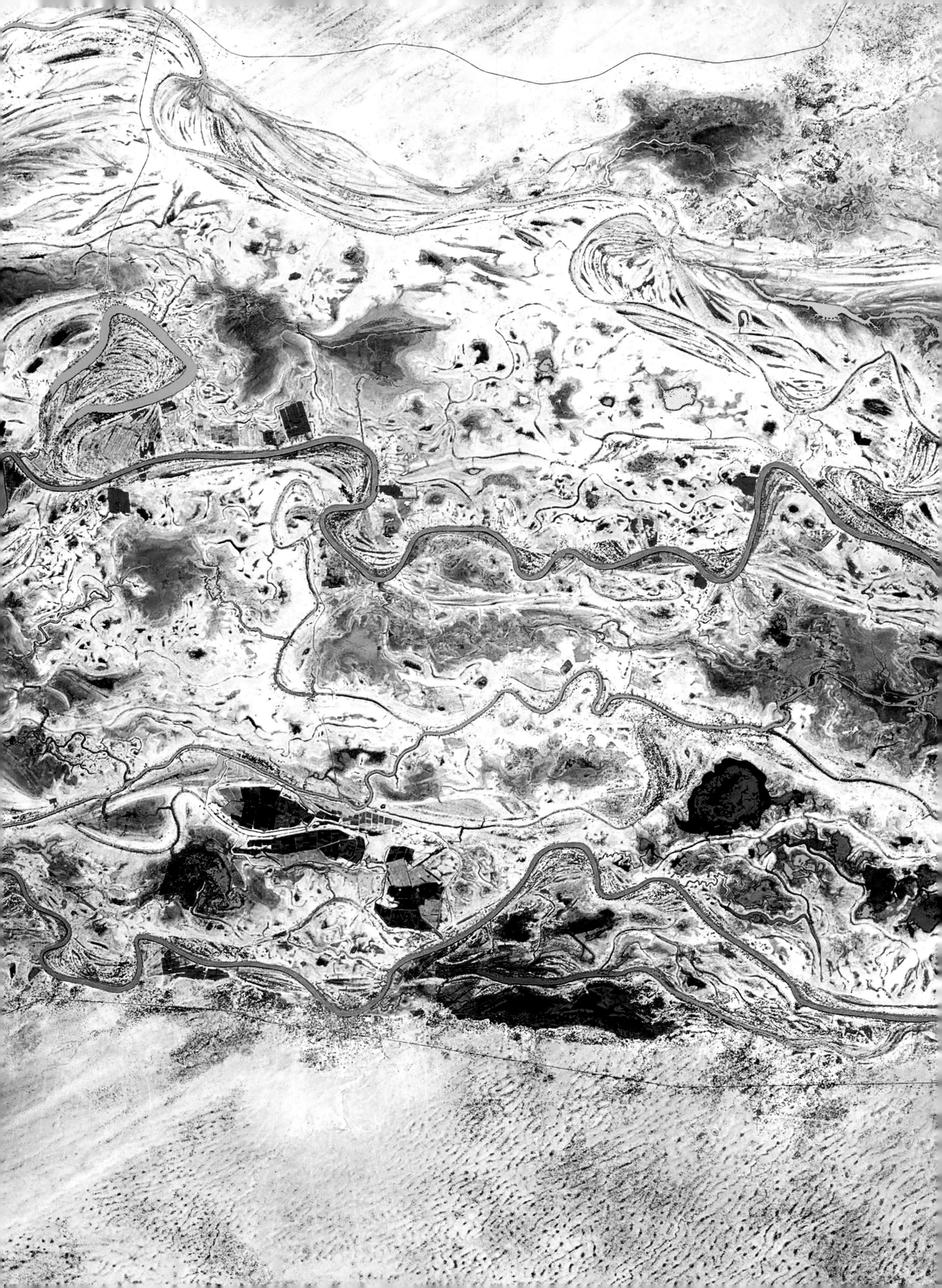

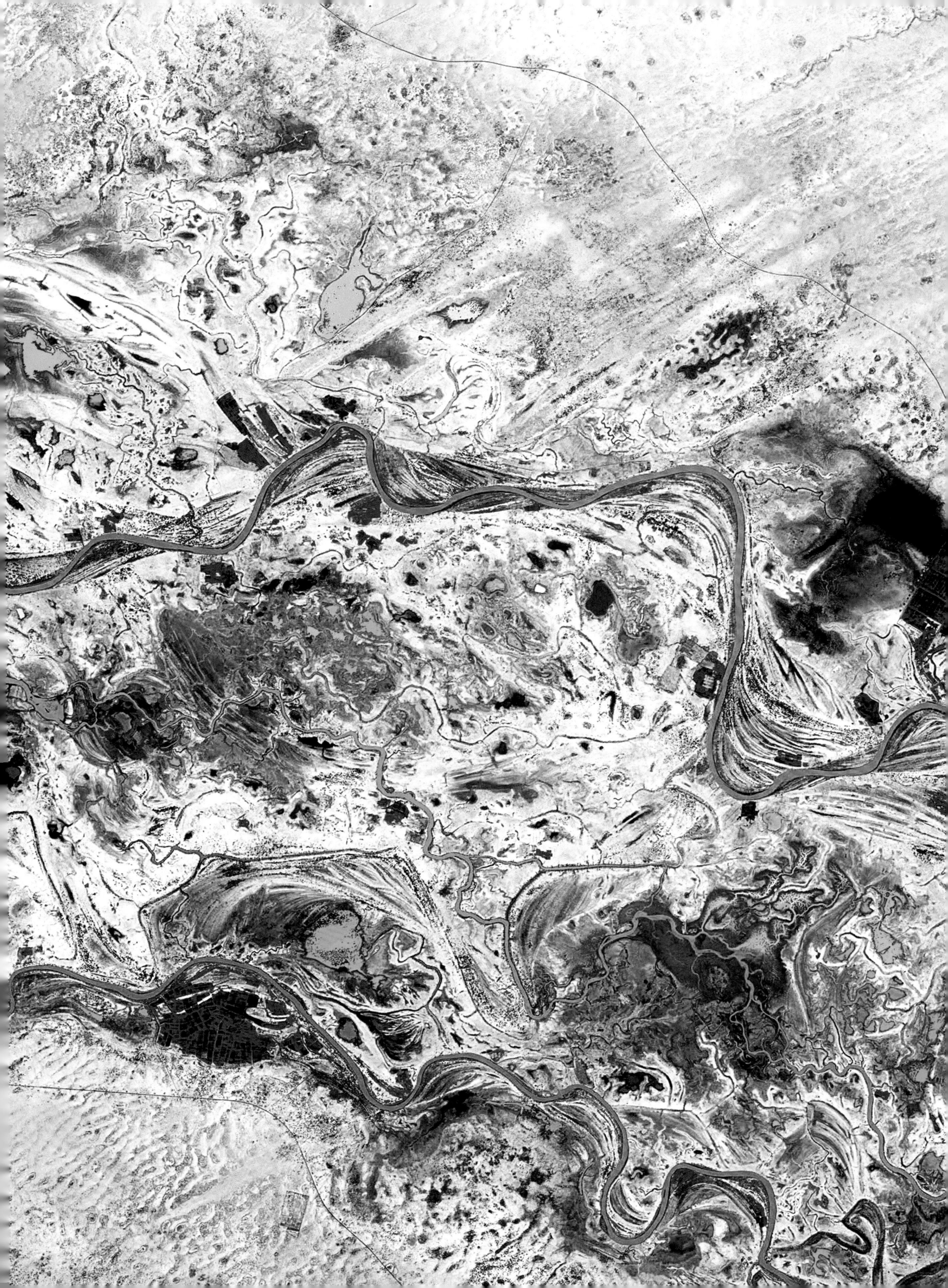

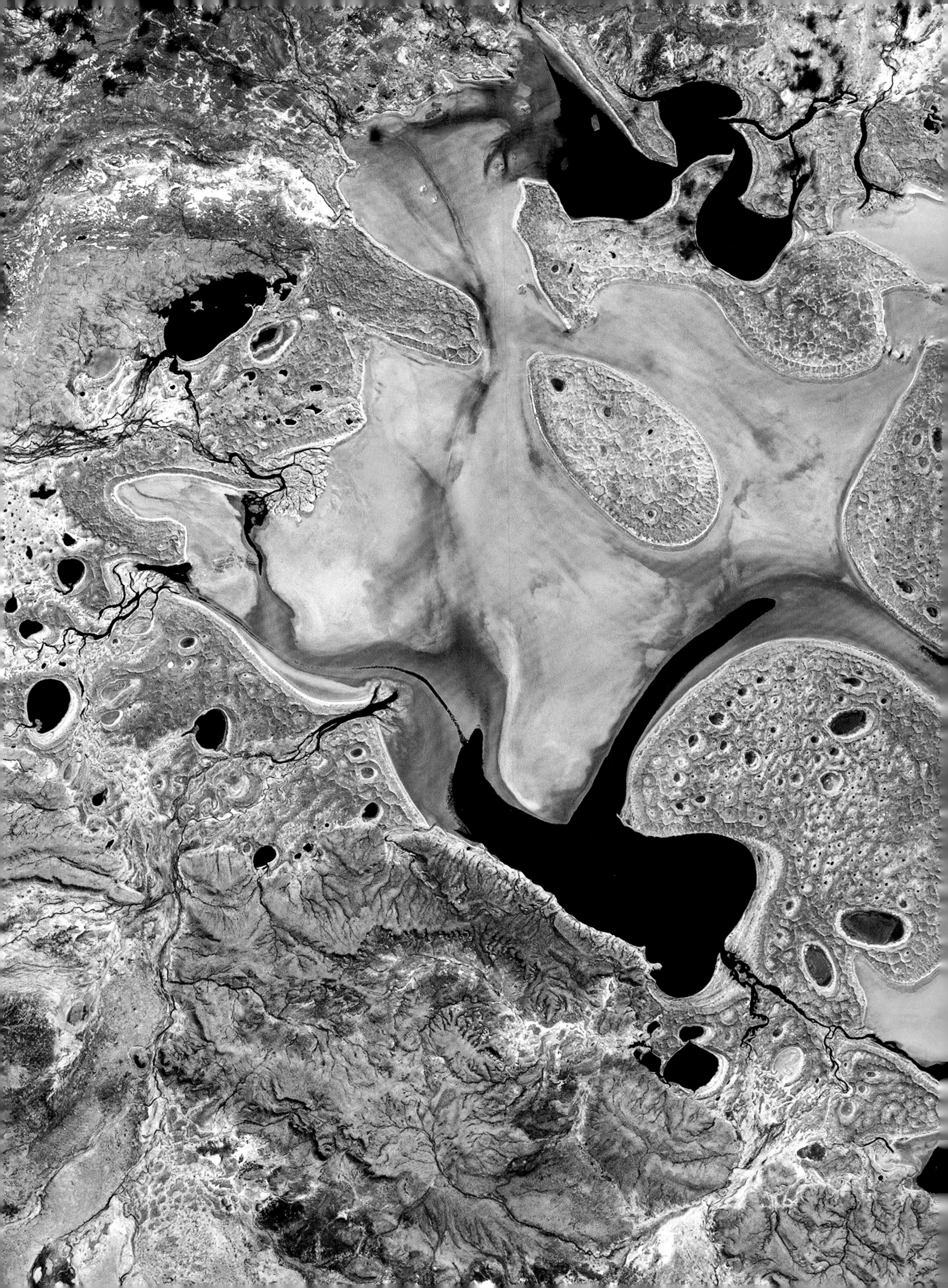

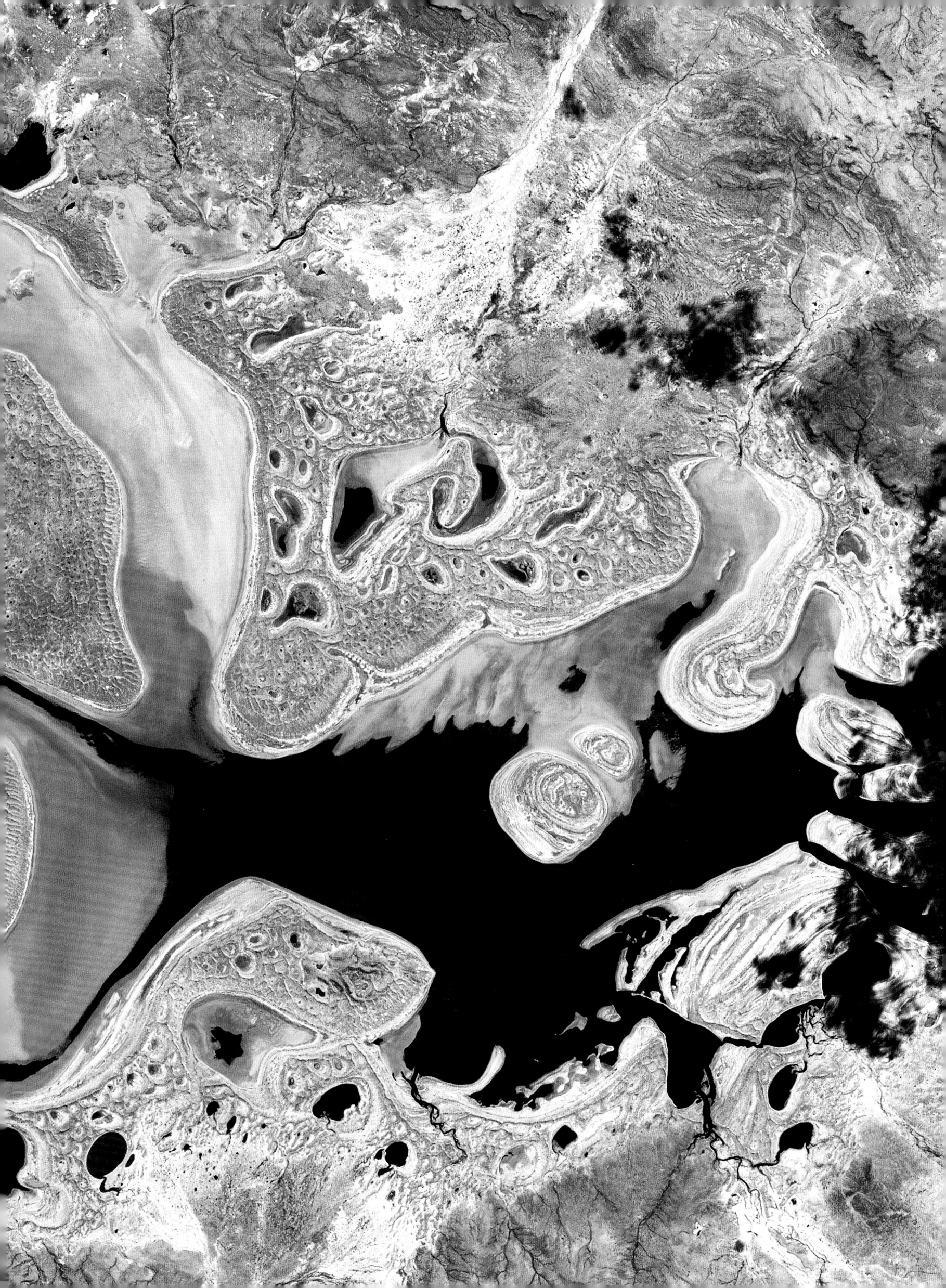

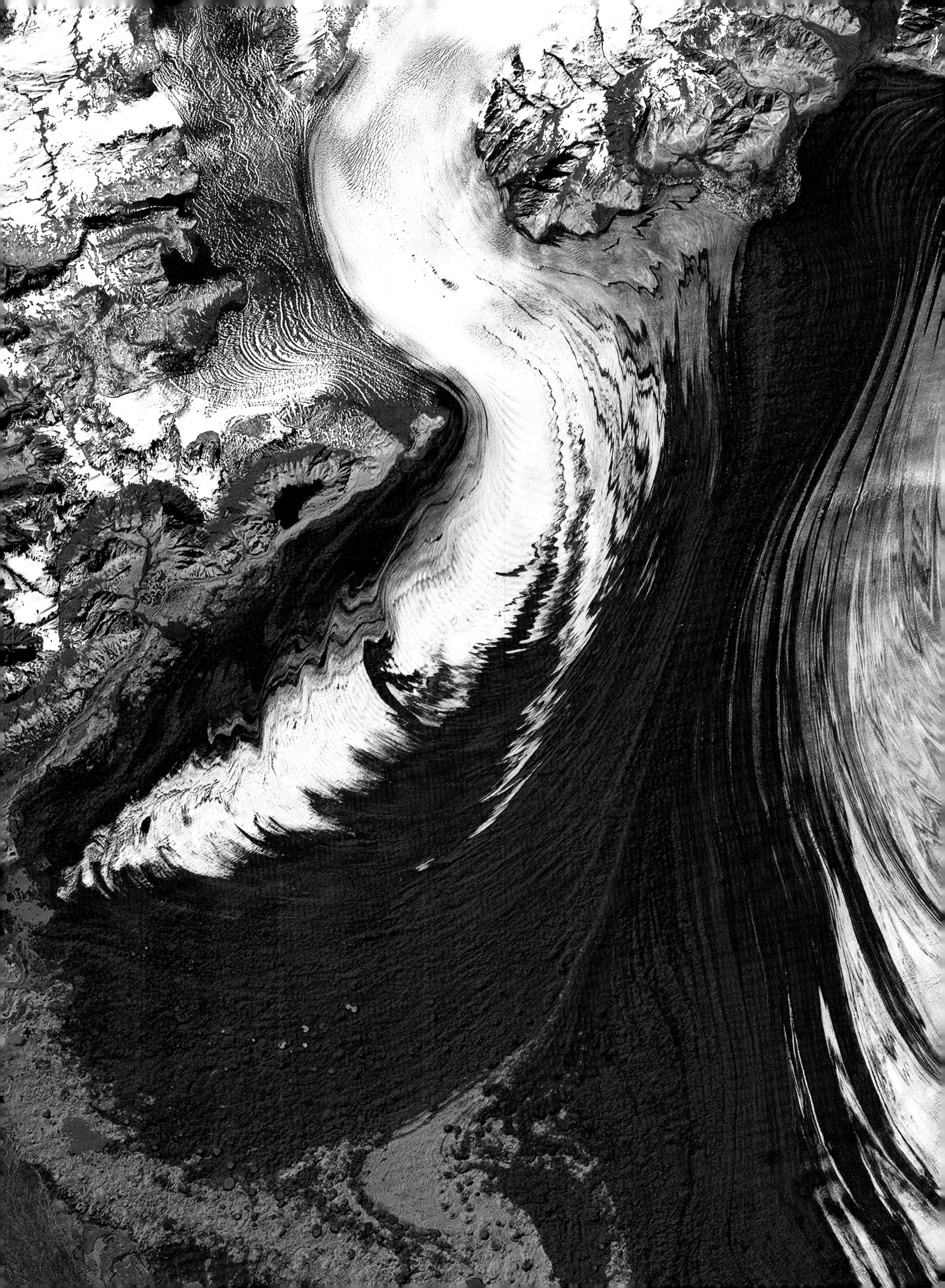

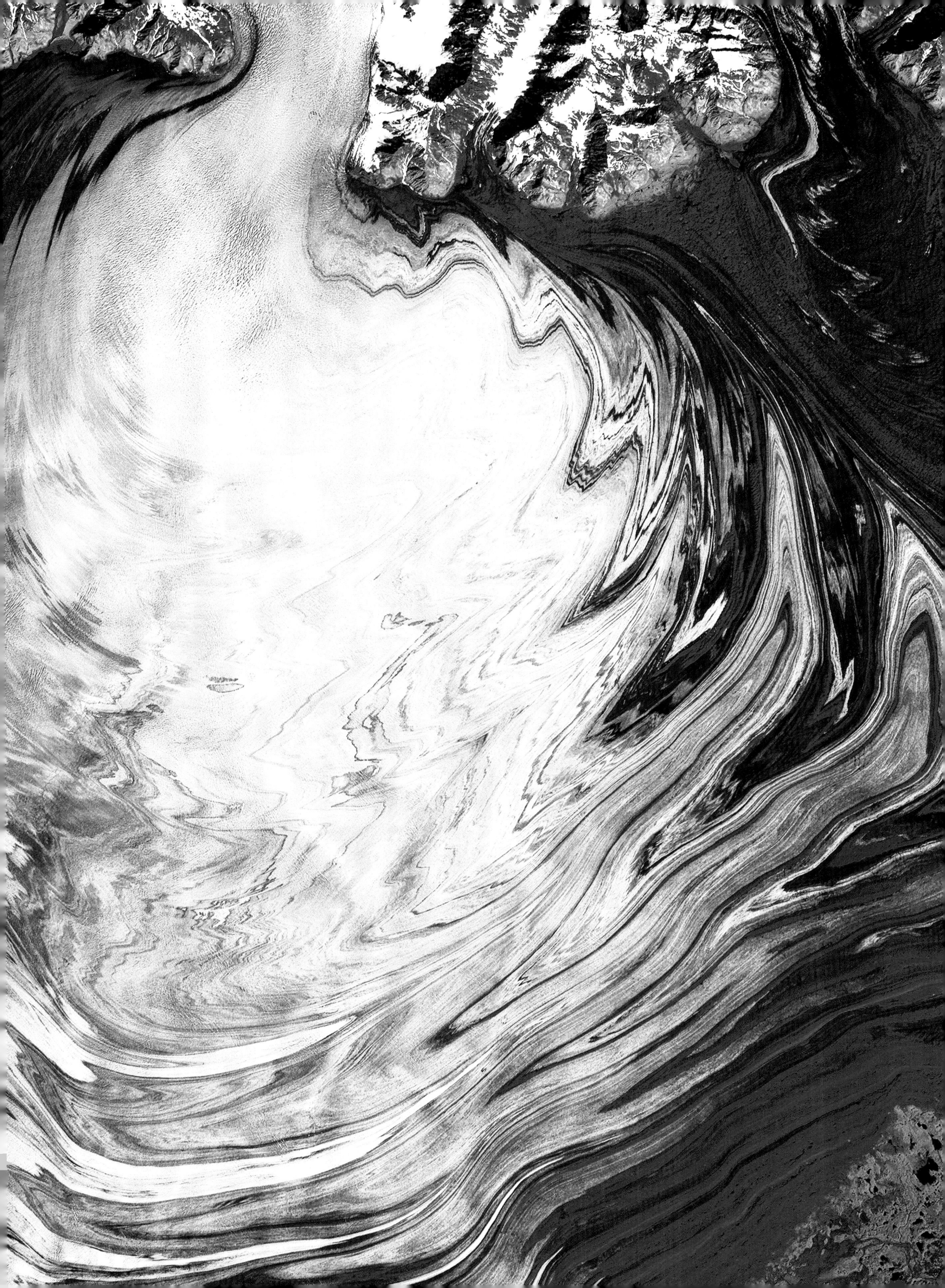

DESIGNED AND DIRECTED BY
OLIVIER BLOND
EDITORIAL DIRECTOR
OF GOODPLANET FOUNDATION

GOODPLANET EDITORS:
ROXANNE CROSSLEY
BENJAMIN GRIMONT
CÉDRIC JAVANAUD
JULIEN LEPROVOST

The GPS coordinates given in the captions are approximate.

FRENCH EDITION

ART DIRECTION
VALÉRIE GAUTIER

GRAPHIC DESIGN & LAYOUT
ANNE-MARIE BOURGEOIS
WITH **MARIE VARÉON**

PROOFREADING
CLAIRE LEMOIN

ABRAMS EDITION

EDITOR
LAURA DOZIER

DESIGNER
SHAWN DAHL, DAHLIMAMA INC

PRODUCTION MANAGER
ERIN VANDEVEER

Library of Congress Control Number:
2013935761
ISBN: 978-1-4197-0962-3

Copyright © 2013 Éditions de La Martinière,
La Martinière Groupe, Paris
English translation copyright © 2013 Abrams,
New York
Photography credits listed on page 256.

Discover the GoodPlanet Foundation's activities
and support it at www.goodplanet.org
Keep up to date with news about the planet at
www.goodplanet.info

Published simultaneously in French under the
title *Espace>Terre*.

Published in 2013 by Abrams, an imprint
of ABRAMS. All rights reserved. No portion
of this book may be reproduced, stored
in a retrieval system, or transmitted in
any form or by any means, mechanical,
electronic, photocopying, recording, or
otherwise, without written permission
from the publisher.

Printed and bound in Italy
10 9 8 7 6 5 4 3 2 1

Abrams books are available at special
discounts when purchased in quantity
for premiums and promotions as well as
fundraising or educational use. Special
editions can also be created to specifi-
cation. For details, contact specialsales@
abramsbooks.com or the address below.

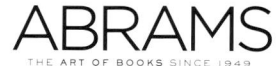

ABRAMS
THE ART OF BOOKS SINCE 1949

115 West 18th Street
New York, NY 10011
www.abramsbooks.com

EARTH FROM SPACE

BY **YANN ARTHUS-BERTRAND**

TRANSLATED FROM THE FRENCH BY
NICHOLAS ELLIOTT

GOODPLANET
FOUNDATION

ABRAMS NEW YORK

PAGES 2-3
MAURITANIA
& SENEGAL
SENEGAL RIVER
16°38' N – 14°25' W

PAGES 4-5
AUSTRALIA
LAKE CARNEGIE
26°09' S – 121°22' E

PAGES 6-7
UNITED STATES,
ALASKA
MALASPINA GLACIER
59°55' N – 140°20' W

OPPOSITE
NASA
EARTHRISE
1968

PAGES 14-15
UNITED STATES,
ALASKA
ALEUTIAN ISLANDS
52° N – 173° W

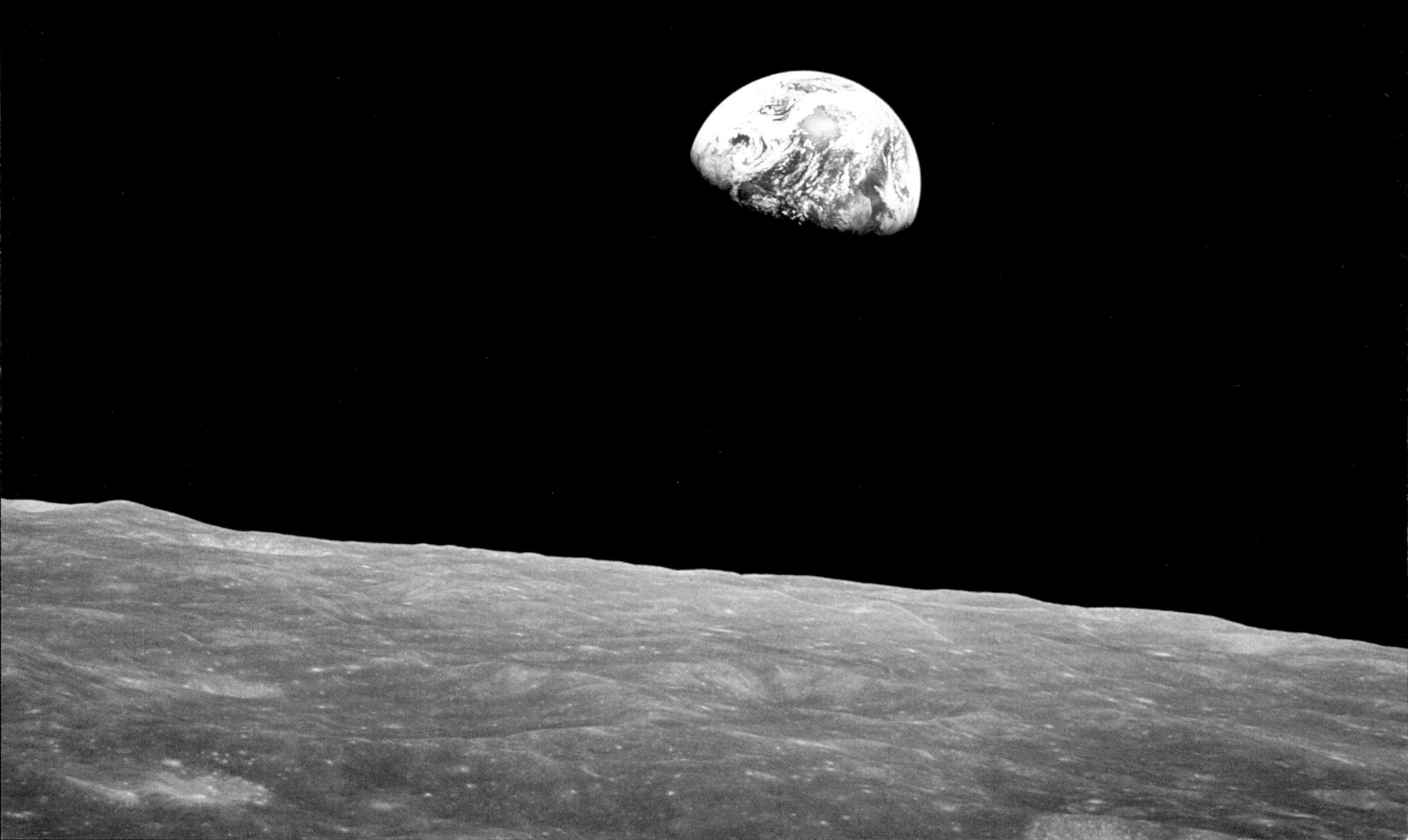

Yann Arthus-Bertrand
President of the GoodPlanet Foundation

Our Earth

One of the most important photographs in the history of humanity was taken on August 23, 1966. It is the first picture of Earth taken from the moon's orbit, an angle that allowed the planet to be captured in its entirety. It showed our planet as we had never seen it before: a little ball isolated in space, vulnerable and magnificent.

All the astronauts who have been lucky enough to see Earth from space with their own eyes are in agreement. No matter their nationality or background, they all describe being overcome by emotion when they contemplated this little blue sphere we inhabit.

Even in the heart of the Cold War, the conflicts dividing us down here often seemed vain up there. The borders separating us lost their meaning: Like never before, these astronauts experienced the feeling of belonging to a single human race and the need to protect this cramped planet we share.

Since then, technology has made incredible advances. The resolution from sensors has been multiplied and computer technologies have been honed. But ultimately nothing has changed since that first photo, which immediately placed space photography in the double perspective of wonder and protection.

Or rather, one thing has changed: These images have also become tools to better understand and change the world. To measure deforestation, track the growth of cities, observe the melting of the glaciers . . . This is why I am delighted to be making this book with Astrium, the European and global leader in the field, which has provided most of the pictures included. For years, we have collaborated on a project to stop deforestation in Madagascar. Astrium's photos have helped us to protect 1.2 million acres (500,000 hectares) of forest and measure our initiatives' impact on the environment.

Personally, I have never been lucky enough to see Earth from space, though it has always been my dream. Aside from a childhood fantasy, in artistic terms it would be a way of furthering the photographic work I began with *Earth from Above*. Or completing it. Or maybe completely reinventing it.

One thing is certain: Whether seen from the sky or from space, no matter the difference in scale, our planet arouses the same sense of wonder by offering us unexpected geometric forms, magnificent structures, and stunning colors—even if the colors in satellite images are often artificial.

While photographing our planet, I discovered the beauty of the world at the same time as the threats weighing on it. I felt myself being transformed and became the environmentalist I am today. I hope that this book and its magnificent photos will have at least some of the same effect on you. And I know that the texts gathered by my foundation's editorial team, headed by Olivier Blond, will be good company along the way.

François Auque
Chief Executive Officer of Astrium

A New Way of Seeing Our Planet

On April 26, 1986, reactor number 4 at the Chernobyl nuclear power plant exploded, causing one of the worst human, industrial, and environmental disasters the world has ever known. A few days later, before the Soviet authorities had released any information, the SPOT-1 LE satellite that we had just launched allowed us to confirm the accident. This was the first time an image from a civilian Earth observation satellite, soon picked up by media around the world, was used to prove that a crisis was taking place and to inform the public.

Space has become a favored location from which to observe the atmosphere, landmasses, and the oceans. A multitude of data is now available to scientists: temperature of the atmosphere, quantity of aerosols, state of the vegetation . . . . Just think: Since 1986, successive generations of SPOT satellites have captured images covering the equivalent of two hundred times the surface of the Earth, with constantly improved accuracy!

Often vital for organizing rescue efforts after major disasters, satellite images have also become essential for protecting the environment and for sustainable management of natural resources. Today they are accessible to everyone online. They also offer every single one of us the opportunity to see the planet from a different perspective and, perhaps, to become aware of its fragility.

Our technology and expertise allow us to build, put into orbit, and operate satellites, then gather space images that can be used in the fields of meteorology, climatology, oceanography, and Earth observation. We carry out these missions with our partners and clients, thereby contributing to the emergence of an even more sustainable society.

As a socially engaged company, it is our duty to make our technology and expertise in space observation available to the collective effort to protect our planet. It was with this in mind that, in 2007, Astrium launched initiatives to support NGOs working to fight climate change with committed partners such as GoodPlanet.

We are delighted to be working with Yann Arthus-Bertrand, a collaboration that began with his project *Earth from Above* and continues today with our contribution to this volume.

Beyond the beauty, occasional strangeness, and even richness of the information held in these images, we hope the spectacle they offer will lead the reader to look at the planet in a new way—the better to know, understand, and protect it.

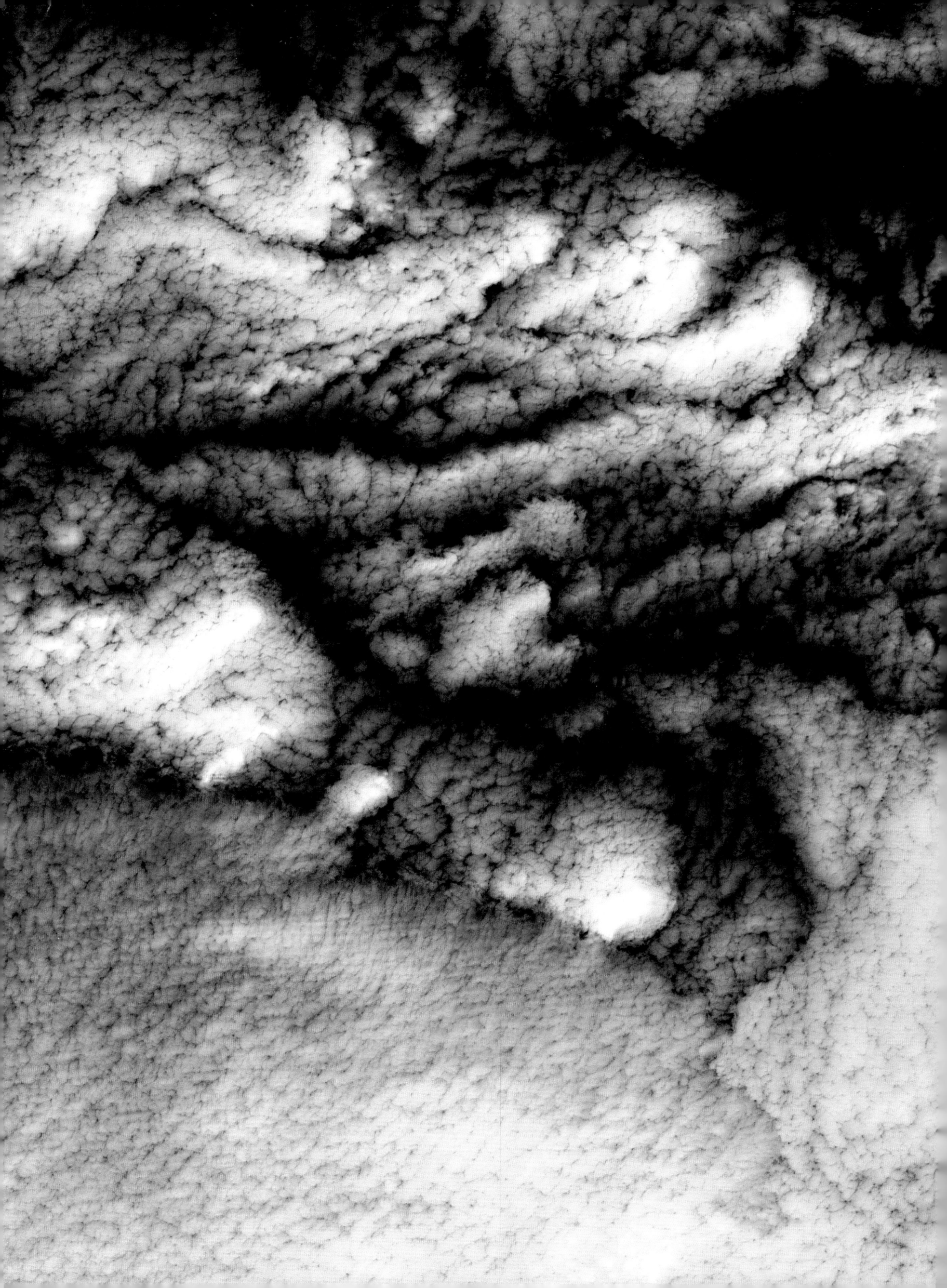

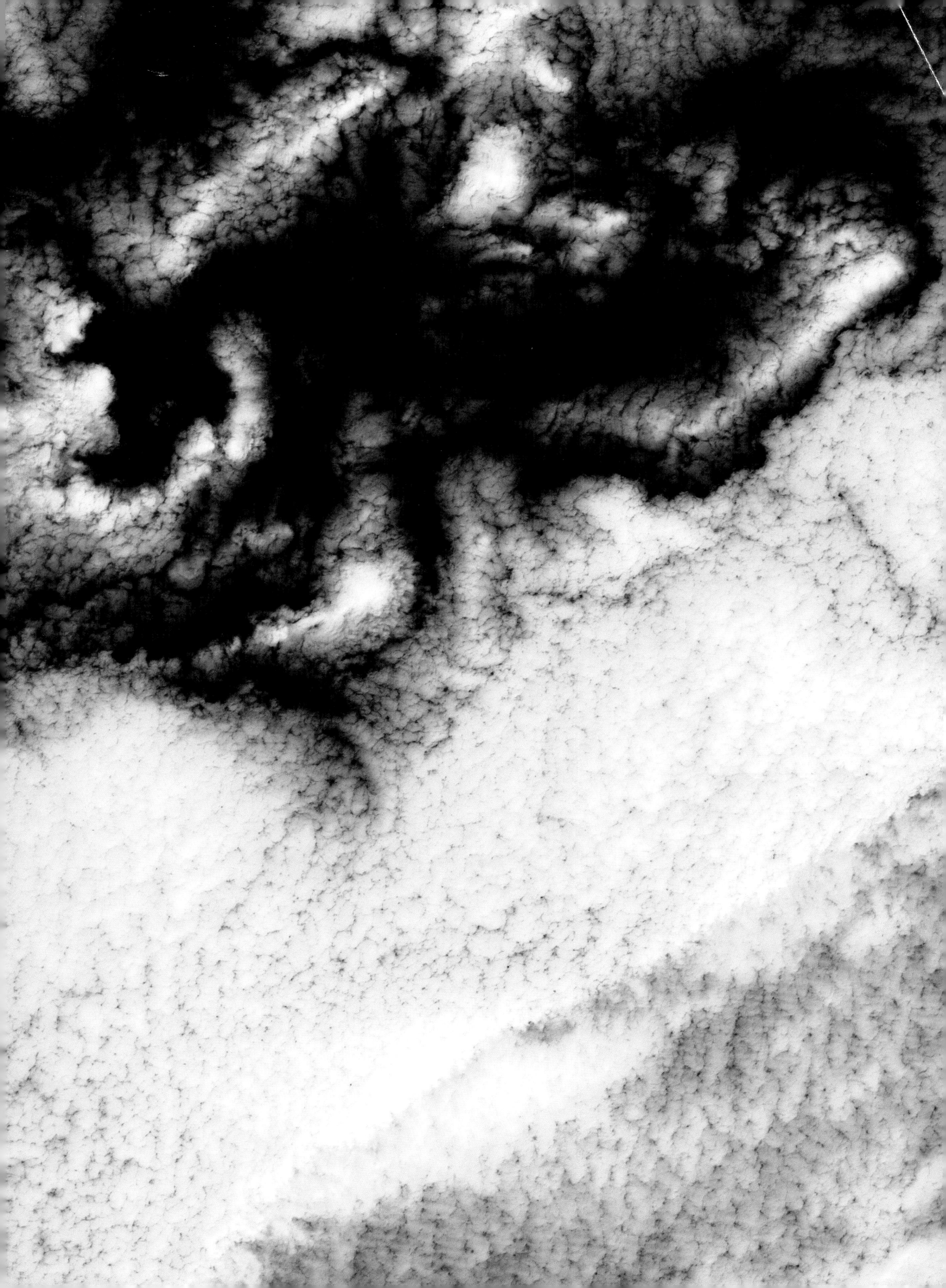

What does Earth look like from outer space? The first astronauts saw a marvelous little blue ball, alone and fragile. But with their increasingly precise devices, satellites perceive incredibly fine details, which are used in the most varied ways—from espionage to environmental protection.

____

SEEING THE EARTH

____

"The view was magnificent. But I did not see the Great Wall of China," said Yang Liwei, China's first astronaut, adopting a nearly regretful tone upon his return to Earth in October 2003. That the thousands of miles of the Great Wall, China's national pride, were not visible from space went against a widely held belief often confirmed by Chinese schoolbooks. But the Great Wall is only about 30 feet (10 m) wide. From space, at a very low orbit of 37 miles (60 km), it is completely invisible; from the moon, trying to make out the Great Wall of China from 250,000 miles (400,000 km) away would be like wanting to look at a hair from a distance of 10,000 feet (3,000 m). Yet what is invisible to the naked eye is detectable using other means, particularly satellites, which have photographed the Great Wall and the entire surface of the planet from every angle imaginable.

THE COLD WAR

The satellite saga, and the space adventure in general, began in 1957 when the Soviets launched Sputnik into space. The satellite's little *beep-beep* was heard around the world, setting off a furious space race. The United States, surprised by the Soviets' achievement, soon launched their own first satellite (Explorer 1) in 1958. They followed, in 1961, with the Apollo program, sending the first men to the moon. The first French satellite, called Astérix, was launched in 1965.

In the half century that followed, approximately six thousand satellites have been launched into space. They have been greatly improved upon and significantly diversified.

The first observation satellites were used as military intelligence satellites during the Cold War. Launched by the Americans (the Corona series, starting in 1959), they were equipped with photographic film that, following exposure, was ejected in capsules to the ground. Far from practical, this system was later replaced by the remote transmission of images.

Weather satellites (the first was TIROS-1, launched in 1960) also have a military legacy, providing useful information about weather on the field of operations. Even the first civilian satellite for observing the environment, Landsat 1, was a product of the Cold War: Its mission included evaluating grain harvests in the Soviet Union.

REVOLUTIONS

Aside from observation satellites, other types include telecommunication satellites (which retransmit telephone calls and television programs), positioning satellites (for the GPS system, for instance), and scientific satellites, which are extremely varied and not always pointed toward Earth (such as the Hubble Space Telescope). Starting with Mariner 9, launched to orbit Mars in 1971, some scientific satellites are used to orbit other planets.

These satellites have so radically transformed our society that we sometimes forget how the world functioned without them. In the past, a mountain was sometimes all it took to block television signals; today, parabolic antennae all over the world allow us to watch TV shows produced across the planet, from Dakar to Borneo. For better or for worse, this has led to a profound standardization of global culture.

Similarly, the services provided by GPS are so ubiquitous that the changes they brought are difficult to measure for

OPPOSITE
AUSTRALIA,
NORTHERN TERRITORY
ULURU
25°20' S – 131°02' E

A UNESCO World Heritage site, Uluru (as it is called by the Aborigines) or Ayers Rock (to certain Anglophones) is a geological structure standing 1,141 feet (348 m) tall and 8,200 feet (2,500 m) long. Uluru is located in the heart of a natural park intended to protect it. But the hundreds of thousands of tourists that visit every year disturb both the Aborigines' sacred traditions and the site's fragile biodiversity.

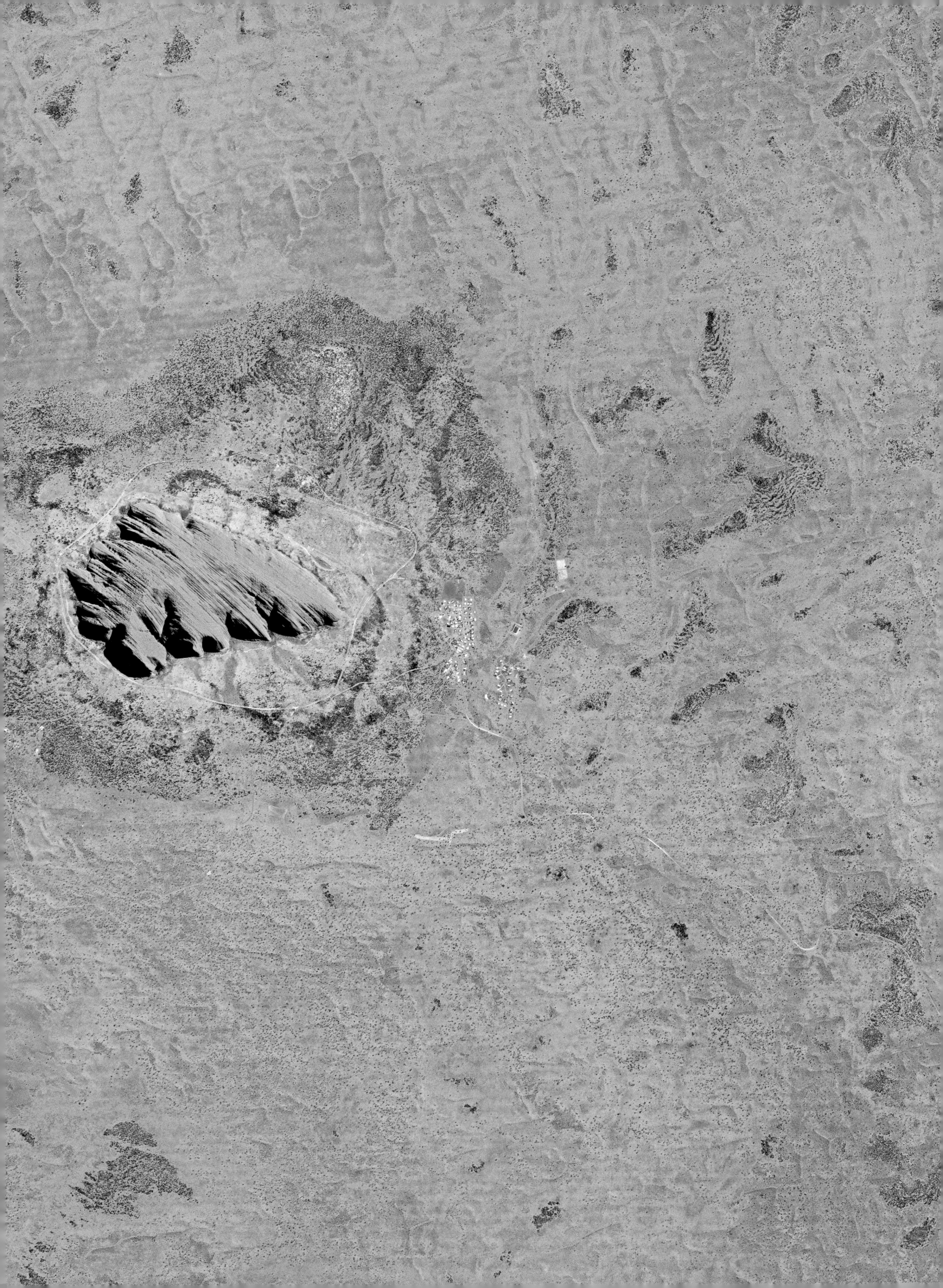

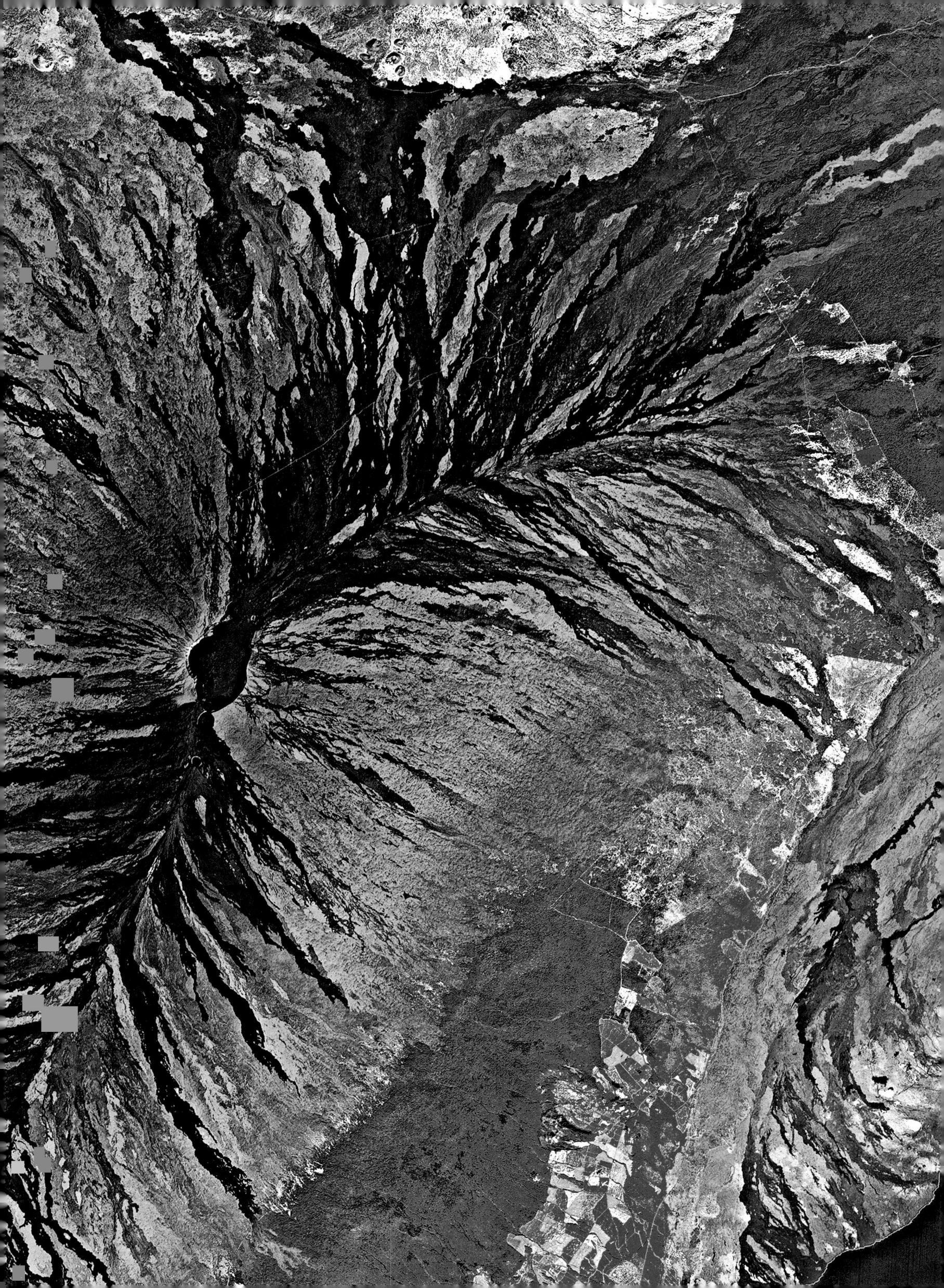

anyone born after their invention. Yes, it was once possible to get lost in a city. GPS positioning and route programming are now part and parcel of daily life in industrialized countries.

DIVERSITY

Satellites' missions are as diverse as their characteristics. For example, telecommunication satellites are often placed in geostationary orbit, over a fixed point on Earth. In order to improve signal transmission, satellites such as the Iridium series are positioned at a relatively modest altitude of approximately 620 miles (1,000 km).

Observation satellites have a sun-synchronous orbit: They rotate around the world in order to be able to take pictures of any part of the planet. The Pléiades satellites travel at a speed of 16,777 miles (27,000 km) per hour and can complete a full rotation in ninety-nine minutes. The lower the satellites are, the better they view the ground. However, the atmosphere provides more resistance to their rotation, which diminishes their life span. The Pléiades satellites are therefore positioned at midrange orbits, from 300 to 500 miles (500 to 800 km). Military intelligence satellites can travel much lower, but have a much shorter life span.

Because they are always used to study the same part of the planet, weather satellites are placed on geostationary orbits. They also orbit at much higher altitudes—22,369 miles (36,000 km) for Meteosat—because they are analyzing the atmosphere rather than the Earth's surface.

Big satellites like Hubble and Envisat can weigh close to 10 tons. Today, the trend is to launch smaller, more practical satellites: The Pléiades satellites weigh 2,160 pounds (980 kg).

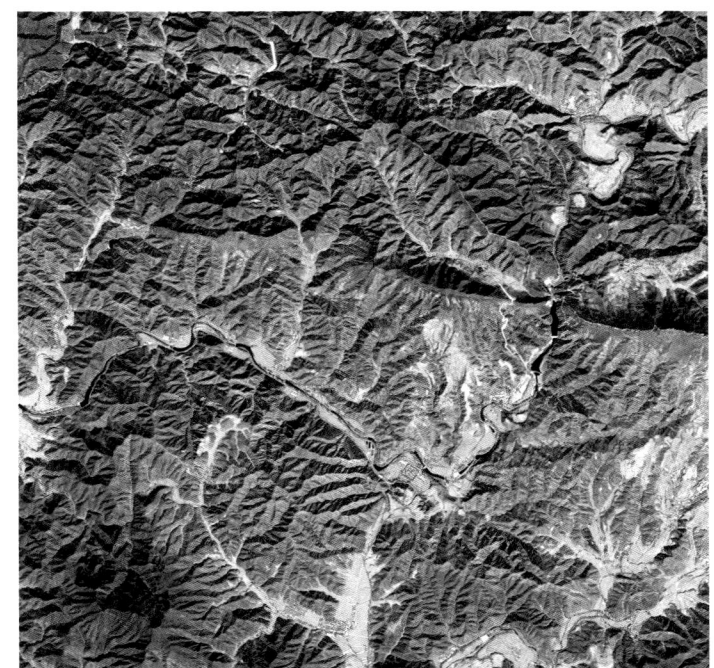

ABOVE
CHINA
THE GREAT WALL, NORTHEAST OF BEIJING
40°40' N – 117°16' E

The Great Wall has been on UNESCO's World Heritage List since 1987. The largest architectural structure in the world, it is 19.6–22.9 feet (6–7 m) tall and approximately the same across—generally, less than 32 feet (10 m). Its length has been reevaluated several times. In June 2012, the Chinese authorities announced their latest estimate of its length as being 13,170 miles (21,196 km)—versus the approximate 4,038 miles (6,500 km) reported just a few years earlier . . .

OPPOSITE
UNITED STATES, HAWAII
MAUNA LOA VOLCANO
19°25' N – 155°33' W

Hawaii's Big Island was formed by volcanoes, including Mauna Kea, which has been extinct for millennia. Its 13,796-foot (4,205-m) peak is an excellent observation point, where several space observatories and a few of the biggest telescopes on the planet are located. At 13,681 feet (4,170 m), Mauna Loa is a hair shorter in size.

The PLÉIADES SATELLITES travel at a speed of
16,777 MILES (27,000 KM) PER HOUR
and can complete a full rotation in NINETY-NINE MINUTES.

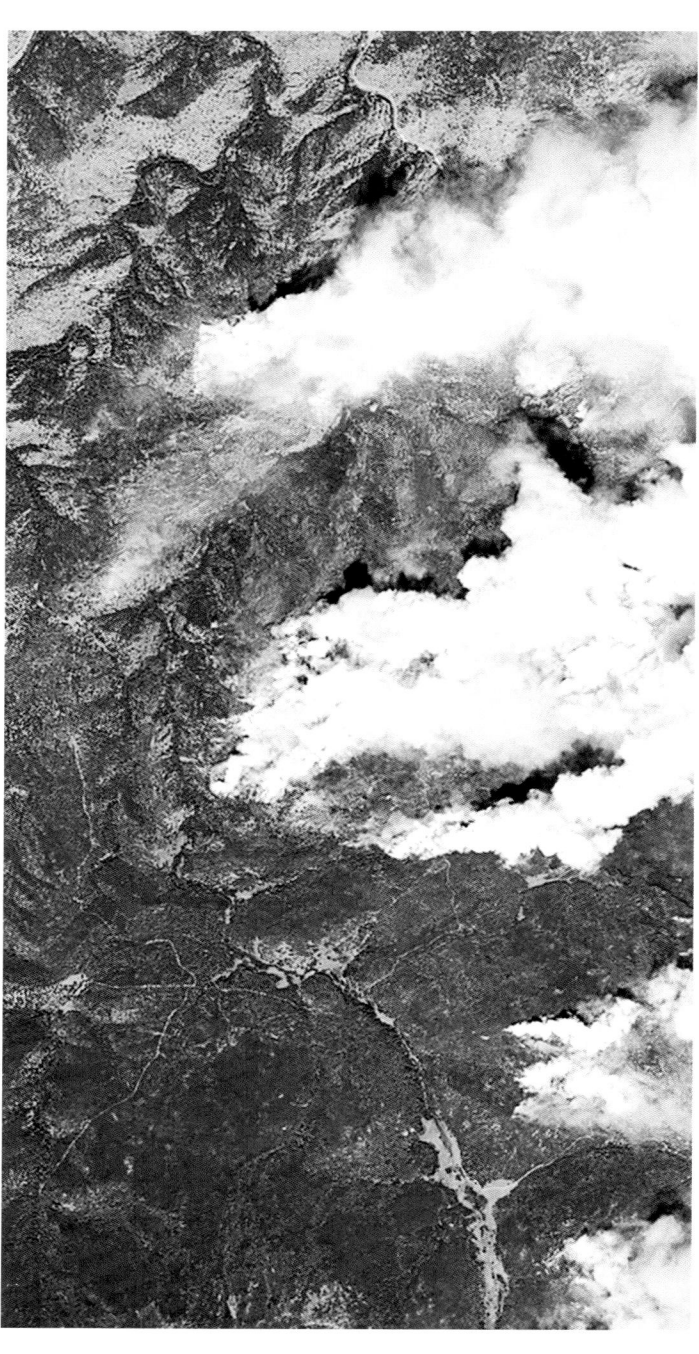

THE BLUE MARBLE

The most famous image of the planet was taken by Apollo 17's crew in 1972. It shows a totally illuminated, round Earth, which has since come to be called the blue marble. This type of image is now achieved by editing together several satellite images laid end-to-end, notably to avoid gaps caused by potential cloud cover. But these images, which are taken in visible light (and therefore similar to what the eye perceives), are only one of the many facets of satellite observation.

The diversity of measurement devices seems to be practically unlimited: Spectrometers, radiometers, interferometers, altimeters, and radars all have a unique way of observing Earth. In fact, most satellites carry several kinds of instruments (called "useful payload").

HURRAH FOR SCIENCE

In just a few years, satellites have allowed us to make major scientific advances. Thanks to them, we have better knowledge of the Earth's shape. While scientists since Newton had known that it was not truly round, they can now say it is potatolike, with hollows and bumps all over its surface. Even the oceans are not truly flat. Aside from waves and tides, of course, oceans can be higher or lower than the average sea level by as much as a hundred feet (or several dozen meters)—for instance, over a chasm or a volcano.

A better understanding of the oceans is fundamental for making models of the planet's climate and anticipating phenomena such as El Niño. In just the last few years, high-precision altimetry satellites (TOPEX/Poseidon, then Jason) have informed us about sea levels, which are indispensable

OPPOSITE

UNITED STATES, COLORADO
FOREST FIRE
39°35' N – 105°30' W

Probably sparked by lightning, this fire ravaged
the wooded area west of Denver for several weeks
in 2012. More than five thousand firefighters
were mobilized to contain it, while it was fanned
by violent winds such as the ones seen in this
picture. The fire consumed more than 77 square
miles (200 sq km) of forest, killed one person, and
destroyed two hundred buildings.

BELOW

ALGERIA
IN AMENAS
28°03' N – 90°34' E

The deadly hostage crisis perpetrated by Islamist terrorists
on the Algerian gas plant of Tiguentourine took place 25 miles
(40 km) west of the city of In Amenas. This picture, taken by
the SPOT-6 satellite on January 8, 2013, about one week before
the crisis, is an example of satellites' potential use for military
or security intelligence. On the left of the shot, you can see
the storage tanks of a former oil refinery. To the right, you can
distinguish the Sonatrach company's residential base, outlined
in black, and farther east, the Y-shaped airfield. The two landing
strips, 1.2 miles (2 km) long and 1.8 miles (3 km) long, respectively, provide a sense of scale.

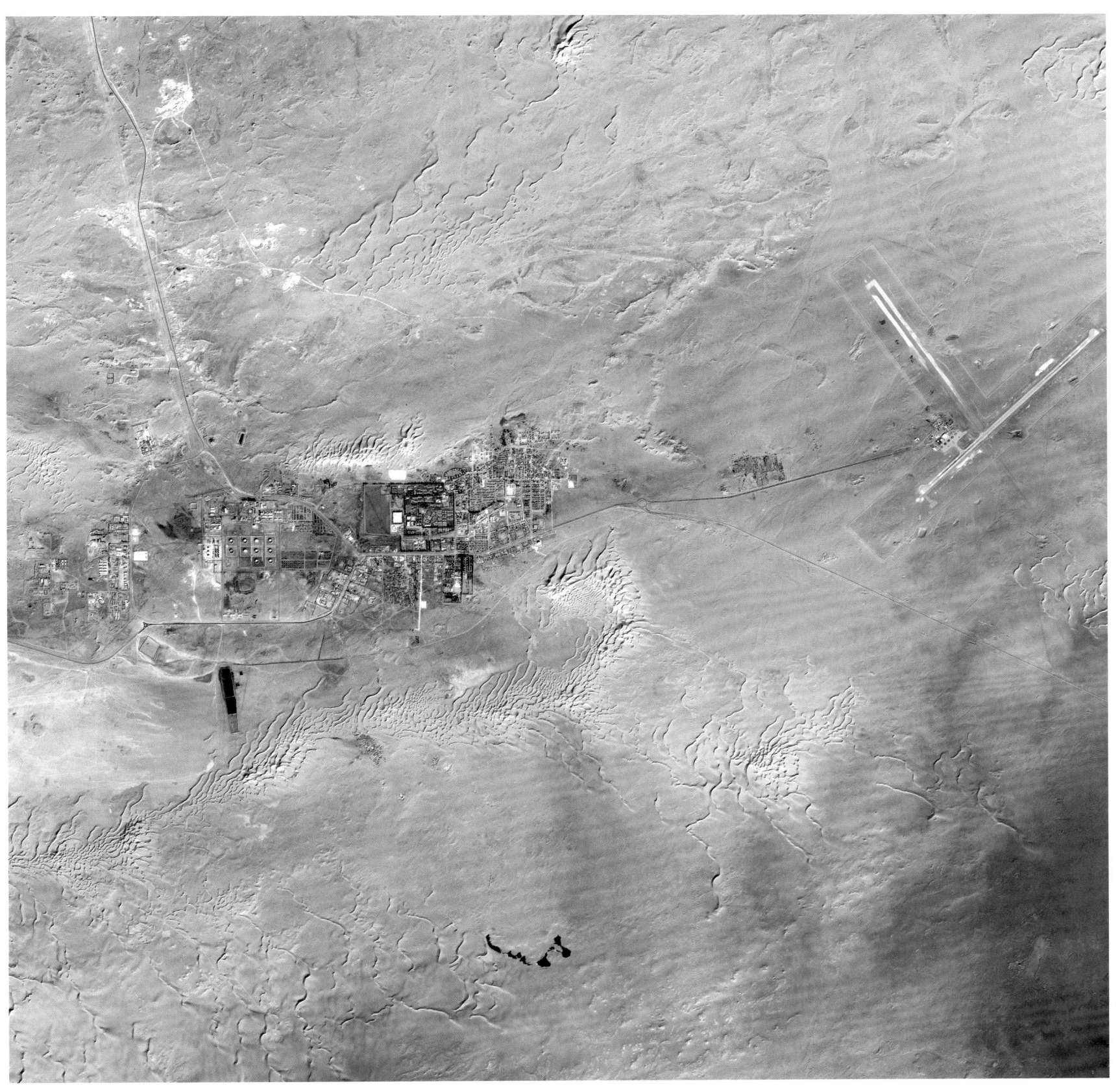

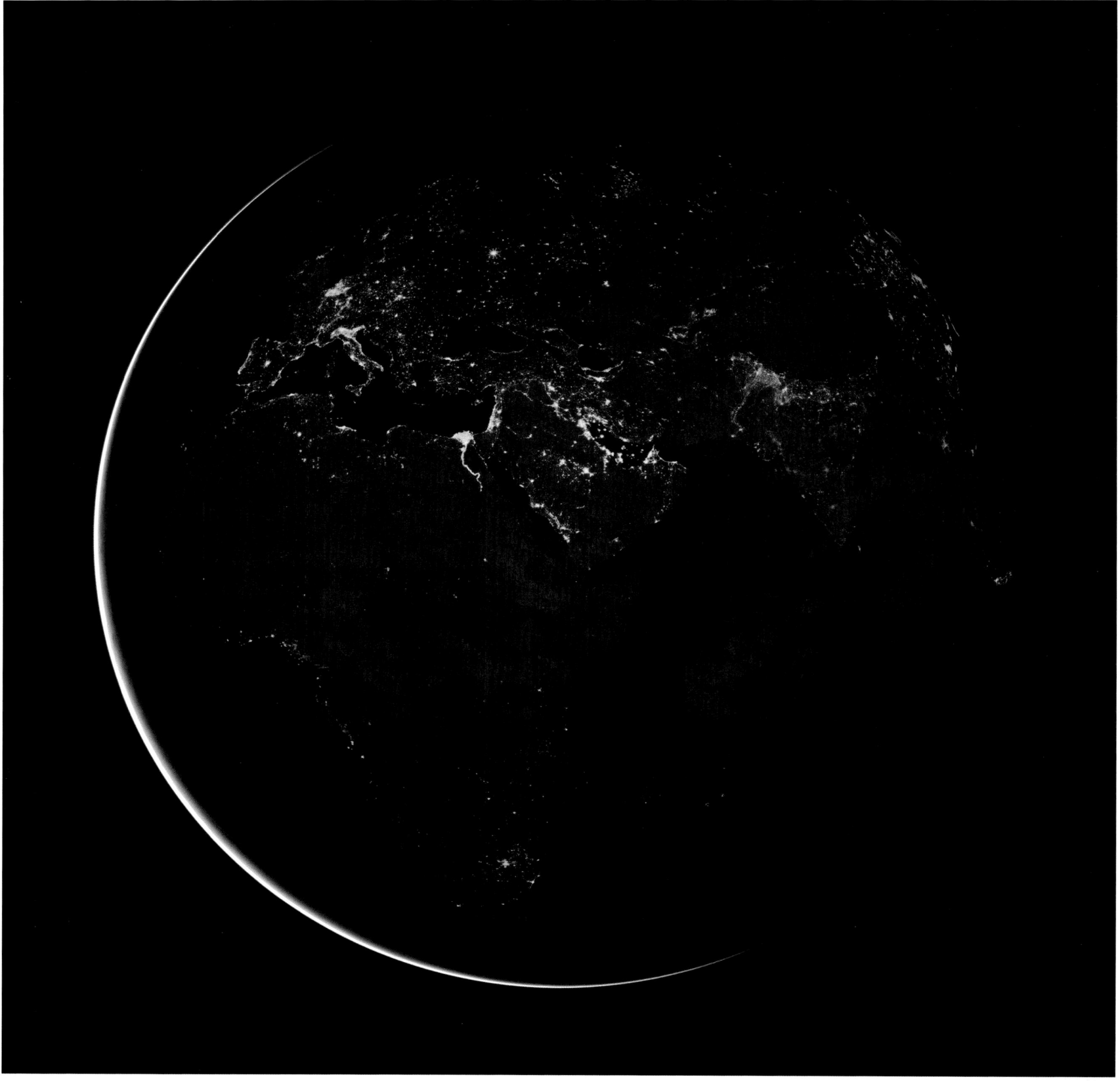

NASA
EARTH BY NIGHT
2012

This image is the product of a combination of multiple photographs taken by NASA from April to October 2012. The orange spots show urban lighting: On this side of the Earth, western Europe, the Mediterranean coasts, certain areas of the Arabian Peninsula, the Indian peninsula, and China appear brightly lit, but a large part of Africa remains dark. Half of the world's population now lives in cities, but close to 1.4 billion people, principally located in Africa and Asia, do not yet have regular access to electricity.

CNES
SPOT 5, SPOT 6, AND PLÉIADES
2002–2012

The SPOT-5, SPOT-6, and Pléiades satellites were built by France in partnership with other European countries and launched in 2002 (8.2 ft/2.5 m resolution), 2012 (4.9 ft/1.5 m resolution), and 2011 (2.2 ft/0.7 m resolution), respectively. In fifty years, satellite resolution has improved by 100 percent.

Today's observation satellites have
a RESOLUTION of 20 inches (50 cm)—and
LESS THAN 8 INCHES (20 CM)
for MILITARY INTELLIGENCE SATELLITES.

for tracking the rise of water levels related to global warming. Seas are currently rising 0.11 inch (3 mm) a year—and the trend is accelerating. Satellites also help us to estimate the volume of ice melting at the poles and glaciers.

Along with other information provided, the above data is increasingly precise. For example, in the 1970s, observation satellites' spatial resolution was 262 feet (80 m); today it is 20 inches (50 cm). In the case of military intelligence satellites, whose resolution has improved by more than 100 percent, it is less than 8 inches (20 cm). Other devices have improved in similar proportions.

"I SEE EARTH"

Let's return to the Great Wall of China. The naked eye cannot make out much from outer space: the continents, the major cloud masses . . . But this limitation is not particularly relevant, given that, to date, only some five hundred human beings have been able to observe Earth from space, and often for very short periods. Other living beings have been sent into space, including Laika, the first dog sacrificed to outer space, who was launched aboard Sputnik 2 in 1957 (she died of stress and heat only seven hours after takeoff); the American monkey Miss Baker, who was launched in 1959 and lived until 1984; and the cat Félicette, the French space hero of a fifteen-minute flight in 1963.

Meanwhile, satellites have constantly observed the planet for decades, seeing it better than any animal's eye. This technology has become accessible to many countries: China, India, South Korea, North Korea, Iran, Israel, and recently even Nigeria.

These satellites send us images of the major terrestrial structures, such as the Great Barrier Reef, spanning about 135,000 square miles (350,000 sq km); the Grand Canyon, at 277 miles (450 km) long and more than 3,280 feet (1,000 m) deep; and Uluru (Ayers Rock), which is 1.8 miles (3 km) long and stands in the heart of the Australian desert. Aside from these natural masterpieces, the great structures of mankind are also visible, including the Great Wall of China, the Giza pyramids, Venice, and the palm-shaped artificial islands in

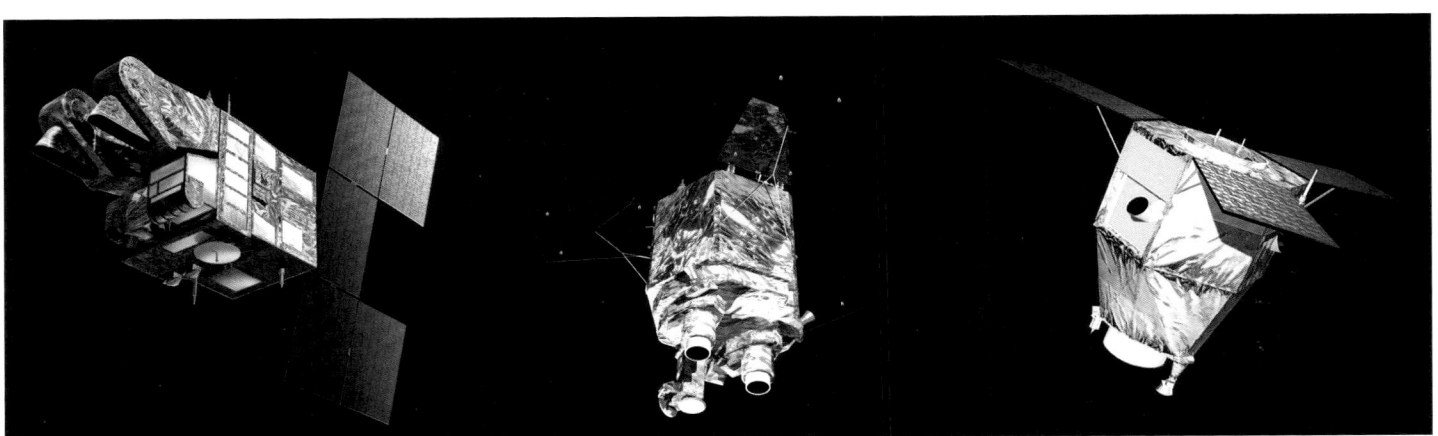

"I SEE EARTH! It is so BEAUTIFUL!"
said Soviet cosmonaut YURI GAGARIN on April 12, 1961.

Dubai. Even a network of night-lit roads can be easily distinguished from space. Man's constant activity spins a web over the planet and leaves a mark on it.

A connection has been established between technologically wondrous observation satellites and the marvels and masterpieces they allow us to observe. The Angkor Wat temples of the Khmer and the Tulum pyramids of the Maya are monitored from space in the context of a UNESCO program designed to better protect them.

It is no small achievement that space technologies born of military rivalries serve to protect humanity's cultural heritage, even when it is their secondary purpose. For sustainable development does not only consist of protecting endangered species. It also and perhaps primarily aims to allow us to better live together, which requires raising awareness of our differences and intertwined histories and making determined efforts to protect them.

In 1967, in the midst of the Cold War, the great powers signed a treaty known as the Outer Space Treaty, by which they rejected the appropriation of outer space and its military use, banned the placing of weapons of mass destruction in orbit of the Earth, and promoted international cooperation. The treaty was ratified by 107 countries. One hopes it will never be called into question.

"I see Earth! It is so beautiful!" said Soviet cosmonaut Yuri Gagarin on April 12, 1961.

What satellites cannot give us is the meaning of what they observe: Only our species can interpret their images and discern within them the fascinating beauty of our world, as well as its incredible fragility.

———

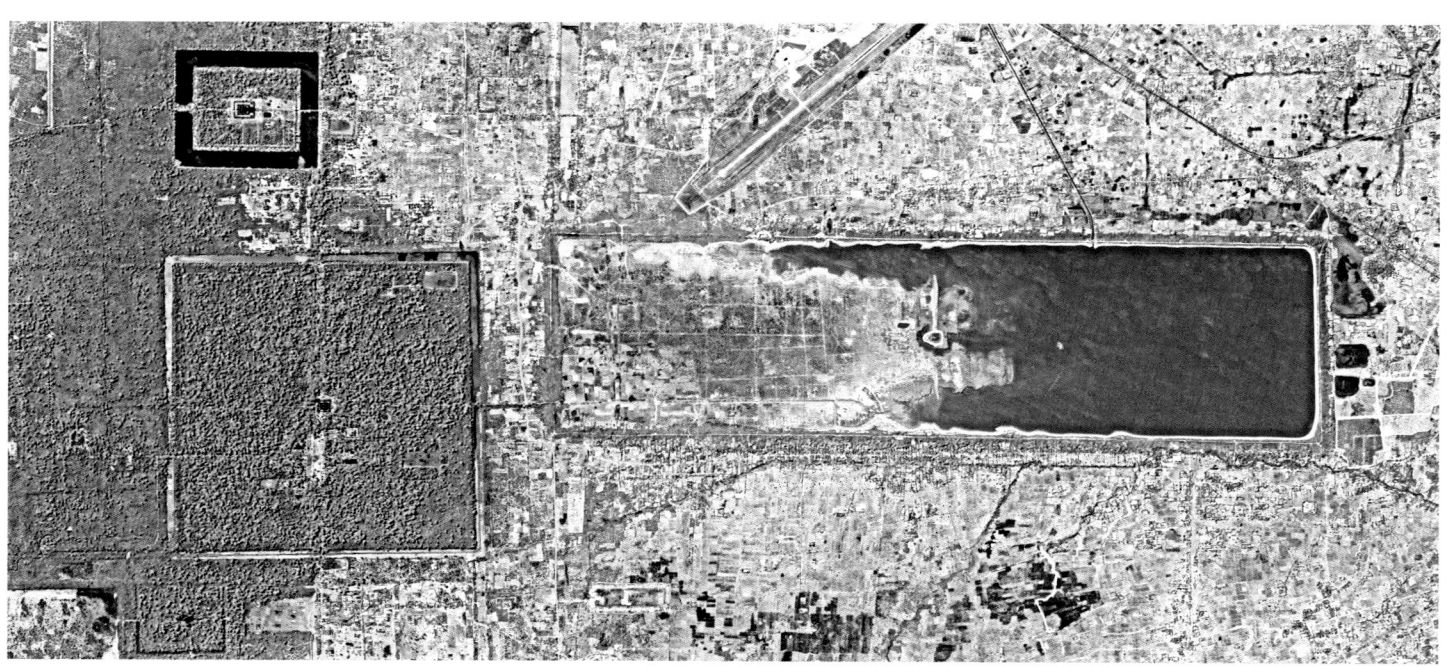

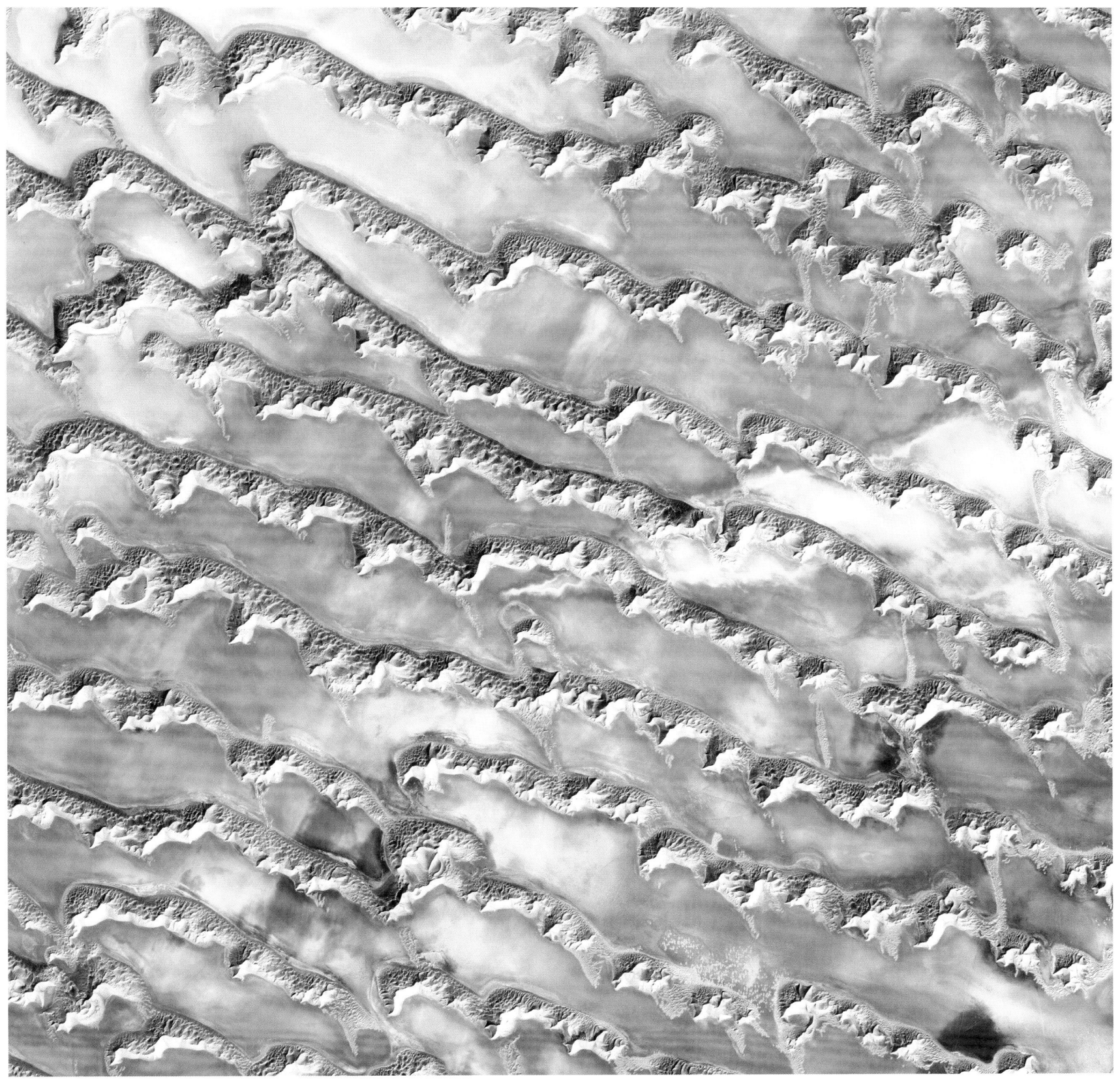

OPPOSITE

CAMBODIA
ANGKOR WAT
13°25' N – 103°51' E

The city of Angkor Wat was the capital of the Khmer kingdom from the ninth to the fifteenth century. Today, the site is on the UNESCO World Heritage List and includes the remains of temples and a vast network of canals, ornamental lakes, and dykes stretching over approximately 150 square miles (400 sq km). A NASA study conducted in 2007 used aerial sensors and satellites to show that, at its peak, Angkor extended over more than 380 square miles (1,000 sq km) and had more than one million inhabitants.

ABOVE

SAUDI ARABIA
RUB' AL-KHALI
20°42' N – 54° E

Appropriately, the name Rub' al-Khali means "the empty quarter." It is one of the largest sand deserts on the planet, forming a rectangle roughly 600 miles (1,000 km) long by 300 miles (500 km) wide, which is located principally in Saudi Arabia, but also in Yemen, Oman, and the United Arab Emirates.

OPPOSITE
INTERNATIONAL SPACE
STATION
AUSTRALIAN AURORA
ALTITUDE 217 MILES (350 KM)
APRIL 29, 2010

As the first French female astronaut, Claudie Haigneré is among those rare human beings who have seen the Earth from outer space with their own eyes, an unrivaled spectacle.

Talking with Claudie Haigneré

The Magical Living Spectacle of Earth

Astronauts often describe the overpowering feeling they experienced the first time they saw the planet from space. Why is sharing this feeling with others important?

Having had the luck to participate in several space missions, I experienced genuine awe every time I discovered the magical living spectacle of the Earth standing alone in the dark of the cosmos. Filled with incredible emotion, I would look through the space station's porthole and contemplate the Earth as often as I could.

Like a privileged spectator of life, I could take in our entire "big house" in one glance and observe the diversity of its landscapes, the variety of colors and movements. As a witness to this incredible, nearly fantasylike beauty, I also grasped our planet's vulnerability as I observed it. My fascination became solemn. Telling people about these experiences is an essential part of this magnificent human and scientific adventure.

What role can this feeling play in raising awareness of our planet?

From space, Earth looks like a beautiful living planet with limited resources, and its protective layer, the atmosphere, seems so thin and fragile . . . From that distance, you realize how important it is to live in constant harmony with it, as well as the responsibility we "earthlings" have to share this awareness.

Seeing from this explorer's position is a great privilege: This helps to develop a more reasonable, responsible, informed relationship to our environment. The most beautiful space conquest of the last fifty years is probably the rediscovery of our Earth.

What is satellite imagery's main contribution to sustainable development?

By taking the place of the human eye, space technologies now allow us to monitor and predict radical changes on our planet, to understand how they function, make models of the consequences, and therefore to control the planet's future and our own.

By scanning the atmosphere, the oceans, and the continents every day, space imagery devices are like our "guardian angels," they make us aware of the globalization of environmental problems, the interaction of the different components, and the necessity of acting collectively, beyond borders and civilizations, to protect the Earth.

In keeping with a quote attributed to Socrates, which states that "man must rise above the Earth, to the top of the atmosphere and beyond, for only thus will he fully understand the world in which he lives," I hope that this observation of Earth from a distance will allow us to feel increasingly close and concerned with preserving its viability and beauty.

Claudie Haigneré was the first female French astronaut. She has completed scientific missions on the Mir station and the International Space Station. A former delegate minister for France's Research and New Technologies, she is now the chair of Universcience, an organization that brings together two Paris museums, the Cité des Sciences and the Palais de la Découverte, with the goal of making the sciences accessible to all.

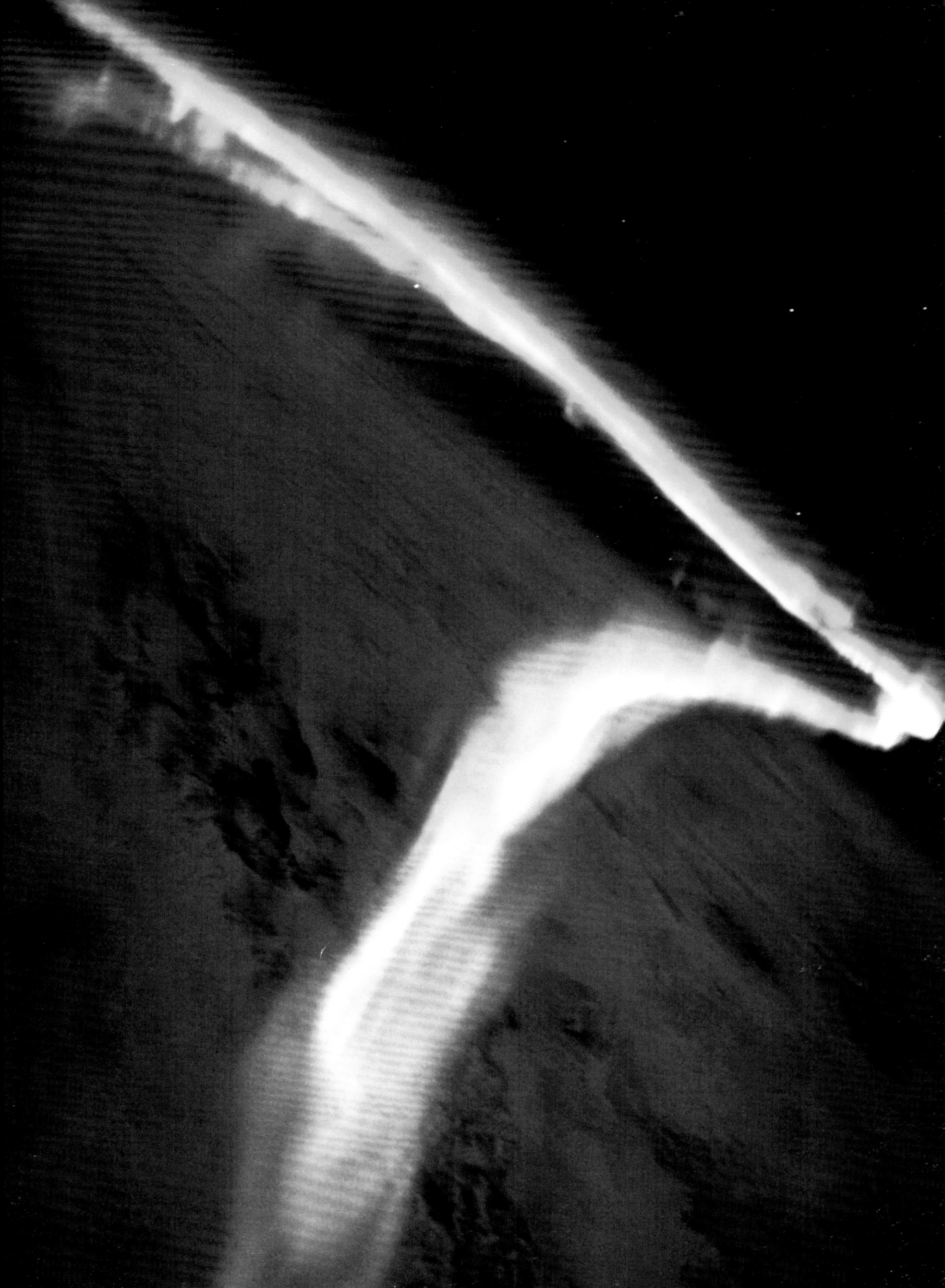

OPPOSITE
NORTH KOREA
WEST KOREA BAY
39°27' N – 125°19' E

PAGES 30-31
FRANCE
MONT-SAINT-MICHEL BAY
48°40' N – 1°30' W

Borders do not exist in space. And even the most closed dictatorship, North Korea, cannot hide its activities from satellites—particularly its political prison camps.

Talking with Christoph Koettl

The Gulag Archipelago

Why do you have to use satellite images to prove the existence of prison camps in North Korea?
Human rights organizations are systematically forbidden to enter the country, and there is no national independent media, opposition party, or independent civil society. Any criticism of the government is punished by incarceration in political prison camps or detention centers. Satellites at least give us access to the country from a distance. They are the only sources of information we have about camps in North Korea, along with accounts from people who have fled the country.

How do you estimate the number of prisoners?
We think there are six camps for political prisoners in the whole country. However, in October 2012, it was reported that Kwan-li-so No. 22 (in Hoeryong, in the province of North Hamgyong) was closed after Kim Jong-un came to power; nothing is known of what became of the twenty to fifty thousand prisoners who were held there.

We cannot determine the exact number of people in the camps with satellite images alone. But based on the size of the buildings, our very rough estimate of the total number of prisoners in North Korea is more than a hundred thousand.

What do we know about these prisoners and their living conditions?
Based on accounts from former inmates and prison guards given to Amnesty International in 2011, the working conditions amount to slavery, and public executions are common. Violations of human rights are severe, systematic, and widespread: forced labor, malnutrition, torture, and other cruel, inhumane, and degrading treatment. Most prisoners are only being held because members of their family are considered hostile to the regime.

Today anyone can see these images on Google Earth. Does that change anything?
Tools such as Google Earth are embarrassing for the North Korean authorities, who deny the existence of prison camps. Thanks to these images, more and more people around the world are aware of this reality.

Have satellite photos been used to prove or denounce the existence of other prison camps?
They are particularly useful for revealing detention centers in places where access is extremely limited or even refused to independent observers. We used Google Earth to expose the horrible Insein Prison in Myanmar. Long-distance images were also very useful to identify and observe detention centers in Eritrea.

Christoph Koettl is the emergency response manager at Amnesty International USA.

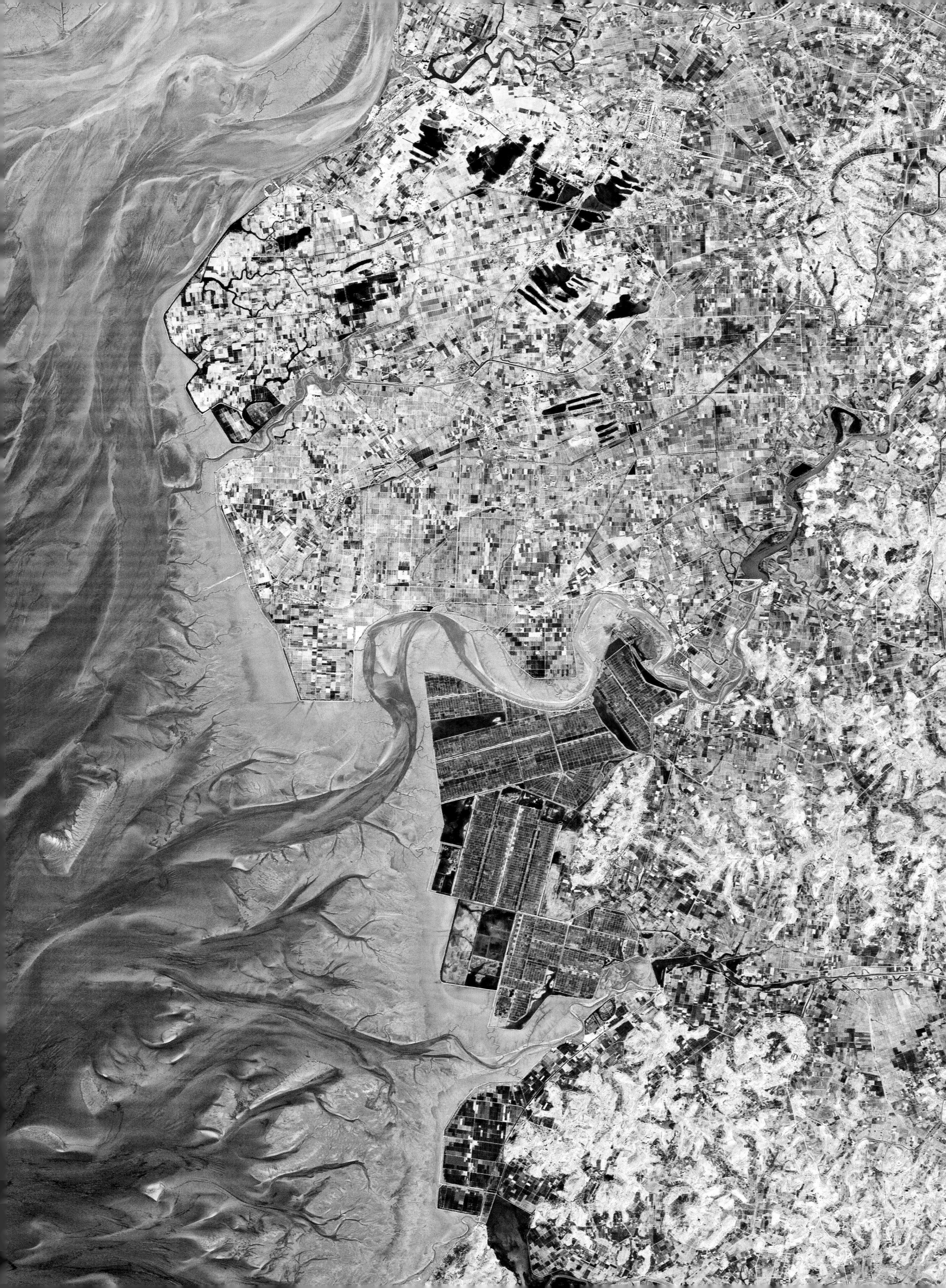

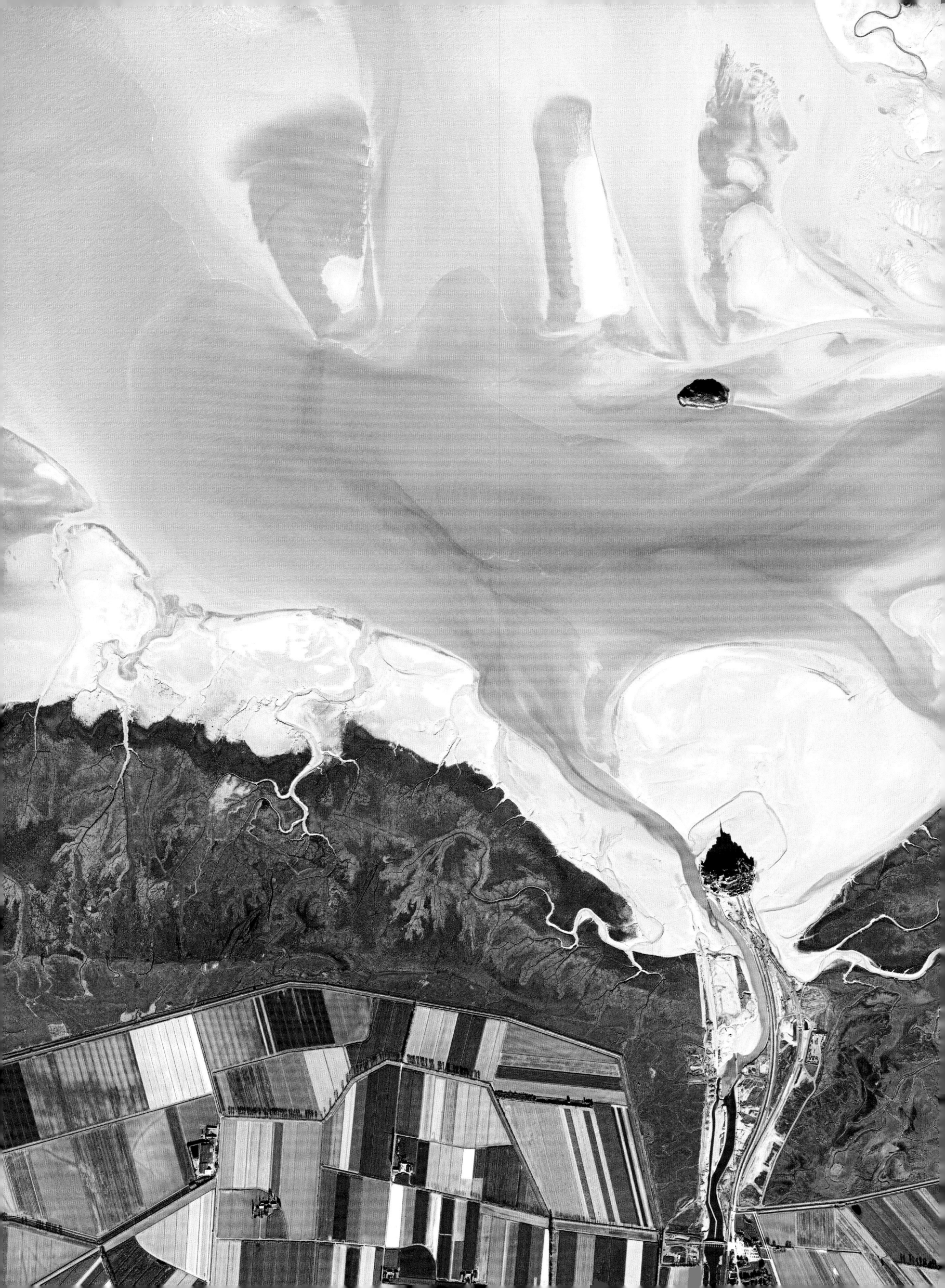

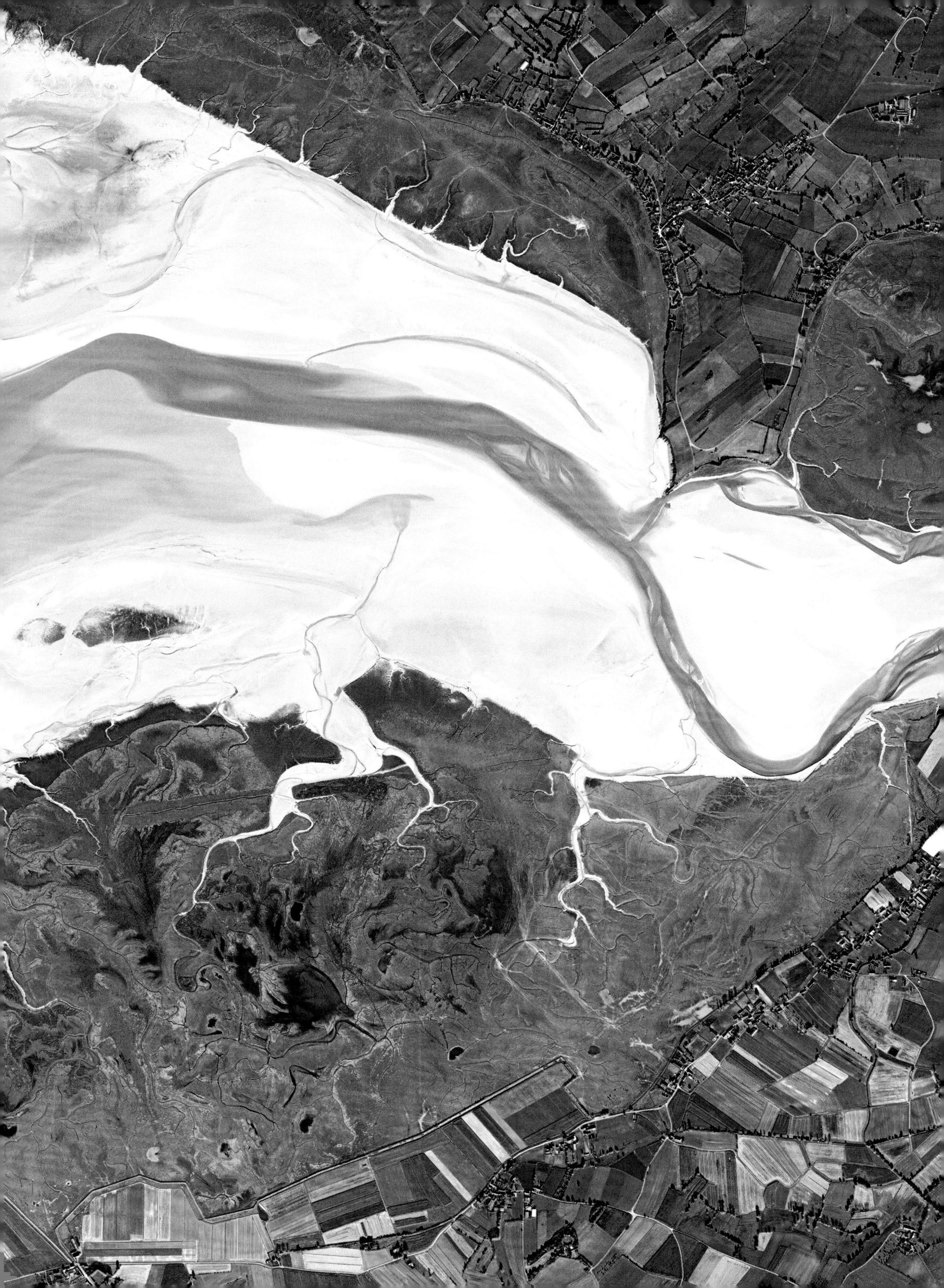

Nearly one billion people around the world go to bed hungry every night. Meanwhile, the global population is expected to grow by several billion in the next decade, and improved standards of living in Asia are increasing the demand for food products. The question of agriculture has hardly been resolved—on the contrary, it has become a major issue of our modern era.

___

FEEDING THE WORLD

___

In one of his notebooks, *Le Sottisier* (also known as the Leningrad Notebook), Voltaire wrote: "In proper political mode, we have found the secret to starving to death those who feed others by working the land." Two and a half centuries later, the situation has not improved much. Many of those who go hungry today are farmers. Other groups who do not have enough to eat live on the edges of urban centers, in shantytowns and slums, and are often descended from migrants who arrived in great numbers, abandoning the countryside, as part of a mass rural exodus.

Yet the global agricultural system produces large amounts of grains, meat, and various other food products. Unlike in previous centuries, hunger is no longer due to shortages, but to problems with distribution or access; in other words, it is a problem of poverty.

Over the last two decades, global food production has increased considerably. Irrigation, fertilizers, pesticides, and the selection of more resistant and productive varieties tripled global agricultural output through the so-called green revolution, which began in the 1950s. Another important, though lesser, growth factor is the expansion of cultivated land, most often to the detriment of forests.

Partially due to the improvement of yields and the increase in cultivated areas, the number of people suffering from undernourishment has diminished over the last twenty years, dropping from 16 percent of the global population to a little more than 13 percent. But over the same period, the population increased, reaching seven billion in late 2011. So the number of people who are undernourished or malnourished has actually increased, growing from 848 million to 925 million in 2010—nearly one in seven people on the planet.

CHANGING AGRICULTURE

Feeding the world is a greater challenge than ever, particularly because it may have an additional three billion inhabitants by 2050, and the population is getting wealthier and modifying its eating habits. Humans are not only consuming more in general; they are also consuming a constantly growing number of products that require a greater amount of farmland. According to the Food and Agriculture Organization of the United Nations (FAO), global food production will have to increase by 70 percent by 2050 to feed the entire global population. Yet at the very moment we need it most, modern agriculture seems to be running out of steam: Yields are now increasing by only one percent a year, slower than the demand. The growth in productivity could even be reversed, for current methods deplete soil and damage ecosystems—the foundation for all of agriculture. Faced with this situation and the constant multiplication of extreme weather phenomena due to global warming, we must rethink existing agricultural models to increase production while preserving the environment. Several ways forward exist, some of which draw on satellite imagery.

OPPOSITE

THE NETHERLANDS
POLDERS
51°42′ N – 3°50′ E

The Netherlands are well named: A third of the country lies 3.2–14.7 feet (1–4.5 m) below sea level. These low-lying areas, known as polders, were reclaimed from the sea, lakes, and marshes beginning in the fourteenth century, thanks to a complex system of dykes, locks, and canals. The polders provide excellent farmland, which has allowed the Netherlands to become the third-largest agricultural product exporter in the world, after the United States and France.

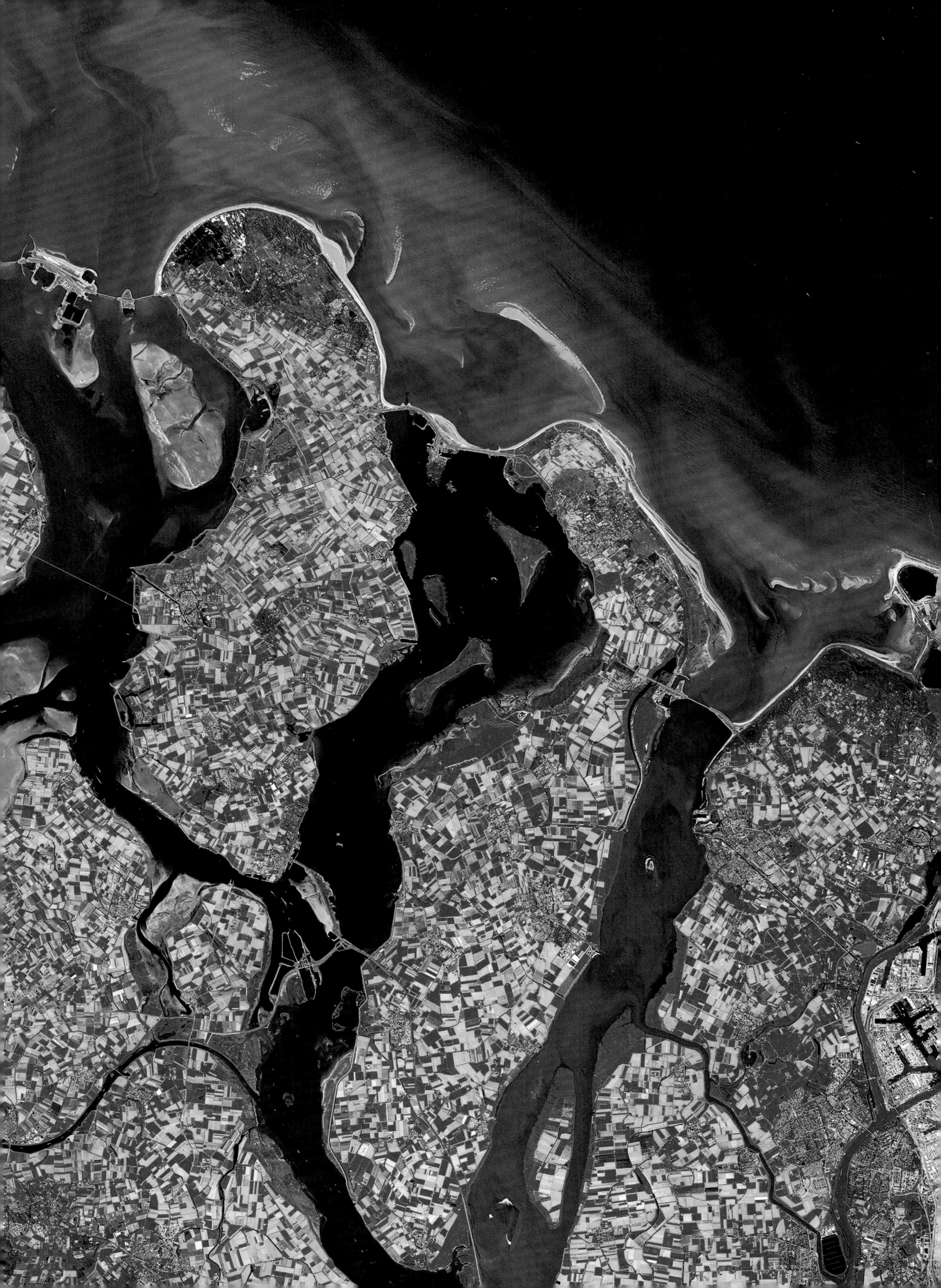

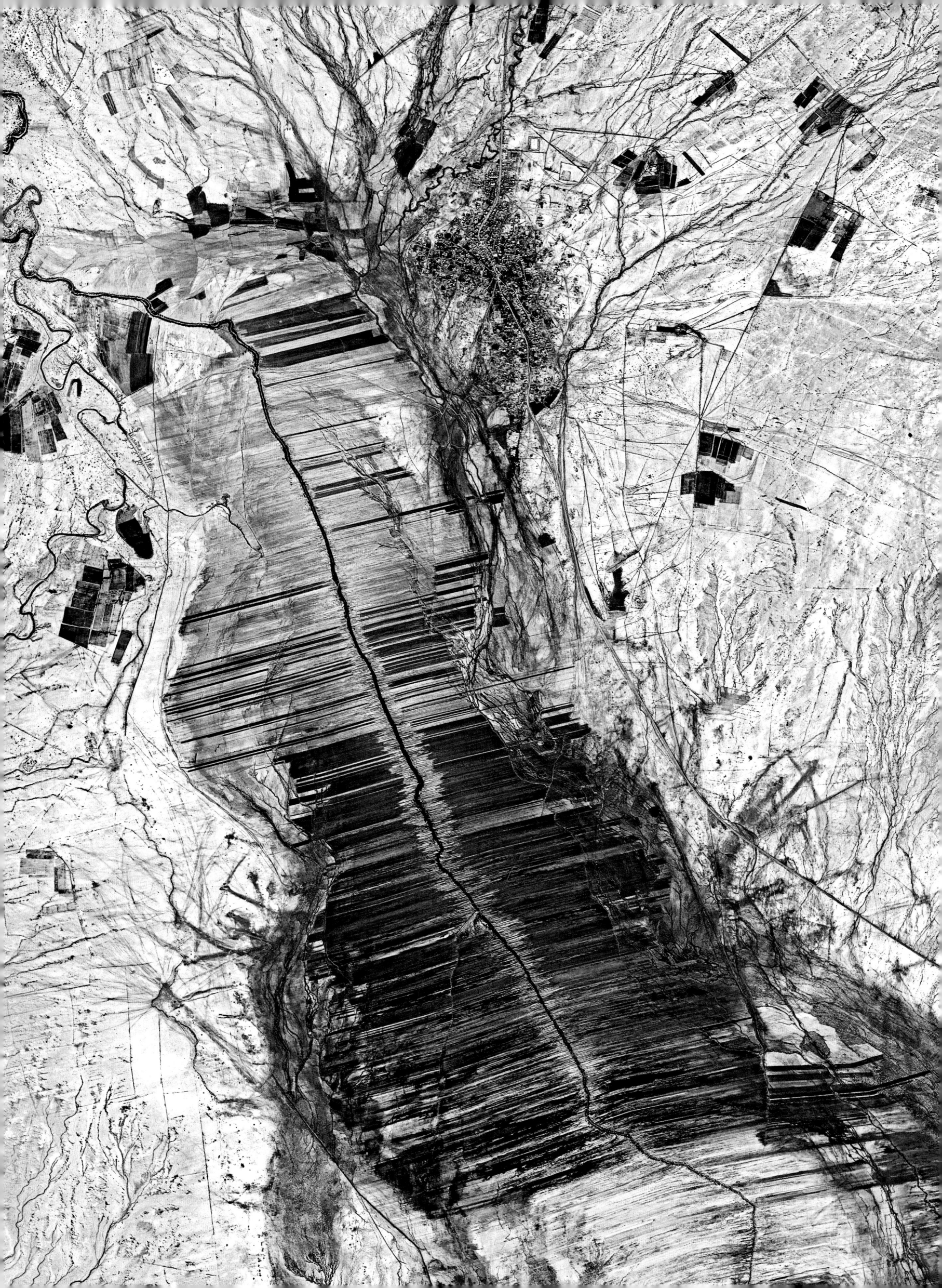

OPPOSITE
ALGERIA
CROPS NEAR BISKRA
34°51' N – 5°42' E

Algeria has tirelessly developed its agricultural production in recent years, particularly
in grains, market gardening, fruit farming,
and viticulture. Agriculture represents
approximately 12 percent of Algeria's
GDP and directly and indirectly sustains
21 percent of its population.

BELOW
SAUDI ARABIA
IRRIGATION NEAR AL-HASA
25°25' N – 49°34' E

These irrigated zones close to the Al-Hasa
oasis have made it possible for humans
to settle here and plant date palms and
lemon trees, among other plants.

PREVENTING CRISES

Satellites are already being used to anticipate famines due
to climate problems, which are among the most tragic forms
of food shortages but are unfortunately only the tip of the
iceberg. Satellites also make it possible to predict the output of farmland for areas as vast as Brazil or China, thereby
anticipating potential shortages that could lead to famines.
Those countries most at risk, notably in sub-Saharan Africa,
are also monitored to identify signs that their agriculture is
failing as early as possible. In 2009, for instance, it was determined one month before harvest time in Kenya that the corn
harvest would yield 15 percent less than estimated due to
drought in the area.

PRECISION AGRICULTURE

But how can the system be improved in the long term? In
the past, farm plots were small and were cultivated based on
the properties and richness of the soil. Today the mechanization of agriculture has transformed these small plots into
huge fields. In order to work these pieces of land as vast as
they are heterogeneous, farmers have gotten in the habit of
using uniform methods that are most often poorly adapted
to the land as a whole.

    Over the last several years, however, farmers have
returned to a more targeted, precise agriculture, using satellite images to manage their fields. These images provide a
wide variety of data on their farm plots, including the crops'
chlorophyll content, rate of growth, and maturity as well as
the soil's temperature, mineral content, and rate of humidity. Combined with meteorological data and measurements
on the ground, the information gathered can contribute
to adopting the appropriate agricultural methods. These
include only spraying areas that require it, intervening with
emergency procedures at the right time, and saving on fertilizer, pesticide, and phytosanitary products. These measures
help farmers financially but also benefit the environment:
The less chemical product is used, the less winds up in the
crops, the soil, and the water table. Satellites serve as tools
to make decisions for producing more and better crops.

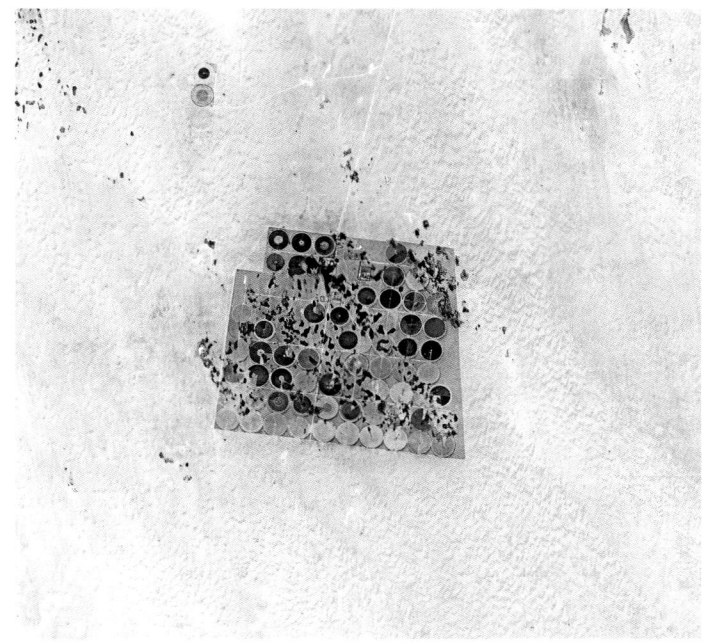

In Africa, for example, SATELLITE IMAGERY is used
to **LOCATE AREAS PARTICULARLY FAVORABLE**
to **CRICKET REPRODUCTION.**

FARMERS IN DEVELOPING COUNTRIES

Nonetheless, intensive grain farming is certainly not an environmentally responsible model for the planet. Additionally, intensive methods are not available everywhere and are not appropriate for the small farmers who make up the majority of agriculturists, particularly in what we refer to as developing countries.

Farmers in the developing world often do not have access to mechanization and have far lower yields than those in the developed world. Yet these farmers have significant room for growth. For them, even a modest improvement in production means a substantial growth in income, which allows them to escape hunger and poverty, send their children to school, and access health care.

Farmers in the developing world can also benefit from services provided by satellites. In some cases they already are. In Africa, for example, satellite imagery is used to locate areas particularly favorable to cricket reproduction, as well as corridors likely to be used by crickets to travel from one area to another. This monitoring makes it possible to anticipate potential infestations and to protect exposed crops. Additionally, a growing number of farmers, notably in Africa, are equipped with cell phones provided by NGOs and governments, on which they receive real-time weather forecasts from satellite data and news about markets to enable them to better conduct their business.

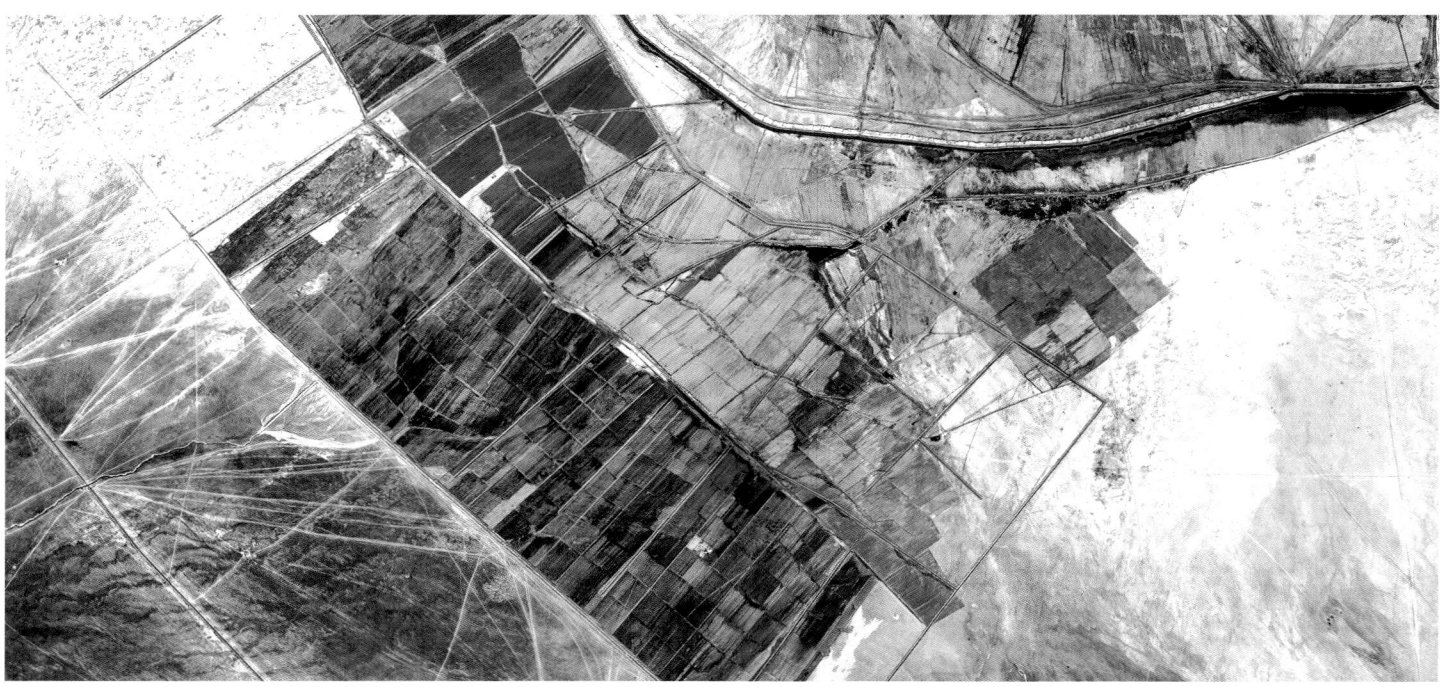

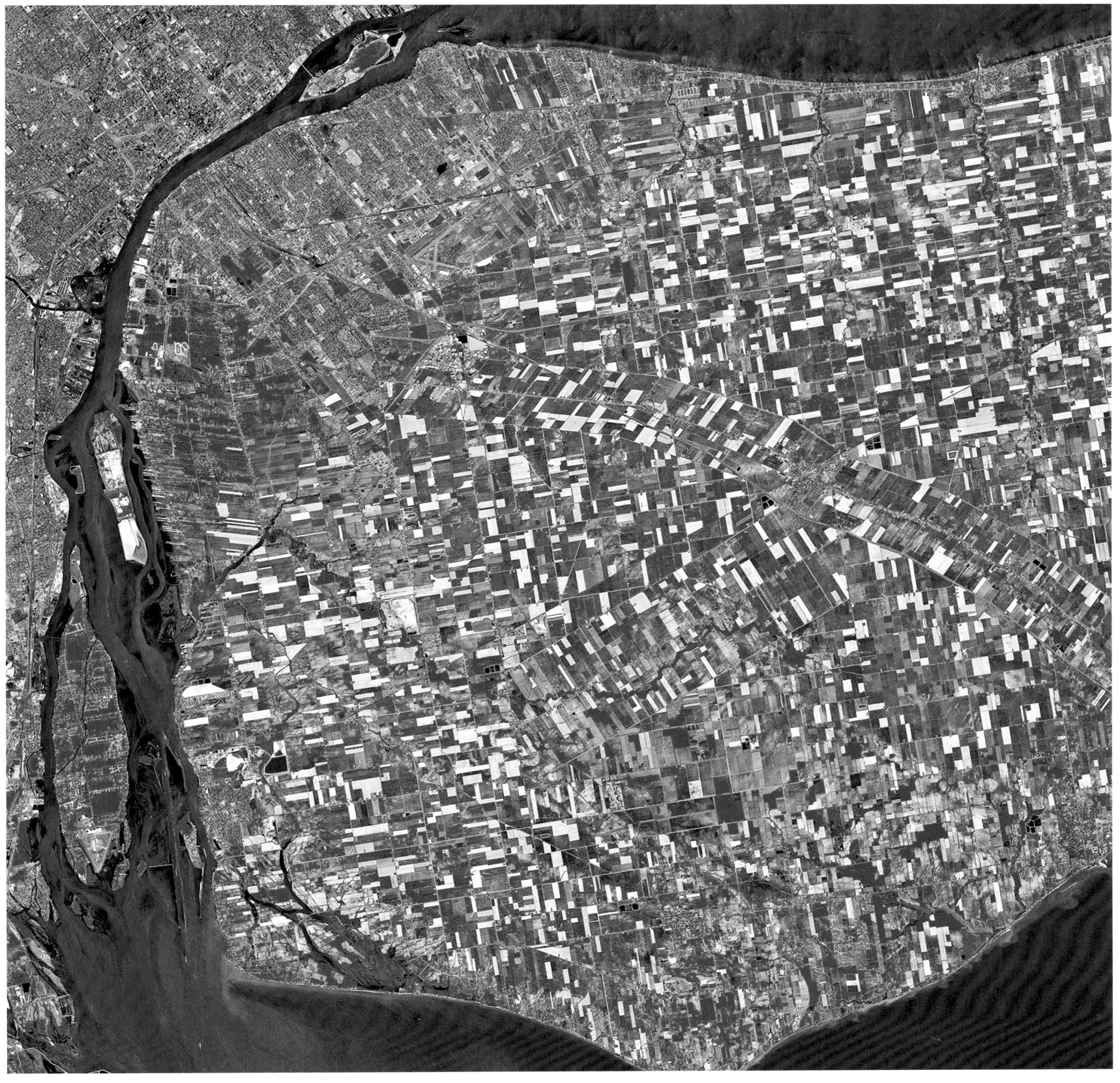

OPPOSITE

TURKMENISTAN
TEJEN OASIS
37°30' N – 60°30' E

Two remarkable watercourses irrigate the cotton fields of the Tejen oasis in the heart of the Karakum Desert, which stretches over more than half of Turkmenistan. Originating in Afghanistan, the Tejen River, or Hari Rûd, flows for 683 miles (1,100 km) and drains into the oasis. The Karakum Canal, the longest irrigation canal in the world, connects the Amu Darya to the Caspian Sea over a distance of 854 miles (1,375 km). Completed in 1988, it made it possible to cultivate large areas of dry land, as can be seen through satellite observation.

ABOVE

UNITED STATES, MICHIGAN
DETROIT
42°18' N – 83°03' W

Standing between Lakes Huron and Erie, the city of Detroit is at the American border with Canada. In this image, the city's grid system can be seen to the northwest across the strait from the agricultural zone near Windsor, Canada's southernmost city. Space here is divided into large squares known as townships, which are allotments inspired by the farms founded in the area by French settlers, starting in 1749.

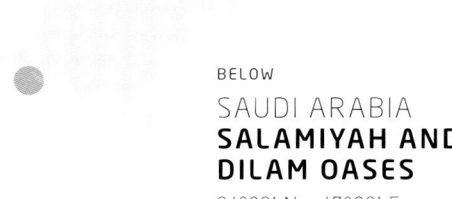

BELOW

SAUDI ARABIA
**SALAMIYAH AND
DILAM OASES**
24°03' N – 47°09' E

Approximately 62 miles (100 km) southeast of Riyadh, the Salamiyah and Dilam oases combine traditional and modern irrigation methods. The remnants of former palm groves are visible, concentrated right next to the road, surrounded by the circles typical of aspersion irrigation zones: Pivots with ranges of approximately 650 feet (200 m) provide water to grain plots over an area of approximately 15 square miles (40 sq km).

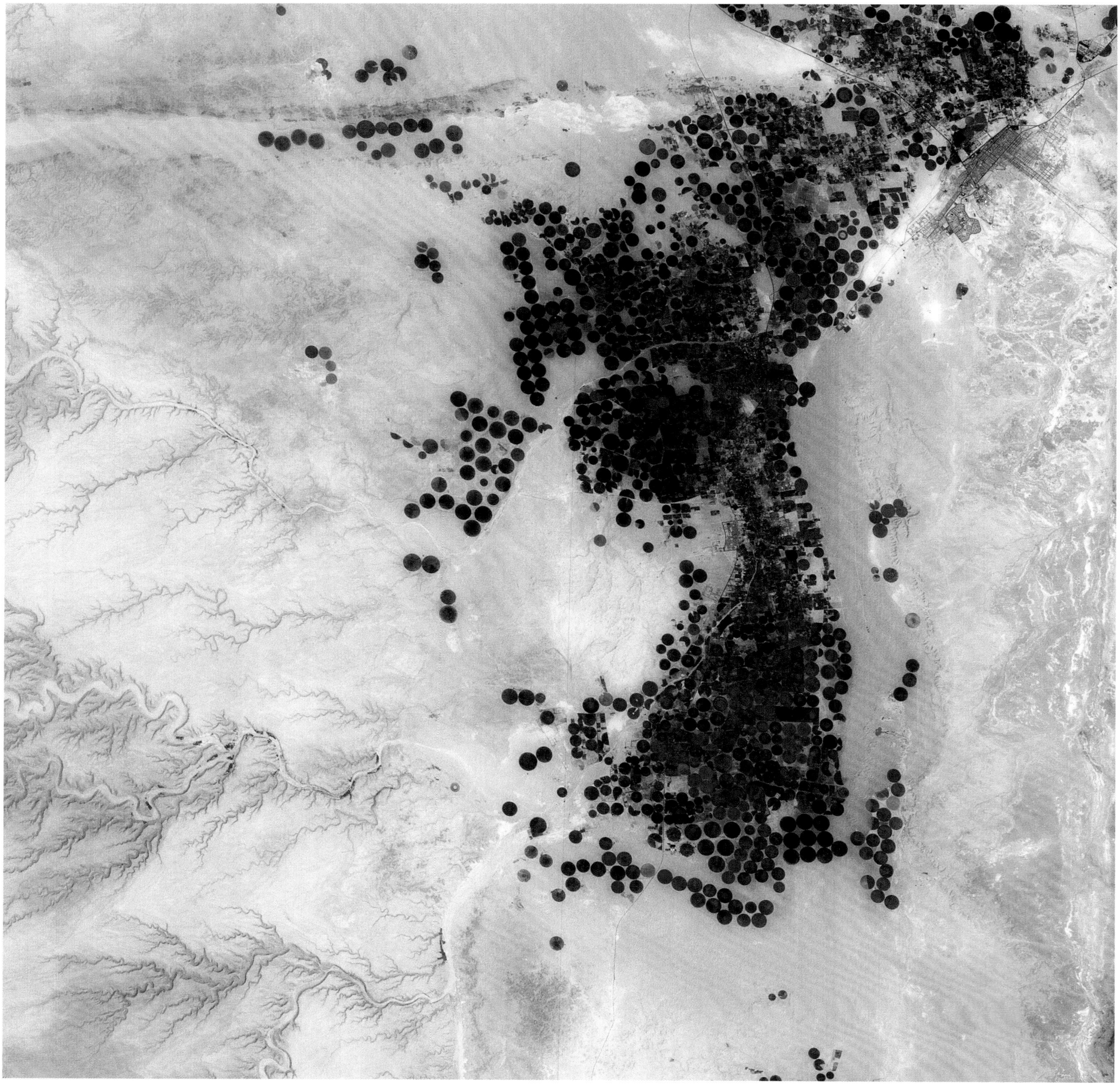

Each year, A THIRD OF THE FOOD produced for
human consumption around the world—

MORE THAN ONE BILLION TONS OF FOOD—

IS LOST OR WASTED.

EXPANDING AGRICULTURAL AREAS

Aside from the improvement of yields, the other major scenario being explored to increase global food production is the expansion of agricultural areas. This is an extremely controversial solution. During the second half of the twentieth century, cultivated land expanded significantly, often to the detriment of forests. Three plants—the soybean, oil palm, and sugarcane—are single-handedly responsible for most of the deforestation in the world. Given what we know about forests' role in climate regulation and the incredible biodiversity forests hold, along with the numerous services they provide, it is undeniable that acute encroachment on these ecosystems is a problem. We need to search for new arable land elsewhere, but it is increasingly rare.

In recent years, the new phenomenon of land-grabbing has developed on a scale as vast as it is problematic. In Asia and the Middle East, but also in North America, states or companies buy arable land in developing countries. Though the figures remain hard to assemble, it is estimated that several tens of millions of acres are concerned, principally in Ethiopia, Mozambique, and Sudan, which are countries with vast but underexploited land resources, cheap labor, and lax regulations.

This practice is poorly supervised from a legal perspective and has been criticized by some NGOs for being carried out to the detriment of local populations. Satellites do not currently appear to be used for land-grabbing. At most, they may serve to analyze the cultivated areas to determine which plants are best suited to grow there.

EATING DIFFERENTLY

Whether they are used for increasing yields or discovering new farmable land, satellites are powerless against another scourge: food waste. Each year, a third of the food produced for human consumption around the world—more than one billion tons of food—is lost or wasted. And though nearly as much food is wasted in developing countries (630 million tons per year) as in industrialized countries (670 million tons per year), the causes are significantly different. In developing countries, food is mostly lost in the fields and during crop transportation, while most of the waste in developed countries takes place between stores and the consumer's plate. Reducing these losses could have a significant impact on global food security.

A second possibility for feeding the world without damaging the planet's environmental balance would be to reconsider the Western diet. While it is considered a model by the populations of developing countries, this diet is linked to certain pathologies (particularly cardiovascular disease and obesity). And it cannot be adopted by the entire planet for it puts too much pressure on natural resources.

Reducing the proportion of meat and processed foods in our diet would significantly reduce food's ecological footprint on the planet.

If we simply changed our eating habits and reduced waste, we would no longer need to increase our food production by 70 percent by 2050, but by a third of that—which is still no small challenge.

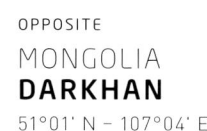
In countries struggling with hunger, the solution is not to import food but to support local small farmers.

Talking with Olivier De Schutter

Hunger Is a Political Problem

What is the situation with hunger in the world today?

In 1992, 840 million people around the world were unable to feed themselves properly. Today there are approximately 925 million people in this situation, about 98 percent of whom live in developing countries. This data should probably be interpreted in relation to demographic growth, which remains strong—every year, the global population grows by 75 million people. But in 2012, a whole 16 percent of the population of developing countries went hungry, a proportion that has only dropped slightly from 20 percent in 1990 and falls far short of the first of the Millennium Development Goals set in 2000. Moreover, in at least one region, sub-Saharan Africa, the proportion of people suffering from hunger has grown, not decreased: Today 30 percent of the region's population lack sufficient caloric intake.

Why haven't we resolved the problem of hunger, in your opinion?

Hunger's causes are primarily political ones. In most developing countries, there was not enough investment in small family agriculture, which is destined to feed local communities. Rural poverty has increased. Millions of small farmers have migrated to cities, while the service and industrial sectors did not grow enough to absorb this workforce: Today 1.3 billion people live in shantytowns on the edges of major cities in developing countries. Based on international institutions' advice, governments have often responded by importing food, which further accelerated the collapse of local networks. This is a vicious circle that is hard to escape.

What measures do you recommend?

We need to massively reorient our efforts in two directions. The first is to rebuild local food production systems, which is the only way to break the current competition between export agriculture and local agriculture—which harms the most marginalized farmers. The second is the spread of agri-environmental methods, in order to separate agricultural production from fossil fuels and reduce agriculture's ecological footprint. These transitions will only be possible if our mechanisms of governance are improved. We are hostages of short-term thinking that affects both political decisions and the market. We have to get out of it by focusing on the long term when making political decisions. We cannot afford the luxury of waiting: Tomorrow is on the line today, because by tomorrow it will be too late.

Olivier De Schutter is the UN Special Rapporteur on the Right to Food. He is also a professor of international law at the Université catholique de Louvain and at the College of Europe.

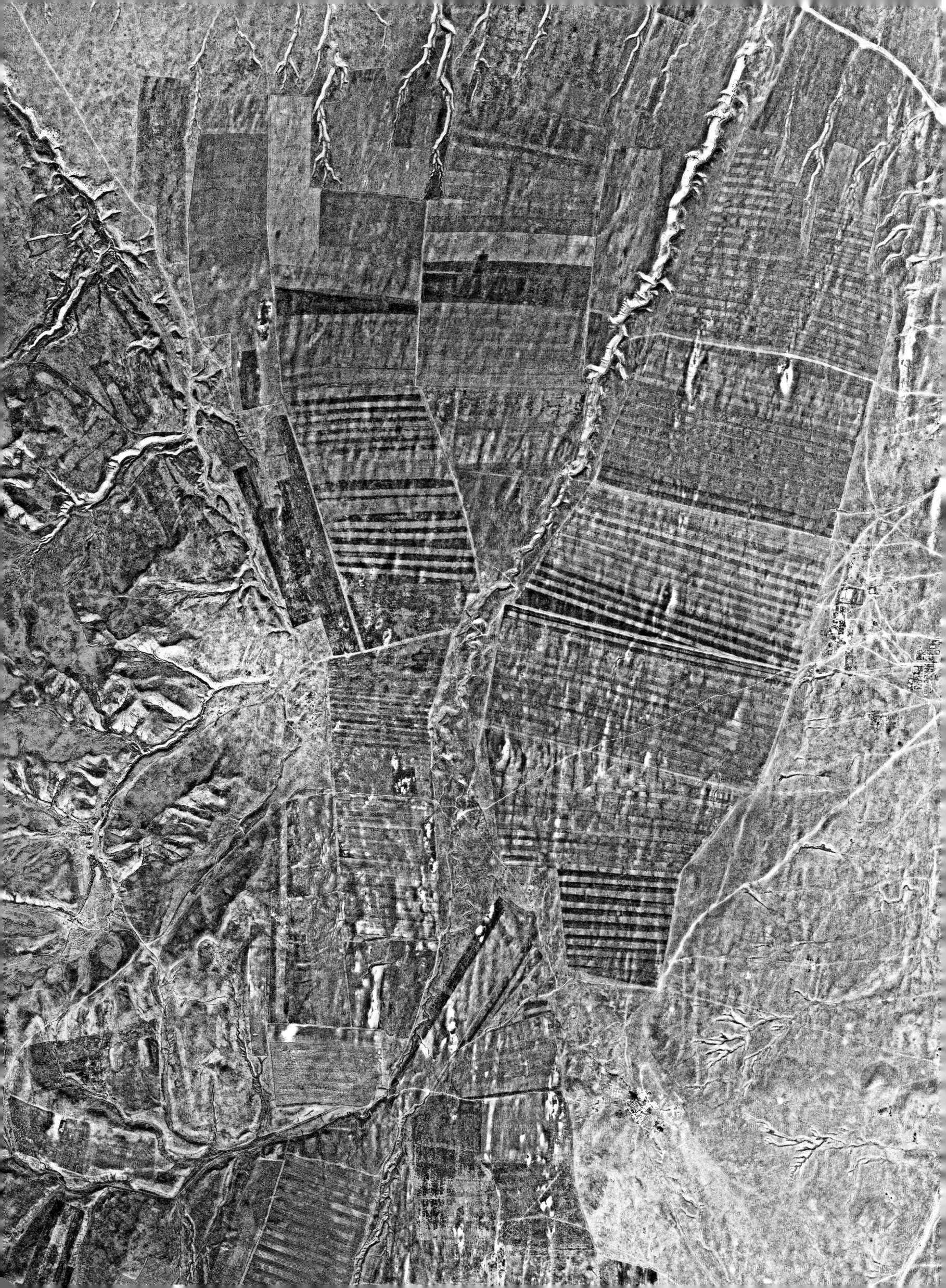

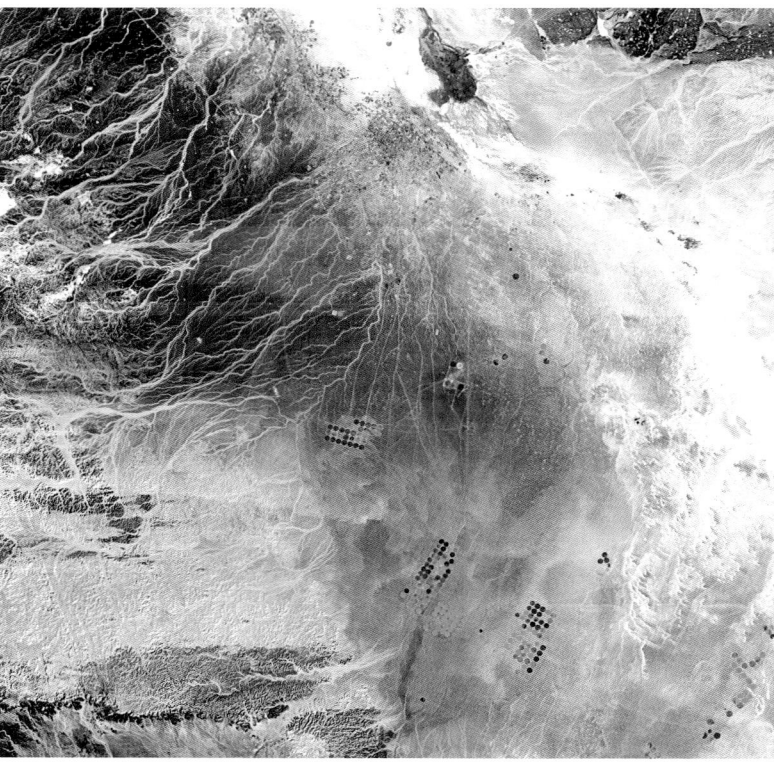

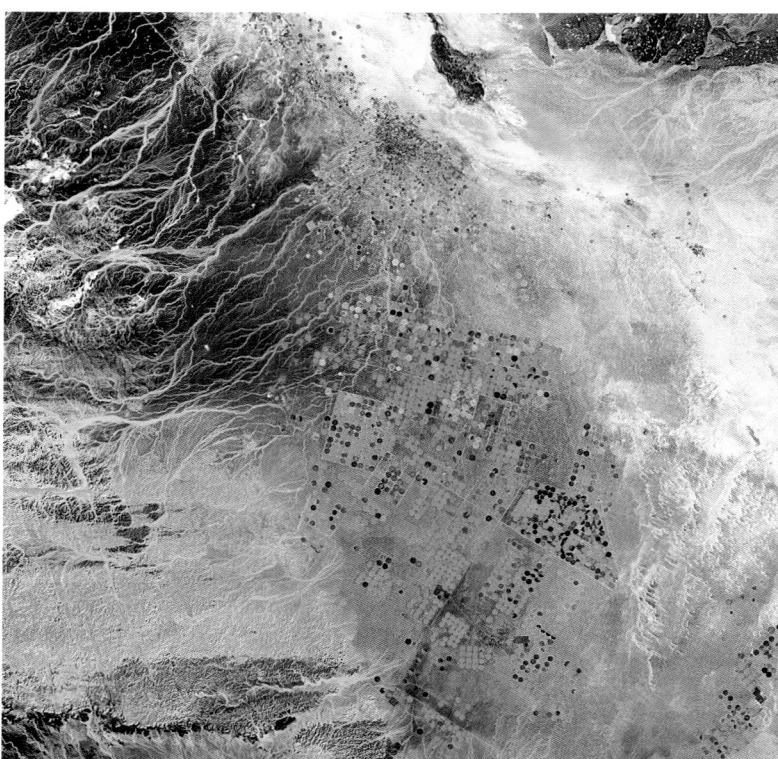

1987

1991

2000

2012

SAUDI ARABIA
WADI AS-SIRHAN BASIN
30°03' N – 38°25' E

Over the last thirty years, Saudi Arabia has drilled its soil in search of a resource far more precious than oil: water. The Wadi as-Sirhan Basin's engineers and farmers have used little-known reservoirs to farm grains, fruits, and vegetables in the Saudi desert. This artificially colored series of images taken in 1987, 1991, 2000, and 2012 shows farming's evolution in the basin. Recently planted surfaces are in green, dry vegetation and fallow land are rust-colored, and dry and bare surfaces (principally desert) are pink and yellow. No one knows how much water lies under the desert; estimates vary from 60 to 208 cubic miles (252 to 870 cu km). Hydrologists have stated that, from an economic perspective, it will only be realistic to pump this resource for a period of fifty years.

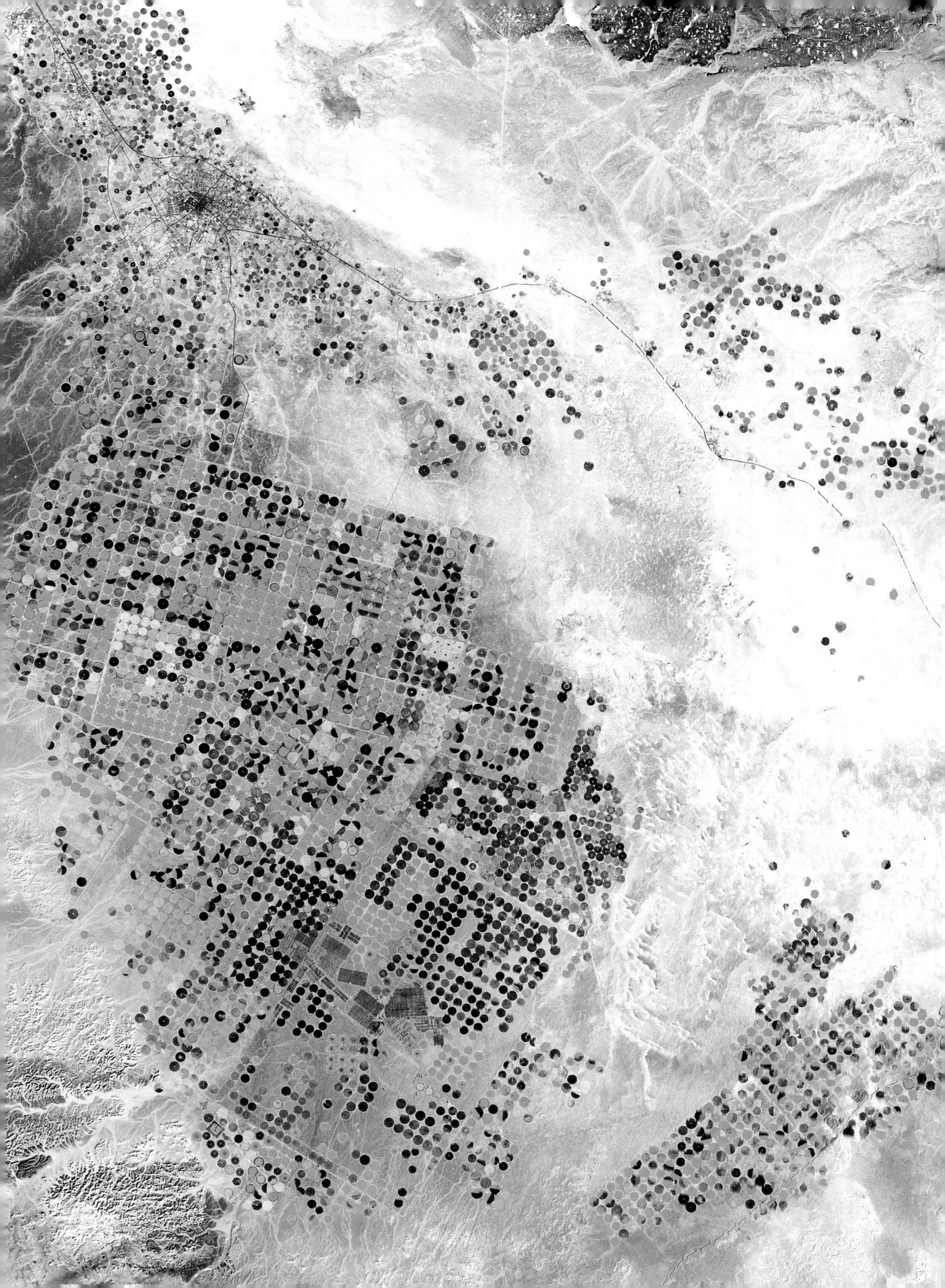

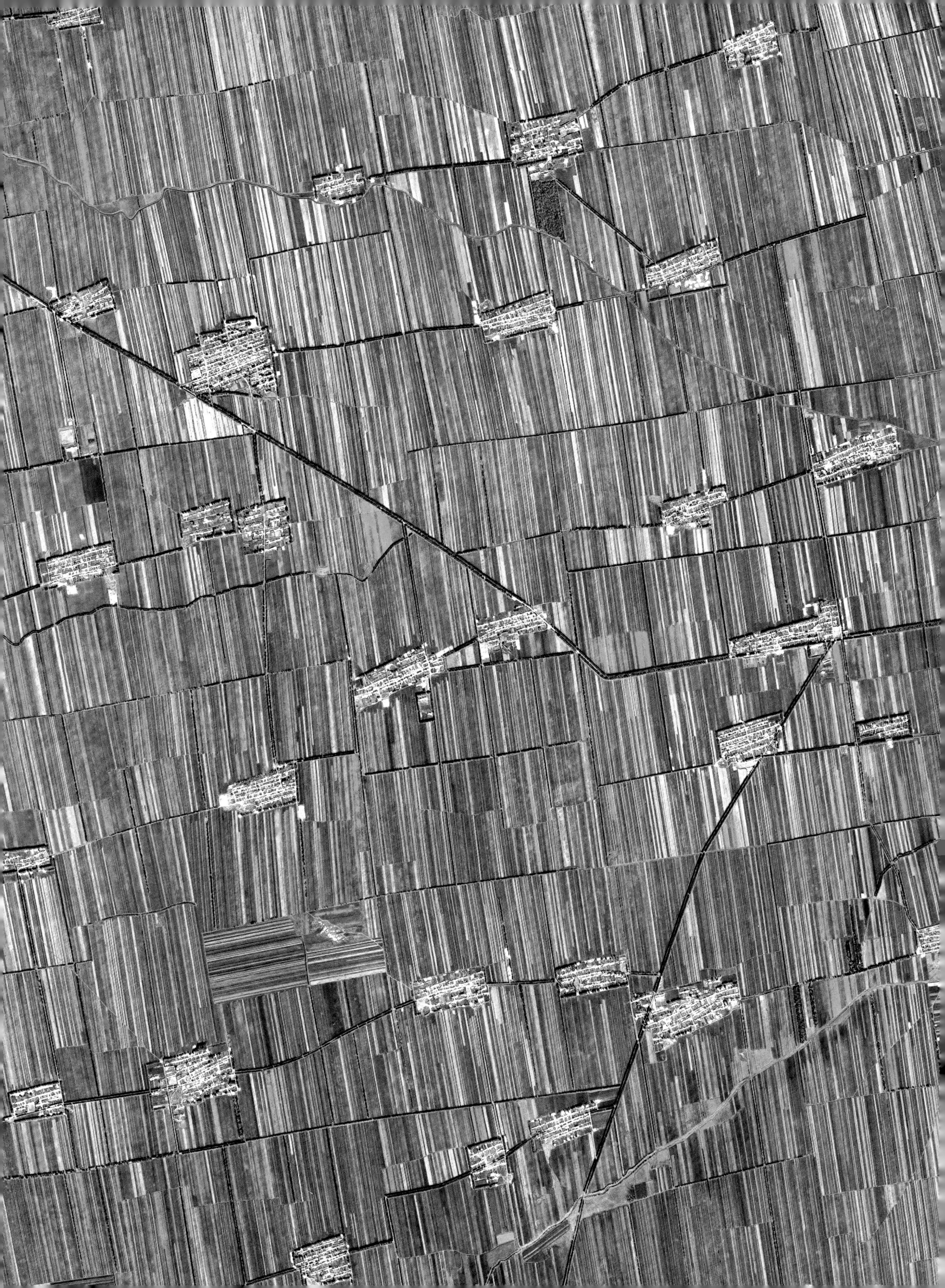

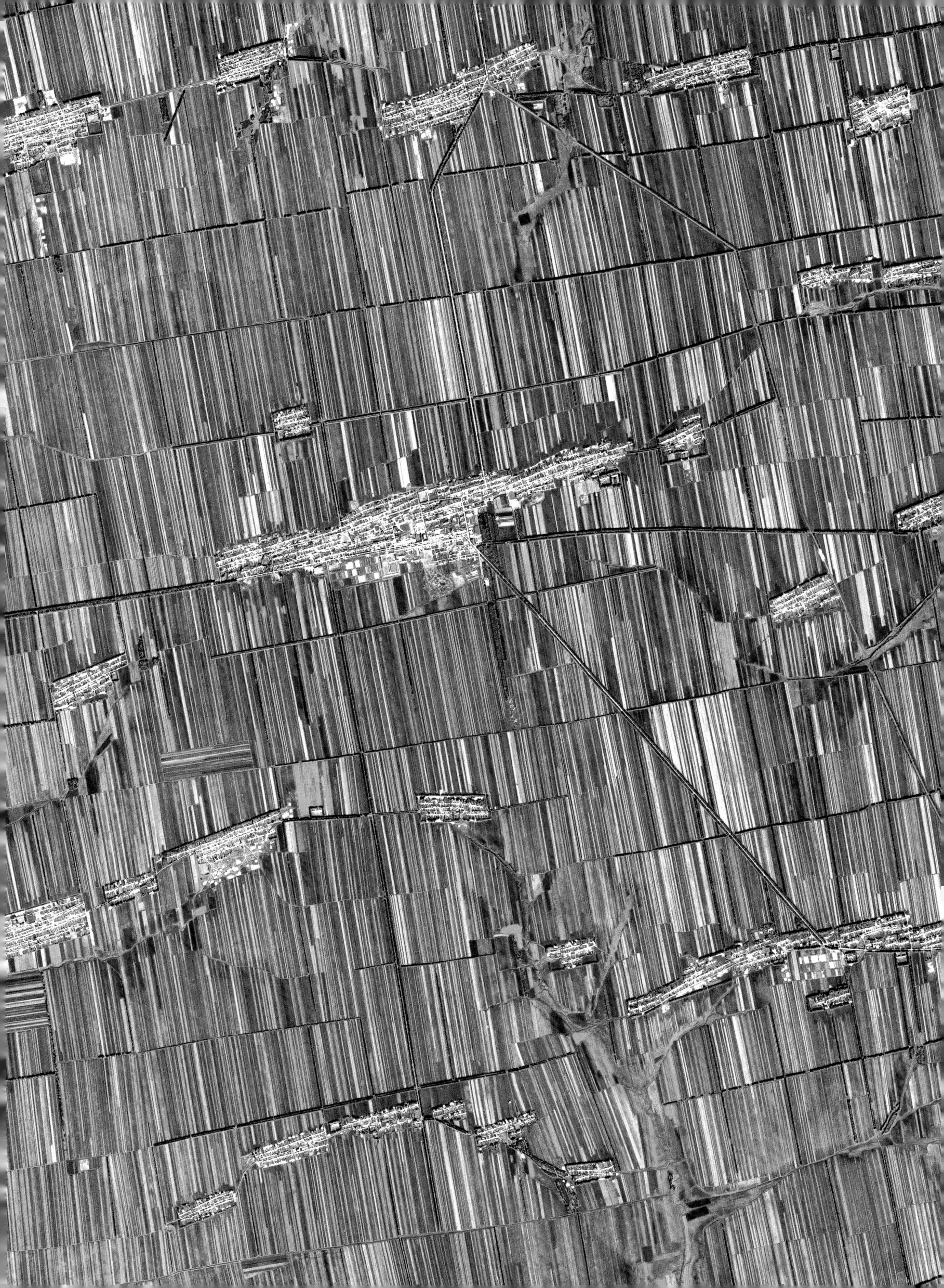

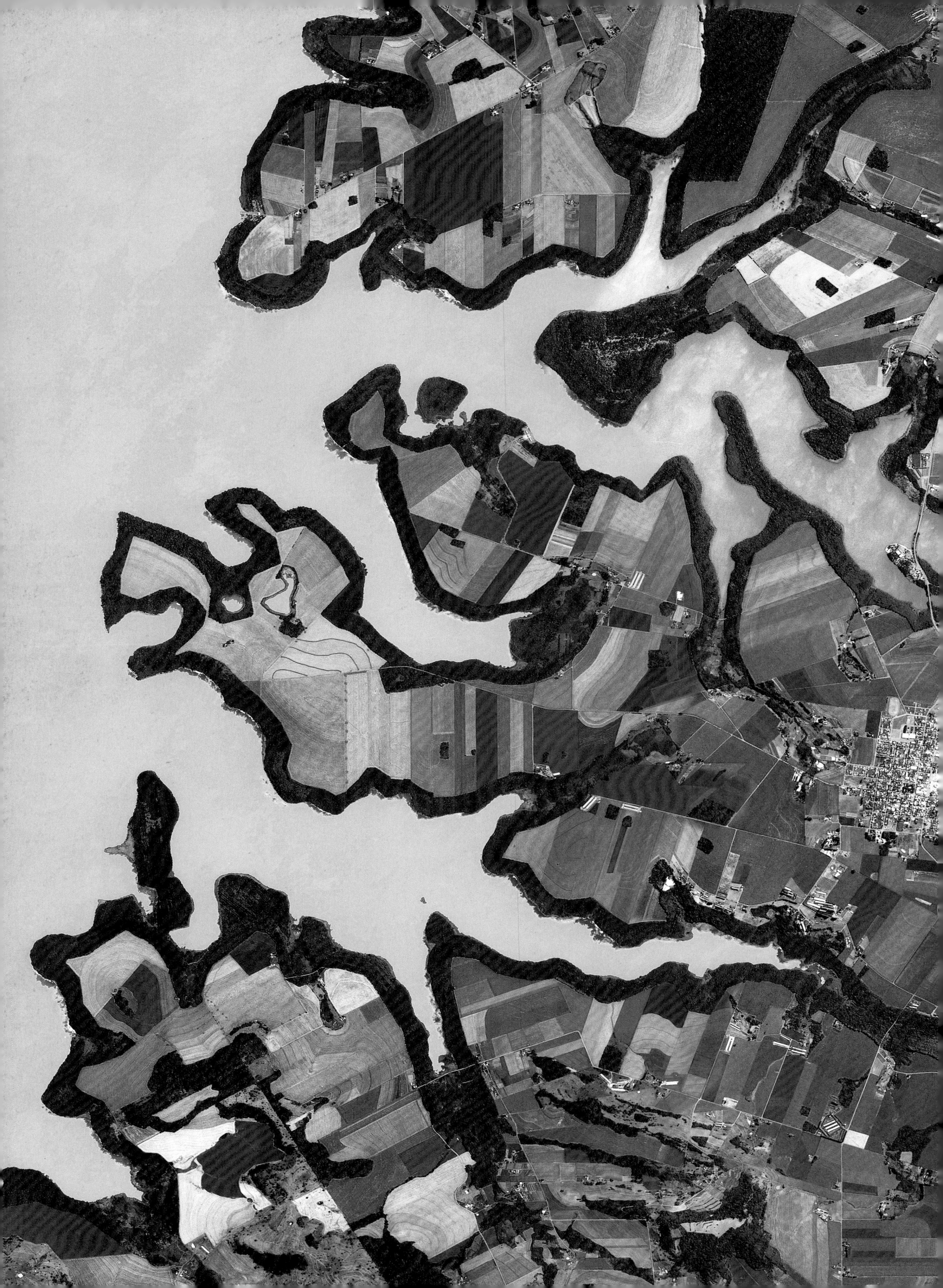

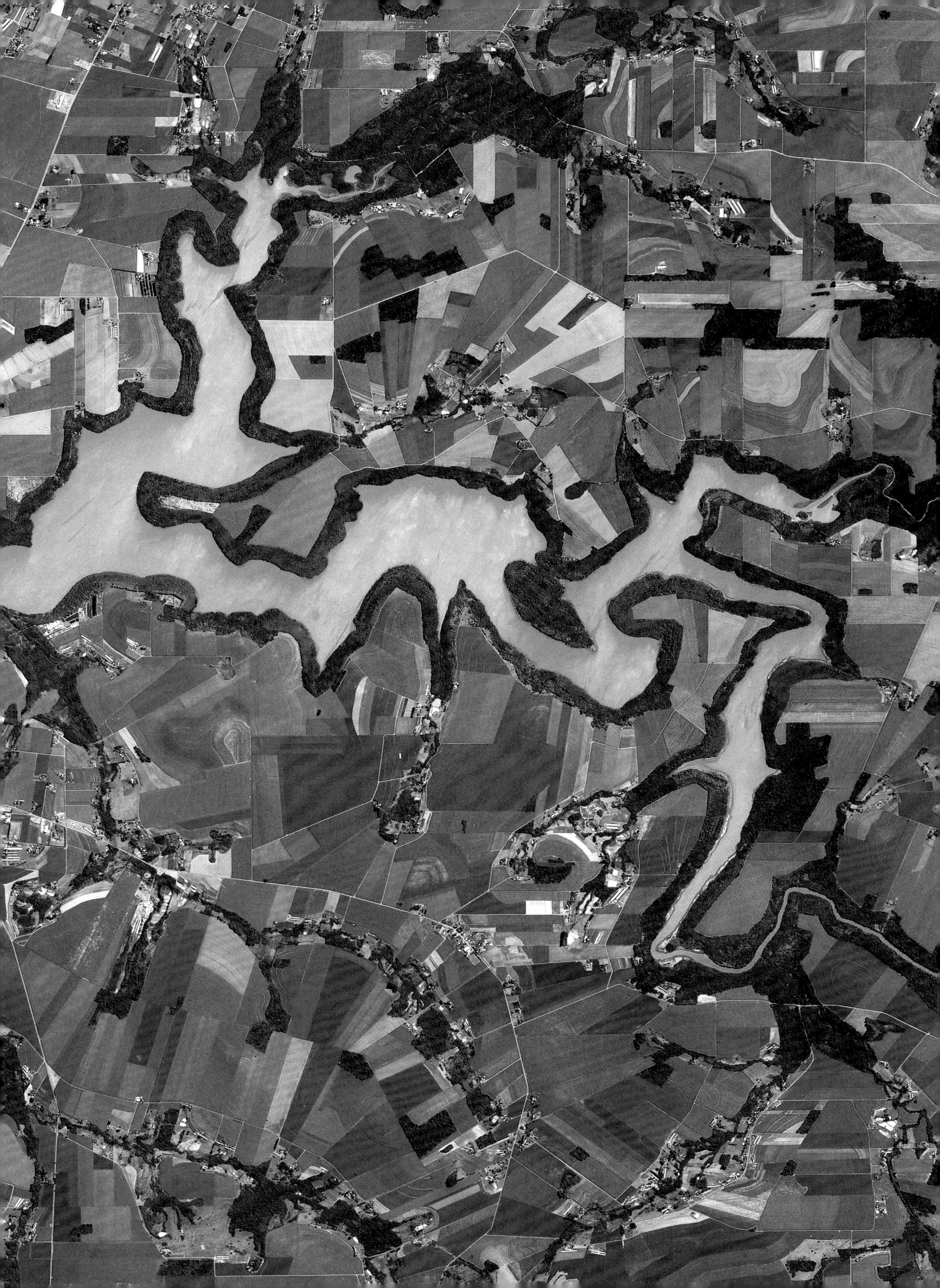

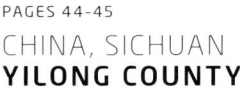

PAGES 44-45
CHINA, SICHUAN
YILONG COUNTY
31°31' N – 106°24' E

Located in China's center, Yilong County is dependent on agriculture, the pillar of the local economy: More than 80 percent of the population in this rural area practices traditional agriculture. The principal agricultural products are rice, oranges, silkworms, and pork.

PAGES 46-47
BRAZIL, PARANÁ
ITAIPU DAM RESERVOIR
25°20' S – 54°30' W

The reservoir of the Itaipu Dam, some of which can be seen here, extends over more than 520 square miles (1,350 sq km) along the border of Paraguay and Brazil. Its immediate surroundings are protected, but the rest of the area is devoted to agriculture.

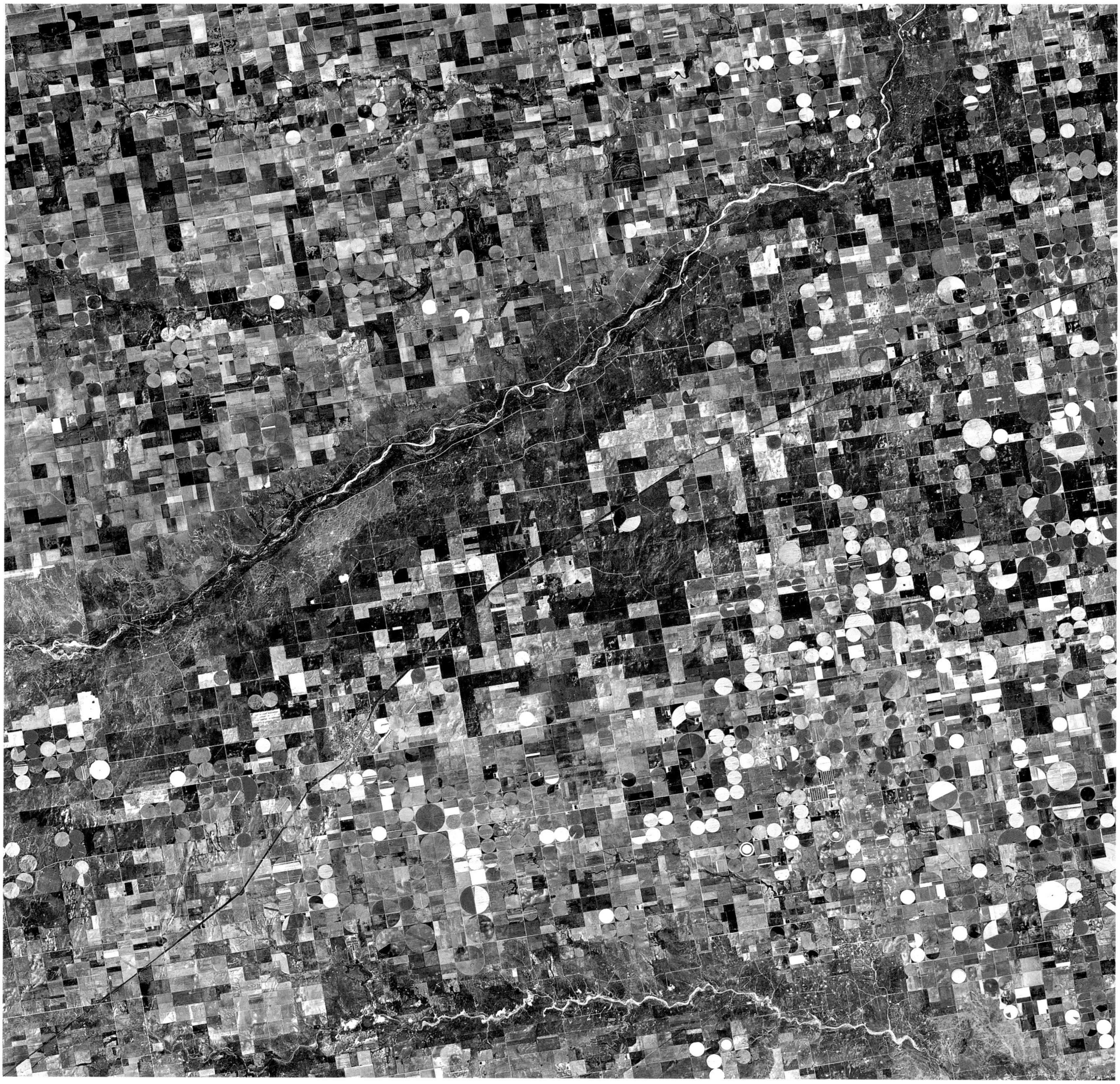

OPPOSITE
UNITED STATES, KANSAS
MORTON COUNTY
37°09' N – 101°45' W

Kansas is part of America's agricultural heartland. The agriculture sector generates $80 billion in this state alone. Most of the farming here is aided by center pivot irrigation (hence the circles in the picture), producing wheat, corn, sorghum, and cotton.

BELOW
PAKISTAN
LAKE MANCHAR
26°30' N – 67°37' E

Lake Manchar is the biggest freshwater lake in Pakistan. The lake's expanse varies according to the season, growing from 135 to 200 square miles (350 to 520 sq km). The red areas, concentrated to the east of the central canal, are farmland.

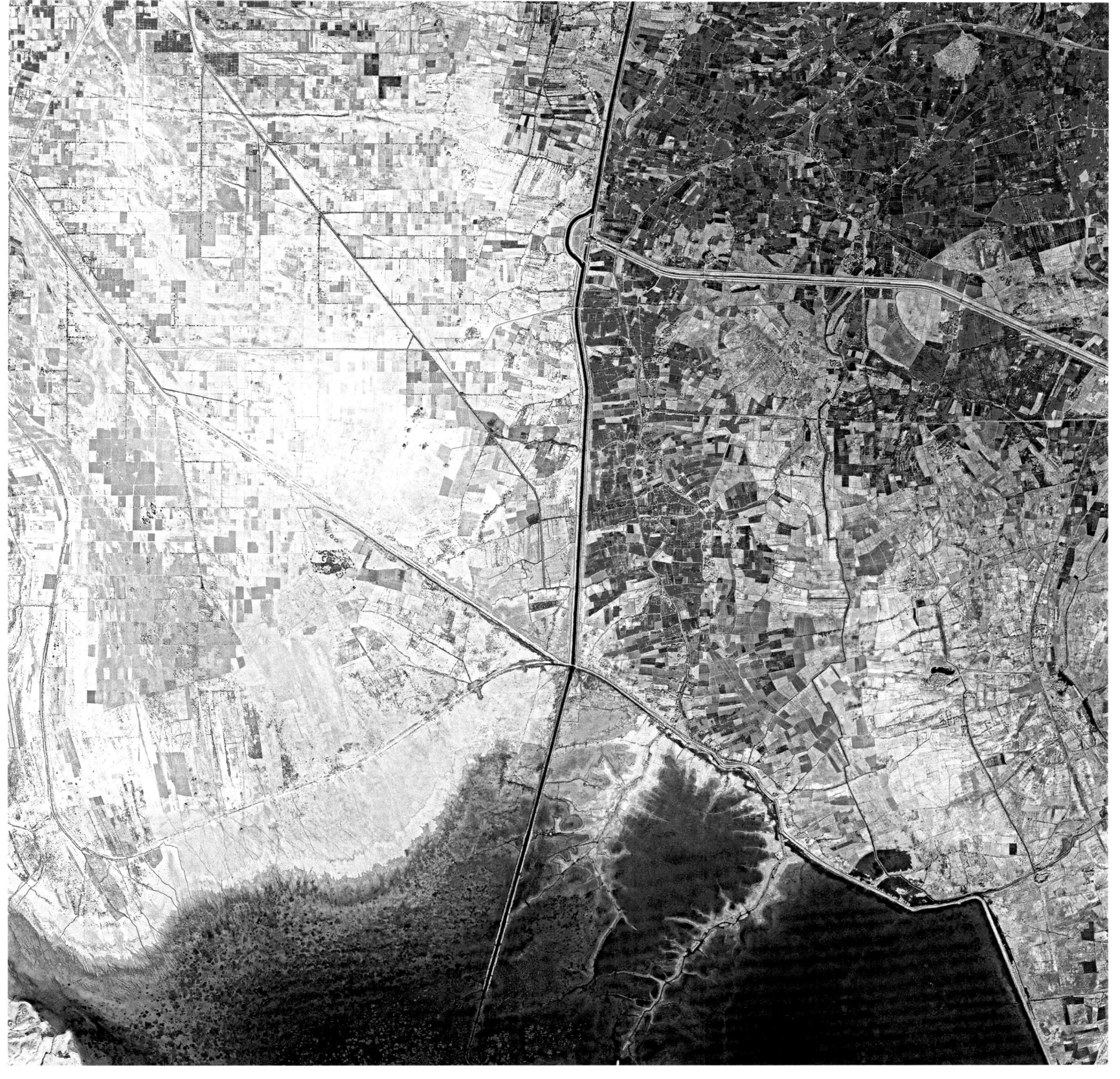

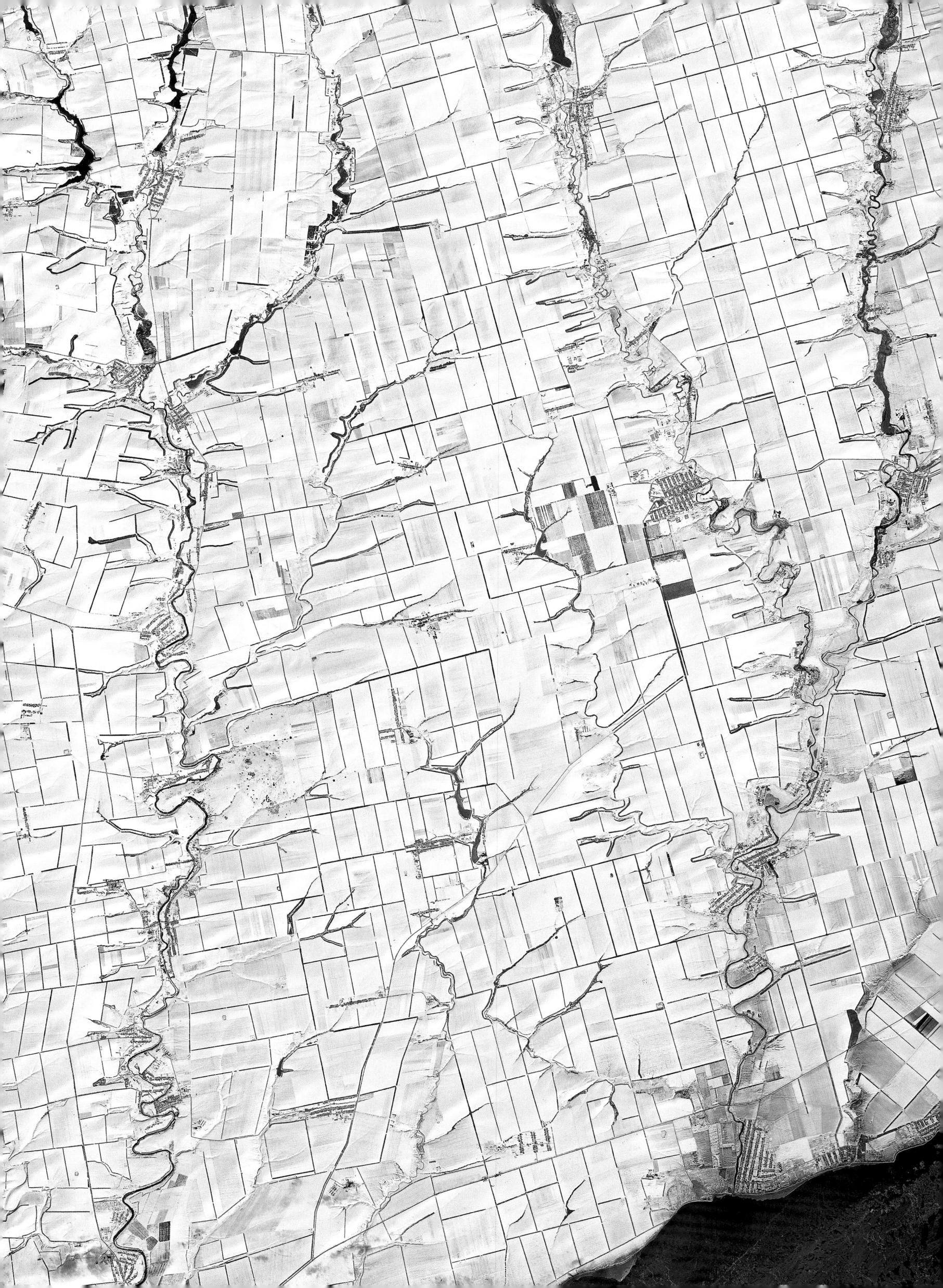

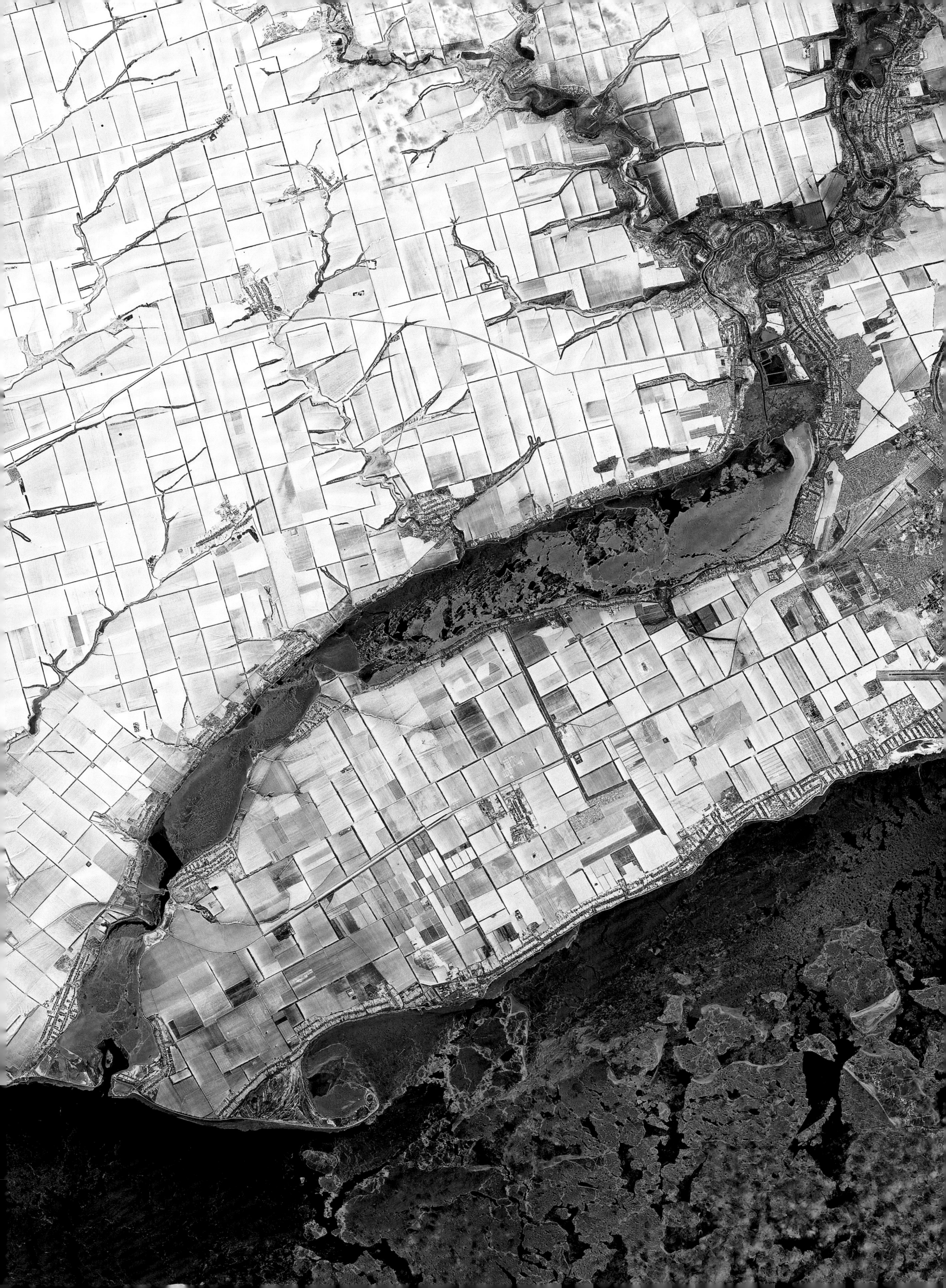

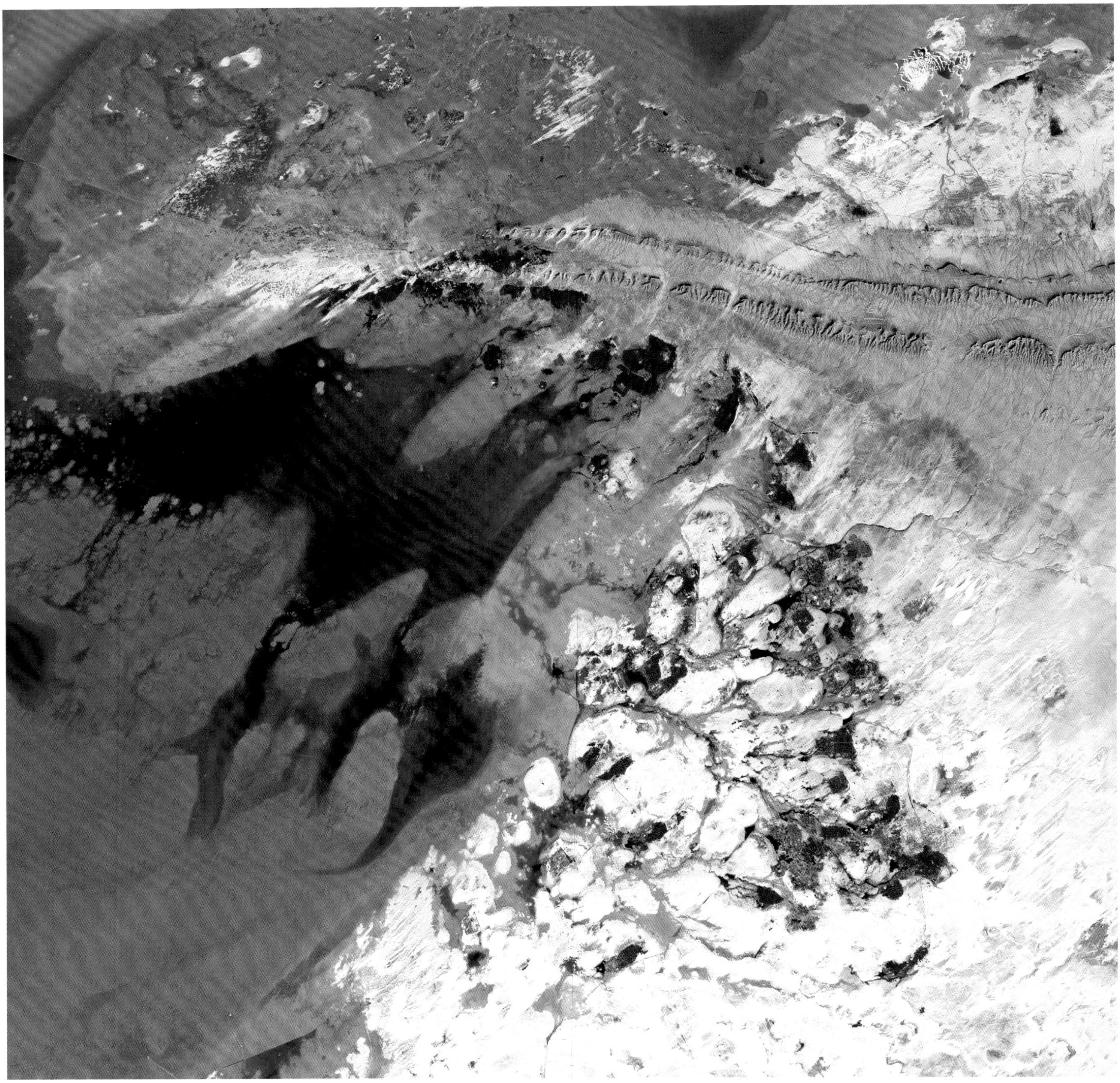

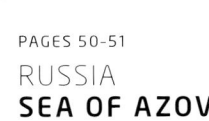

PAGES 50-51

RUSSIA
SEA OF AZOV
47°10' N – 38°42' E

West of the Mius River, old irrigation canals built in the thirteenth century are joined with modest water-courses broken by a succession of small Soviet-era dams. To the east, the river flows into the Sea of Azov. Chernozem, a type of black soil that is among the most fertile in the world, covers eastern Europe's bread basket near the Russo-Ukrainian border. This image shows the area in winter covered in snow.

ABOVE
TUNISIA
CHOTT EL JERID
33°33' N – 8°51' E

South of Chott El Jerid, the humid saline crust appears in black and brown. This is the beginning of the Sahara: The steppe, which extends until Tozeur, gives way to the desert. The shift from white to gray indicates variation in the sand's thickness. The sandy expanse is punctuated by many oases, including the Douz oasis in the center of the image. The palm groves, which receive water from springs, are shown in dark red. The color's intensity is representative of the extent of plant activity in oases, and underscores the contrast between irrigated surfaces and dunes.

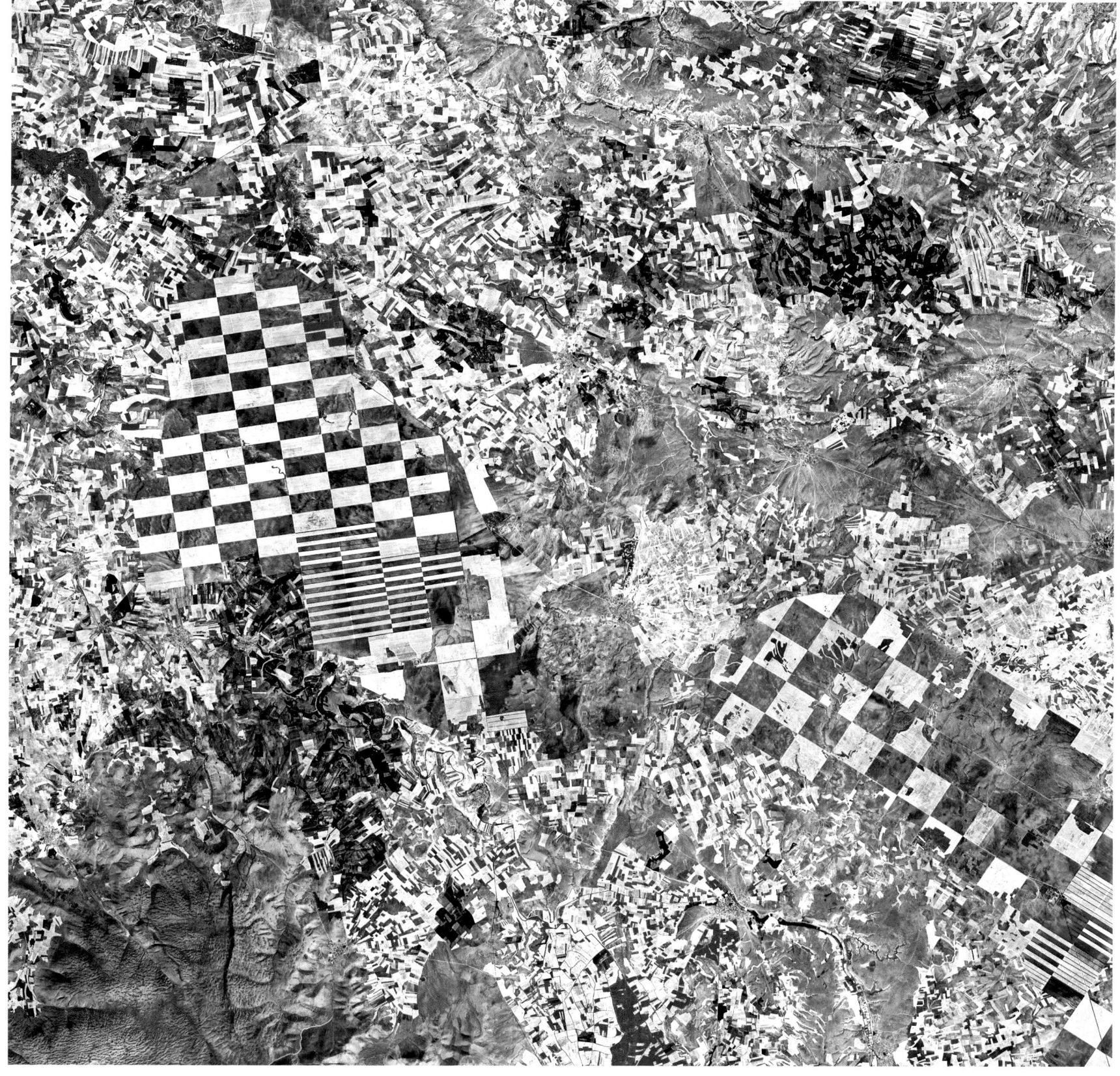

ABOVE

TURKEY
**ANATOLIAN PLATEAU NORTH
OF KONYA**
9°42' N – 34°45' E

Central Anatolia is a large semi-arid plateau
receiving less than 15 inches (400 mm) of precipitation a year. Fields are burned to promote
new growth. Alternating fires on large regularly
shaped plots creates a checkerboard effect, while
traditional slash-and-burn techniques used on
several plots leave irregular black spots.

PAGES 54-55

CHINA
CHONGMING ISLAND
31°39' N – 121°30' E

The alluvial island of Chongming, the third-largest
island in China, is located at the mouth of the Yangtze
River close to Shanghai. While the region was the most
rural zone in Shanghai's surroundings, it is rapidly
becoming urbanized, after the construction of a bridge
and tunnel connecting the island to Shanghai.

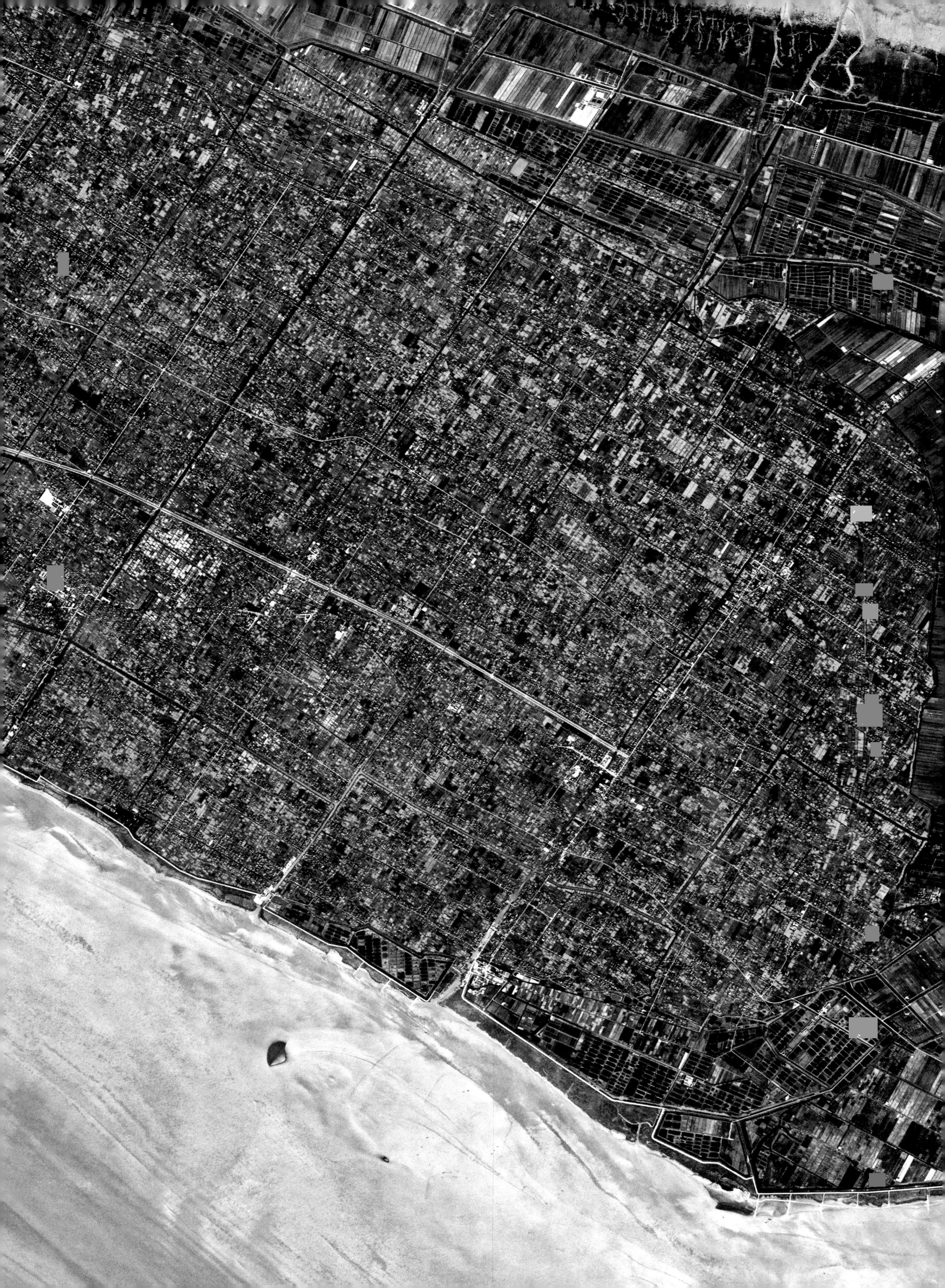

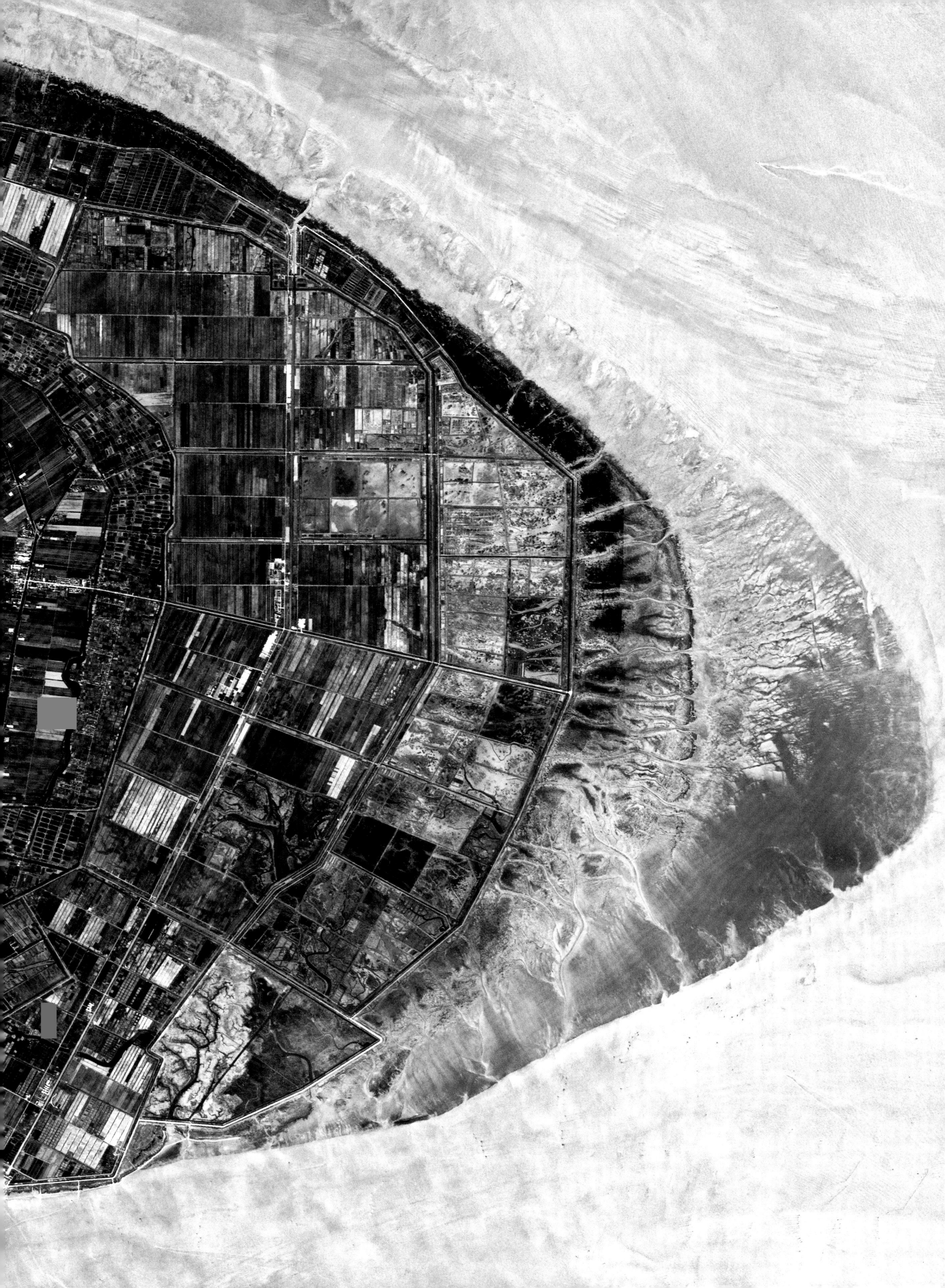

For many people, climate change remains an intangible notion, a distant debate among experts, which stokes doubts about whether this major upheaval is truly happening. Like Saint Thomas, many only believe what they can see. But satellite images do not make guesses or have prejudices; they record facts. Additionally, satellites make it possible to study parts of the world that are difficult to access and to measure developments on a regular basis.

SEEING CLIMATE CHANGE

While the dwindling of terrestrial ice observed in recent decades is measurable, how can one quantify the volume of ice in motion, most of which is inaccessible to man? At altitudes of several hundred, if not several thousand, miles, satellites are equipped to photograph these regions at regular intervals. Their findings are incontestable: The glaciers are shrinking. Or at least the vast majority of them (more than 90 percent), as well as the Greenland and the West Antarctic ice caps.

The images provided by the SPOT program (*Satellite Pour l'Observation de la Terre*, or Satellite for Earth Observation) have revealed significant acceleration in the melting of the Mont Blanc glacier: It shrunk by 3 feet (1 m) a year from 1979 to 1994, then by 10 feet (3 m) a year from 1994 to 2000, and 13 feet (4 m) a year from 2000 to 2003. The same is true of the glaciers in the Andes, which are thought to have shrunk by an average of 30 to 50 percent since the 1970s. As for the Himalayan glaciers, they are diminishing by more than 8 inches (20 cm) a year. Yet these figures conceal very significant regional disparities. Indeed, while the glaciers of the Jammu and Kashmir area are shrinking by 26 inches (66 cm) a year, the Karakoram glaciers are stable. These local differences, which we do not yet fully comprehend, could be due to specific climate conditions related to precipitation and temperature. For instance, the controversy around the melting of the snows of Mount Kilimanjaro has highlighted the key role of factors other than global climate change (see page 58).

OBSERVING THE RISING SEA LEVEL

The massive quantities of water immobilized in the form of ice in the two major polar ice caps (Greenland and Antarctic) and in thousands of terrestrial glaciers around the planet compose what is known as the cryosphere. If all this ice melted and ran into the oceans, it would catastrophically transform our planet's surface. After several millennia of stability, the global oceans have been rising since the 1900s. This rise in the water level, which was of 0.06 inch (1.7 mm) a year throughout the twentieth century, is now accelerating. According to satellite images, it is currently greater than 0.1 inch (3.2 mm) a year. The reason for this worrisome phenomenon is found in the conjunction of two processes: melting of the continental ice sheets and thermal expansion. Warming the ocean increases its volume (as with most solid, liquid, and gaseous bodies), which makes its level rise.

Altimetry satellites such as the TOPEX/Poseidon and the Jasons, which have monitored sea levels since 1993, are the only tools able to accurately inform us about global ocean levels. Their data completes that from the vast network of tide gauges, instruments attached to seafloors to measure the water's height. As with the melting of the glaciers, the rise in sea levels is not uniform and is modulated by currents.

If sea levels were to continue to rise at their present rate, major problems would probably not occur until the end of the twenty-first century. But given the accelerated melting

OPPOSITE

CHINA
EASTERN HIMALAYAS
29°57' N – 93°40' E

The Himalayas are the tallest mountain chain in the world. They include approximately fifteen thousand glaciers, which are currently affected by rising temperatures in the region—+1.8°F (+1°C) since 1970. In the short term, their melting creates a danger of rising waters and floods downstream. But in the longer term, it indirectly threatens populations because the Himalayan glaciers supply the greatest rivers in the world: the Huang He (the Yellow River), the Yangtze, the Mekong, the Ganges, and the Indus, which irrigate areas inhabited by billions of people.

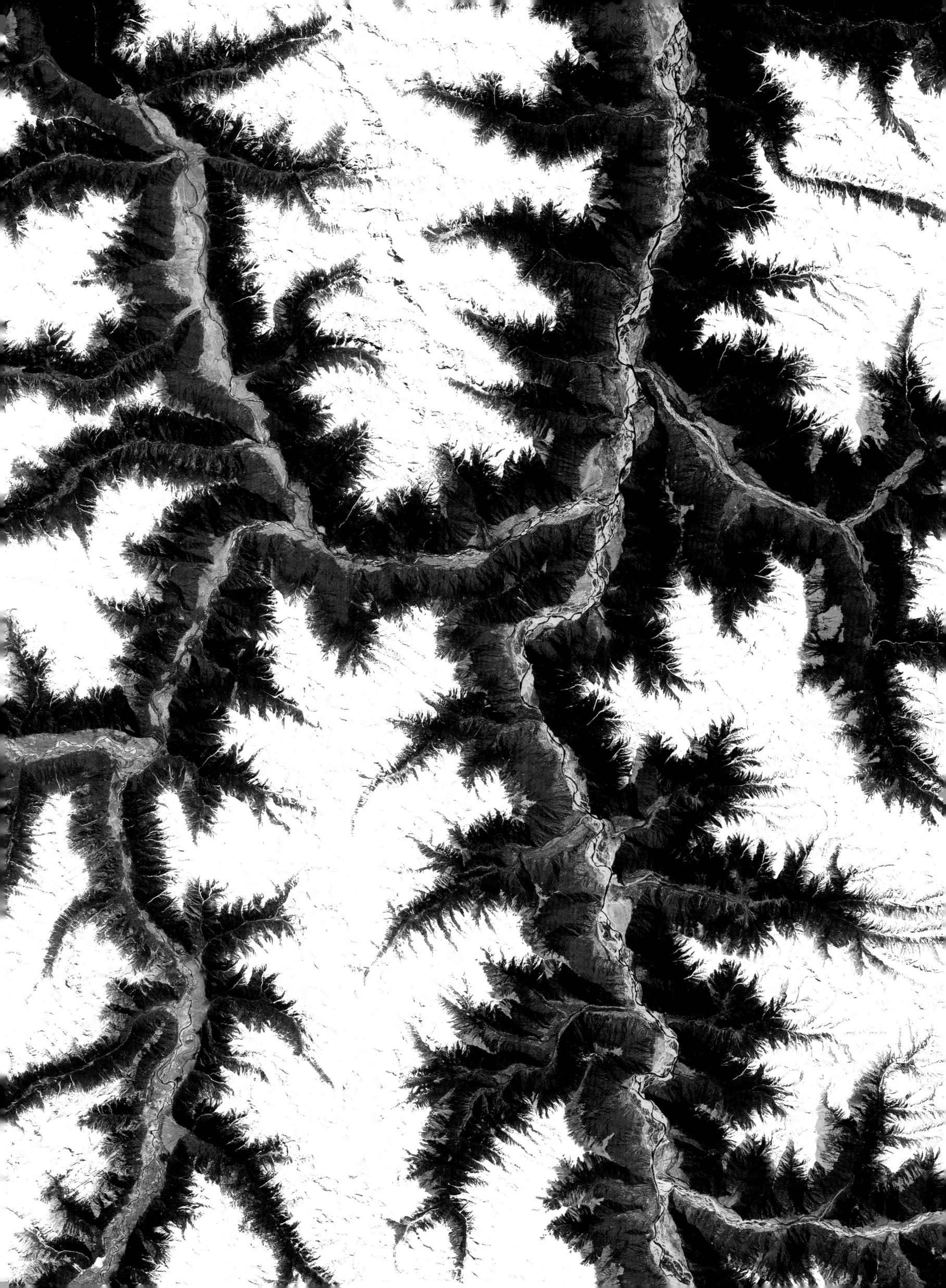

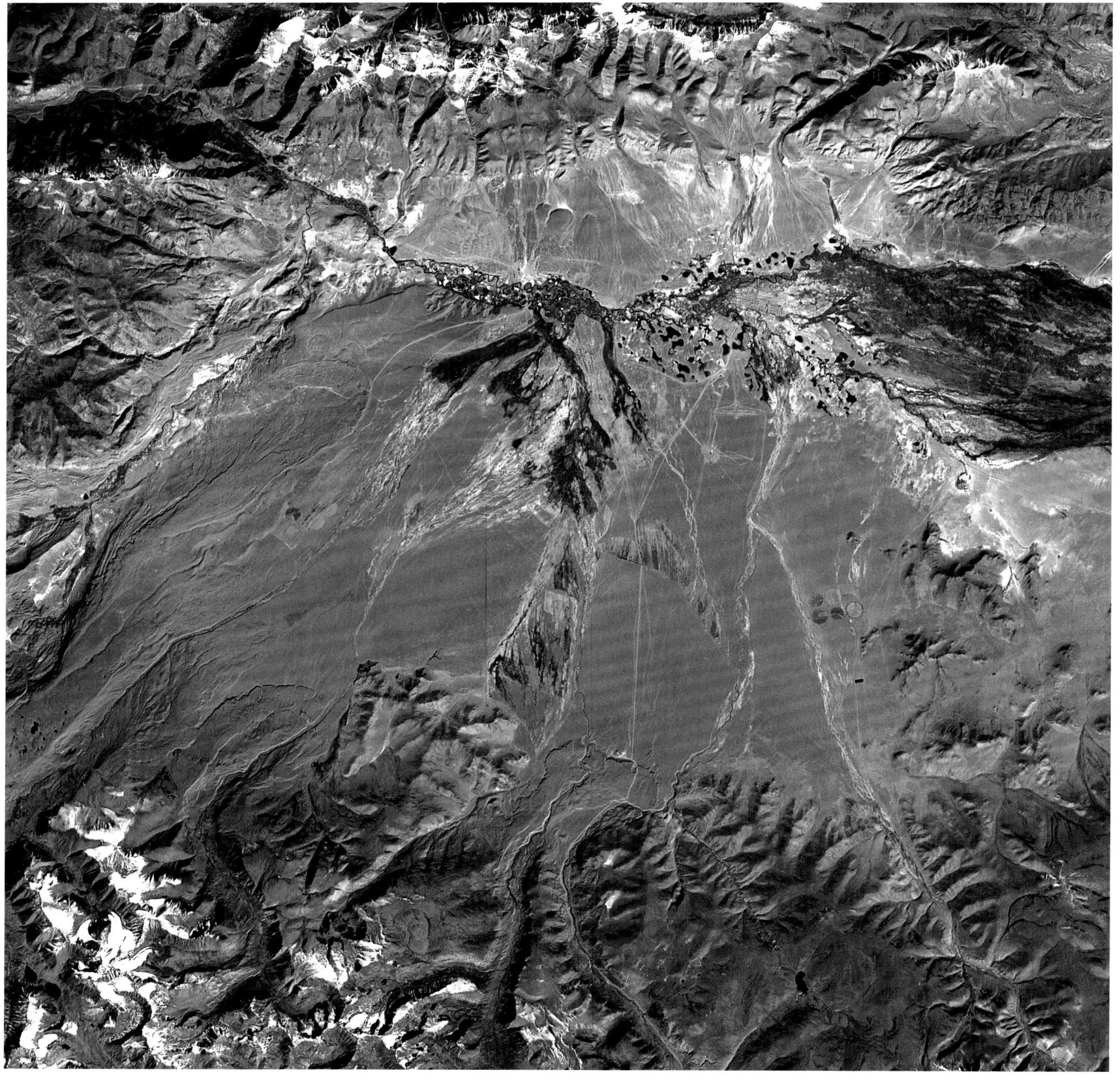

RUSSIA
ALTAI MOUNTAINS
53° N – 92° E

The Altai Mountains are in the heart of Siberia, on the edge of Mongolia and Kazakhstan. The region is home to rare and majestic animals, including the snow leopard. Part of the soil here is permanently frozen, a phenomenon known as permafrost. With global warming, permafrost is melting, endangering certain human structures and archeological sites dating to the first centuries BC and protected by UNESCO.

TANZANIA
KILIMANJARO
3°04' S – 37°22' E

The legendary eternal snows of Kilimanjaro, visible for more than eleven thousand years, could soon disappear. This ice cap, estimated to cover 4.6 square miles (12 sq km) in 1900, has been reduced to 0.7 square mile (2 sq km). The problem here is not the rise in temperatures, but the lack of precipitation, which has been observed locally since the 1950s. Deforestation at the foot of the mountains is the principal cause of the change in rain patterns.

It is believed that if **SEA LEVELS RISE** significantly, millions or **EVEN HUNDREDS OF MILLIONS** of people will have to leave their homes to become "ECO-REFUGEES."

of the Greenland and West Antarctic ice caps, most specialists expect sea levels to rise dramatically: The increase could be over 3.2 feet (1 m) in a century's time.

For example, imagine that the Greenland ice cap melts completely: Water levels would increase by nearly 22.9 feet (7 m) in the Arctic. In Antarctica, the water would rise by 197 feet (60 m)! Thankfully, this scenario is unlikely according to current projections.

Yet even a rise of 3 feet (1 m) could be catastrophic, due to the enormous concentration of wealth and populations in coastal areas where many urban centers have developed. Shanghai, the leading merchandise port in the world, was built at very low altitude, leaving the harbor installations particularly vulnerable to coastal storms. The same is true of New York City, which is notoriously poorly protected against high water levels and was brought to a standstill by Hurricane Sandy in the fall of 2012.

In Bangladesh, where half the population lives in areas less than 16 feet (5 m) above sea level, various scenarios predict that about 20 percent of the territory will disappear underwater and that twenty to forty million people will be displaced. As for Africa, the United Nations Framework Convention on Climate Change (UNFCCC) estimates that "close to 30 percent of coastal infrastructure is at risk" and that the number of Africans threatened by flooding "will grow from 1 million in 1990 to 70 million in 2080." In this sense, the nation of Tuvalu in Oceania is emblematic: Consisting of nine coral atolls whose highest point is 16 feet (5 m) above sea level and populated by ten thousand people, the archipelago is endangered by rising water levels. According to experts, it could disappear entirely by 2050. It is believed that if sea levels rise significantly, millions or even hundreds of millions of people will have to leave their homes to become "eco-refugees."

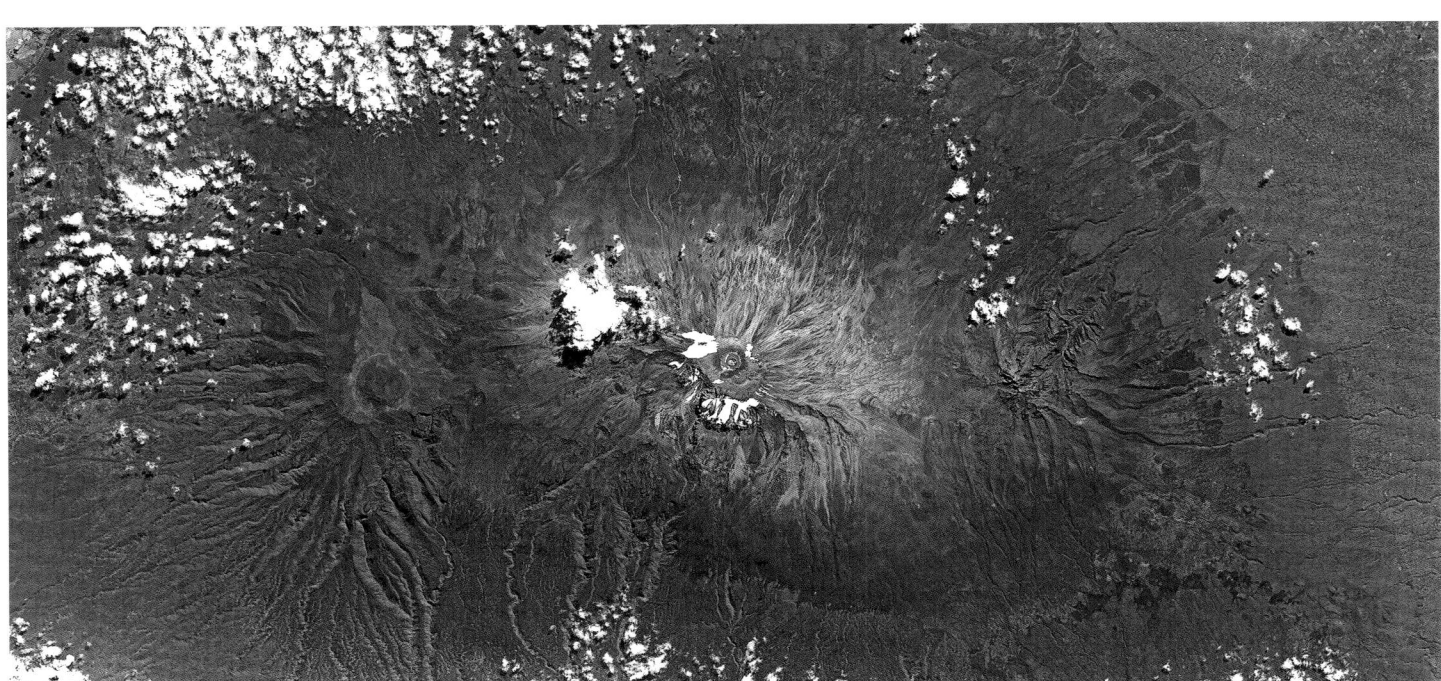

BELOW

ECUADOR
GALÁPAGOS ISLANDS
0°30' S – 90°36' W

When Charles Darwin landed in the Galápagos
Islands in 1835, he noted an incredible
abundance of animal and plant species. Based
on his observations, he developed the theory of
natural selection, which provides a logical and
unified explanation for the diversity and evolution
of life. The archipelago's fifty-five islands are the
emerged parts of volcanic formations, and they
are listed as a UNESCO World Heritage site.

OPPOSITE

ANTARCTICA
LAMBERT GLACIER
71° S – 70° E

At more than 250 miles (400 km) long and 62 miles
(100 km) wide, the Lambert Glacier is the largest
glacier in the world. Because it is 1.5 miles (2.5 km)
thick and hard to access, we resort to satellite images
to study it. This image, taken with infrared sensors,
shows cracks in the ice, which descend more than
1,300 feet (400 m).

Rising sea levels are not the only threat to our coastlines.
Today researchers believe that global warming is causing an
intensification of tropical cyclones and more frequent occur-
rences of extremely high temperatures.

ADAPTING TO THE CONSEQUENCES
OF CLIMATE CHANGE

If we cannot stop climate change, we need to prepare for its
consequences. Complementary strategies need to be implemented: building dykes and dams, moving buildings inland
and adapting them, and the like. Thanks to satellite images,
it is possible to evaluate coastlines' vulnerability and establish maps of weak points—as with the Infolittoral project in
France. The study of parameters such as local intensity of
the rise in sea levels, sedimentary dynamics, progression of
the high-tide line, urbanization, or the rate of deforestation
allows us to measure the consequences of climate change.
In French Guiana, satellite imagery made authorities aware
of the vulnerability of the northern coast, largely due to the
dwindling of the mangroves.

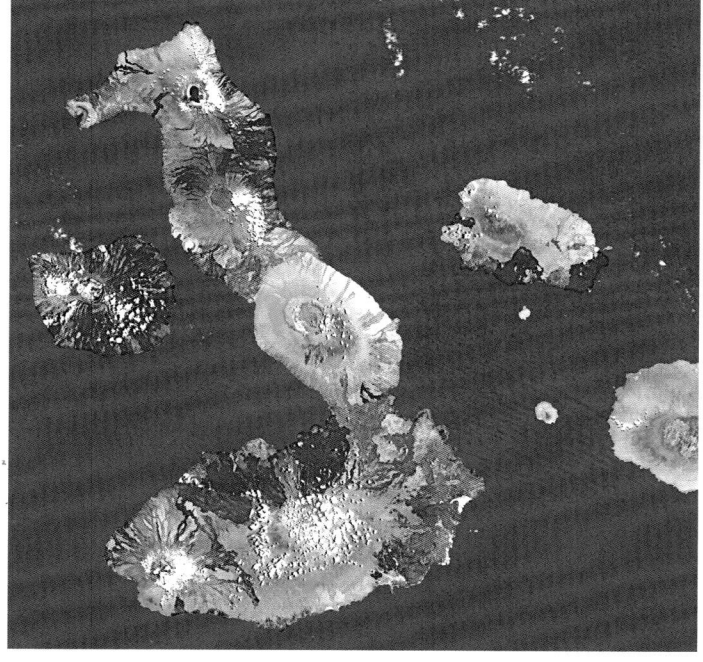

BIODIVERSITY AND RESOURCES OF THE FUTURE

Aside from rising sea levels, the consequences of climate
change are already having an effect on biodiversity. Many
species are migrating to the poles or gaining altitude in
order to recover the natural conditions they are used to.
This is the case with European butterflies: A study of thirty-
five butterfly species showed that 60 percent of them have

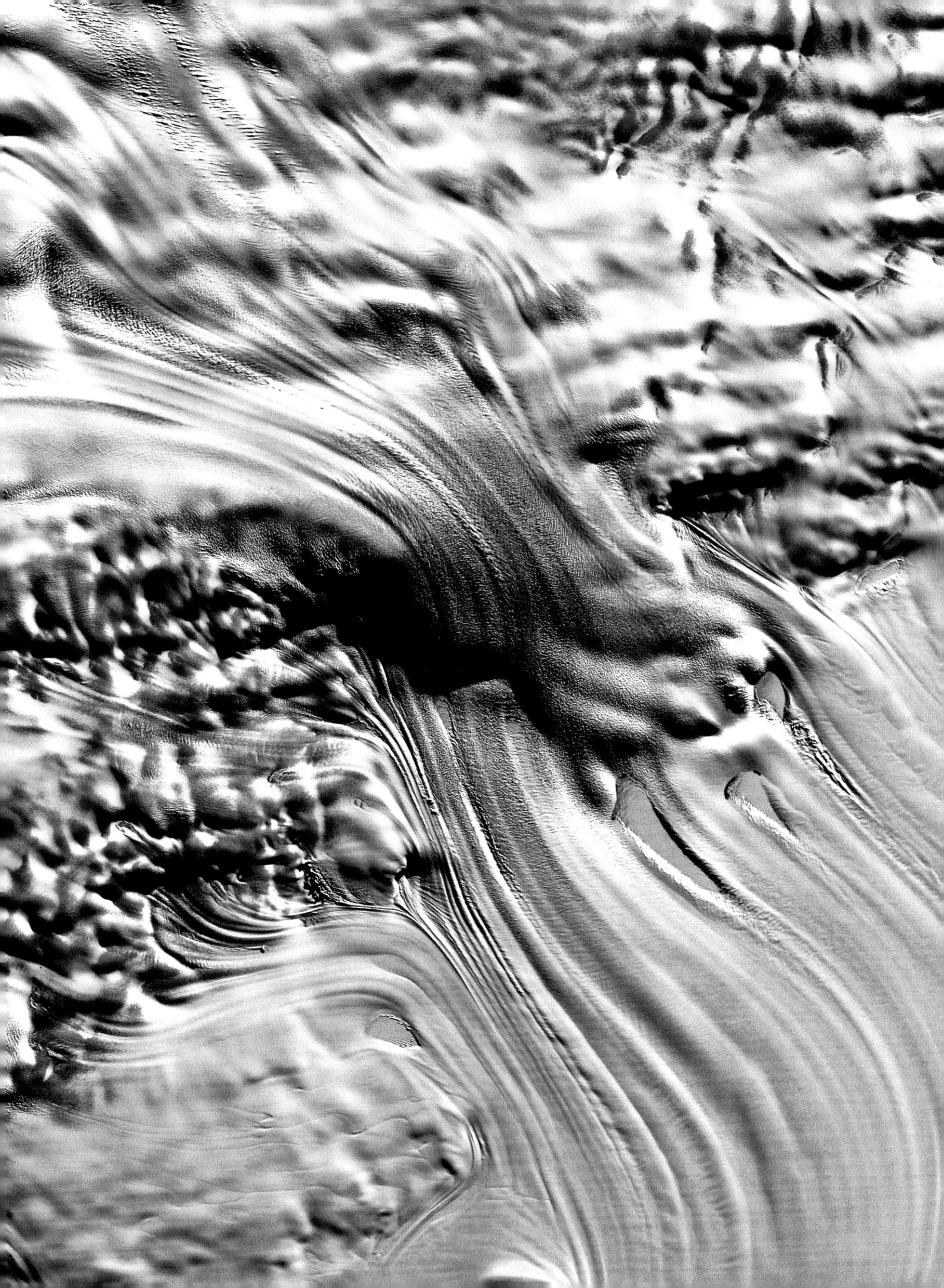

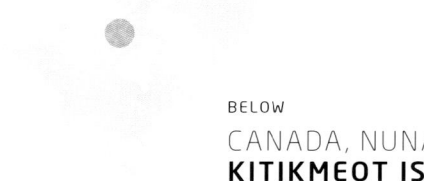

BELOW

CANADA, NUNAVUT
KITIKMEOT ISLAND
67°06' N – 97°30' W

Located in the extreme north of Canada, Nunavut Territory is one of the most remote areas in the world. Its landscape still bears traces of the last glaciation. The blue in this reconstructed image shows lakes, natural basins, and watercourses shaped by the ice's retreat eight thousand years ago. The blue is darker where there are mud sediments and phytoplankton, while red indicates the presence of vegetation: Grasses, mosses, and lichen can grow in this often frozen or snowy ground.

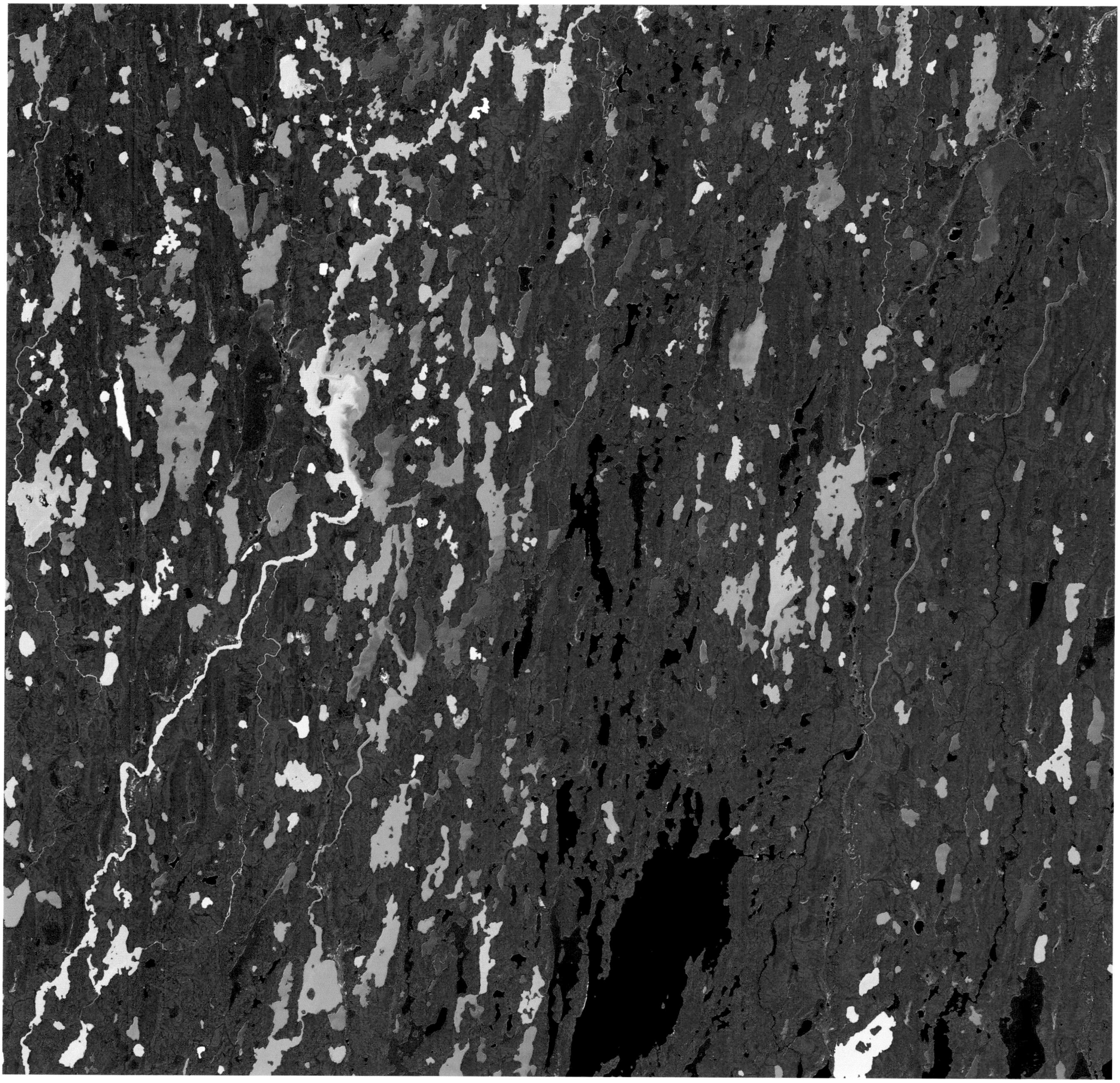

One major current objective is to use SATELLITES' IMAGE SENSORS to accurately describe the **COMPOSITION OF THE ATMOSPHERE** and the **QUANTITIES OF THE PRINCIPAL GREENHOUSE GASES** in it.

relocated themselves to the north; the *Hesperia comma* has traveled north 136 miles (220 km) in twenty years. Similarly, king penguins have modified their food-gathering trips and migratory travels based on variations in the temperature of ocean surfaces. But other species such as tropical coral are sedentary and cannot escape the increase in temperatures: A 1.8°F (1°C) increase over the course of a season bleaches them, while a 5.4°F (3°C) increase would destroy them.

These changes imply not only an increasing scarcity of species but also a considerable modification of their morphology and behavior. The phenomenon is also likely to increase the frequency of disease and reconfigure distribution ranges. These changes can now be observed thanks to technology and the Argos system, which tracks terrestrial and aquatic animals with a satellite beacon. Often used to help manage fisheries, these beacons determine the distribution of phytoplankton and the direction of currents and therefore evaluate the presence and quantity of commercial species such as tuna.

IDENTIFYING CAUSES

Satellite images could become precious tools to identify the causes of climate change and try to remedy it. One major current objective is to use satellites' image sensors to accurately describe the composition of the atmosphere and the quantities of the principal greenhouse gases in it—including carbon dioxide and methane. This is the goal of the Carbon-Sat project implemented by ESA (European Space Agency) and Astrium, which aims to send a satellite dedicated to the study of greenhouse gases into orbit. Multiple issues are at stake: obtaining a global picture of the quantity of greenhouse gas in our atmosphere, locating and identifying the principal emitters, and determining potential positive impacts of the reduction measures that could be implemented.

Yet many questions remain. How can we transform observation, however accurate, into a form of action? The major areas of intensive agriculture will probably be shown to be the principal emitters of methane, and the largest urban areas and coal-fueled power stations will likely appear to be responsible for a significant portion of CO_2 emissions. Once the satellite photo is analyzed and the diagnosis correctly made, what measures will we be able to take? Perhaps we will have to consider creating a sworn international authority for monitoring the climate and inventing methodologies that can make satellite photos serve as evidence. Is an international governing authority for climate still feasible at a time when the great international summits can no longer result in nations making global agreements to preserve our planet?

———

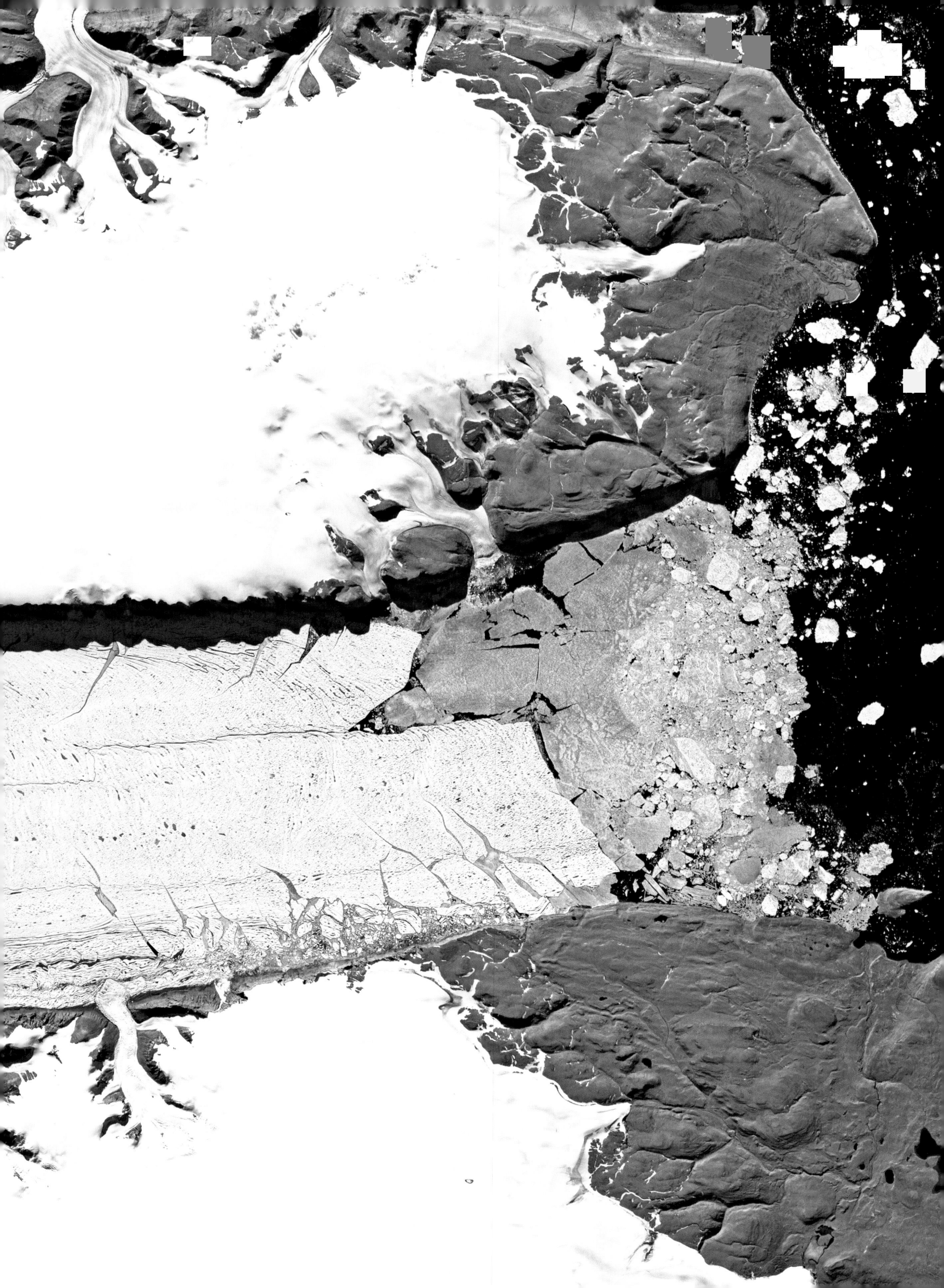

DENMARK, GREENLAND
PETERMANN GLACIER
80°30' N – 59°30' W

Between land and sea, the
Petermann Glacier northwest
of Greenland consists of a
terrestrial component and a
spit extending into the sea.
During the summer of 2010, a
96-square-mile (250-sq-km)
iceberg separated from the
glacier; it was the biggest
piece of ice ever recorded.
In July 2012, the same
phenomenon produced a
smaller iceberg, measuring
50 square miles (130 sq km).
Pictures show the glacier spit
as it becomes more fragile
before breaking off. Arctic
melting accelerates global
warming. Ice reflects light and
the sun's heat, while water
absorbs it. This is one of the
reasons why the Arctic area is
warming faster than any other.

1999
2012

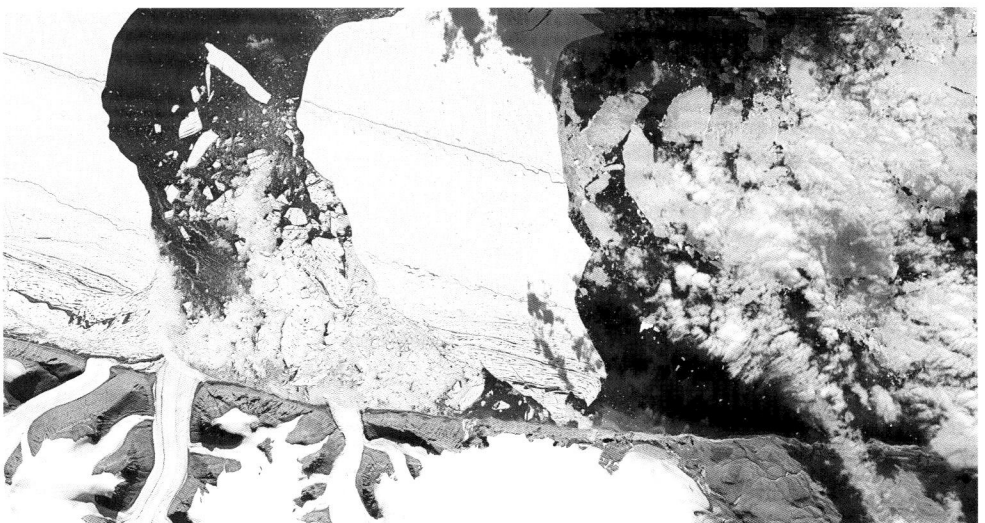

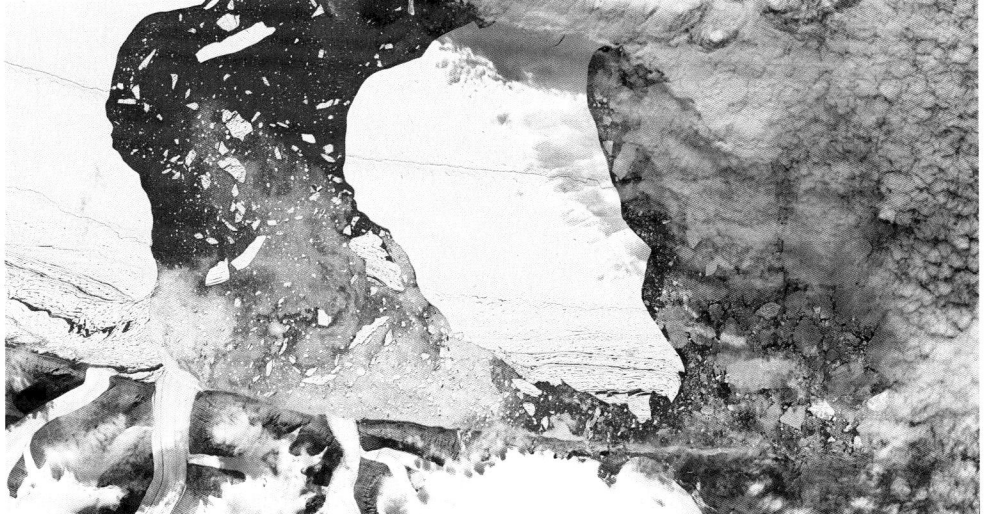

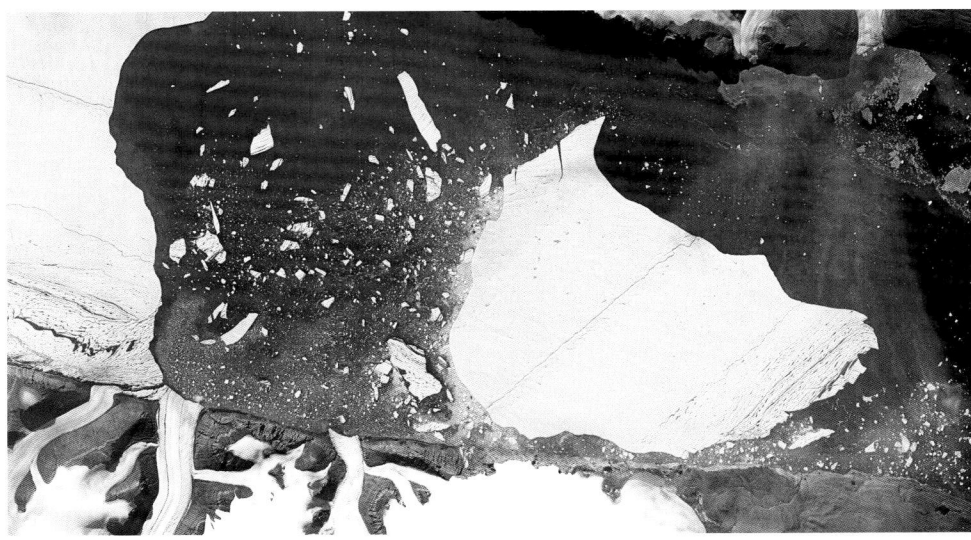

OPPOSITE
DENMARK, GREENLAND
ICE WAVE
62°50' N – 41°45' W

Satellite images now allow us to take global measurements of changes that, as recently as a few decades ago, we were only able to partially perceive.

Talking with Jean Jouzel

Climate Change from a Global Perspective

How have satellites changed our understanding of climate change?

Satellites are excellent tools to grasp the problem from a global perspective. Twenty or thirty years ago, to study climate, we had to go to such and such a glacier to measure the speed at which the ice accumulated or melted. Thanks to satellites, we can now compare the expansion or shrinking of glaciers over entire regions, including in inaccessible areas, year after year.

To measure the growth of sea ice in the Arctic, we used to primarily rely on ships or the few planes that ventured into the area. Now we have detailed maps in real time. This area, which is being warmed at nearly twice the rate of the rest of the planet, has now become the sentinel of global warming. Satellite images also enable us to make a quantitative estimate of the volume of ice melting.

Finally, a visual representation of the phenomenon is particularly important to inform decision makers and citizens.

What has changed the most?

Until the third report by the Intergovernmental Panel on Climate Change (IPCC), in 2007, we considered that, overall, the ice in Antarctica and Greenland was in about the same state. We have since learned that it is melting—and quickly—in Greenland. The situation has probably gotten worse in the last ten years due to global warming. But our understanding of the scope of the problem has also been further refined since the report, thanks to satellite data.

As for the rise in sea levels, the level of precision we have achieved is practically unhoped-for. We can actually measure the phenomenon, though it is only on a scale of 0.12 inch (3 mm) a year. Here, too, we had good data throughout the twentieth century, thanks to tide gauges. But for the most part these devices were only available along coastlines. Now we can study an entire zone at once, including in the heart of the Pacific.

This has allowed us to identify very significant local differences. In some places, the rise in sea level is far above average, while in others it is practically nonexistent.

What issues lie ahead?

In scientific terms, one of the most important issues is to achieve a better understanding of the atmosphere, particularly the interaction between clouds, emissions, and aerosols (suspended microparticles, dust, soot, and the like). Depending on the situation, these interactions can warm or cool the planet, but they remain poorly understood. Satellites will play an essential role in informing us about that.

Jean Jouzel is a climatologist and former director of the Institut Pierre Simon Laplace. He is the vice president of the IPCC scientific group, which was awarded the Nobel Peace Prize in 2007.

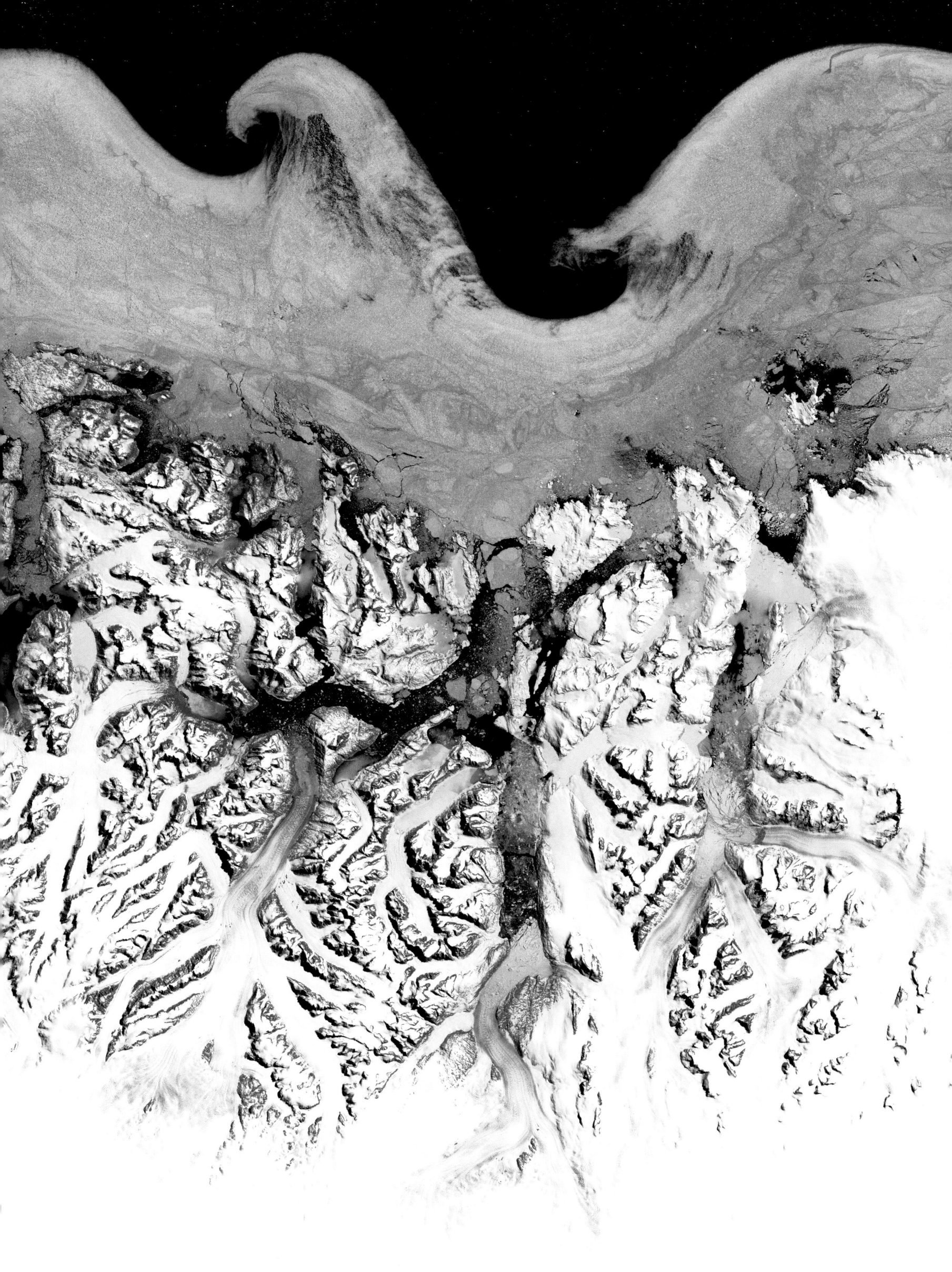

DANGER IN THE HIMALAYAS

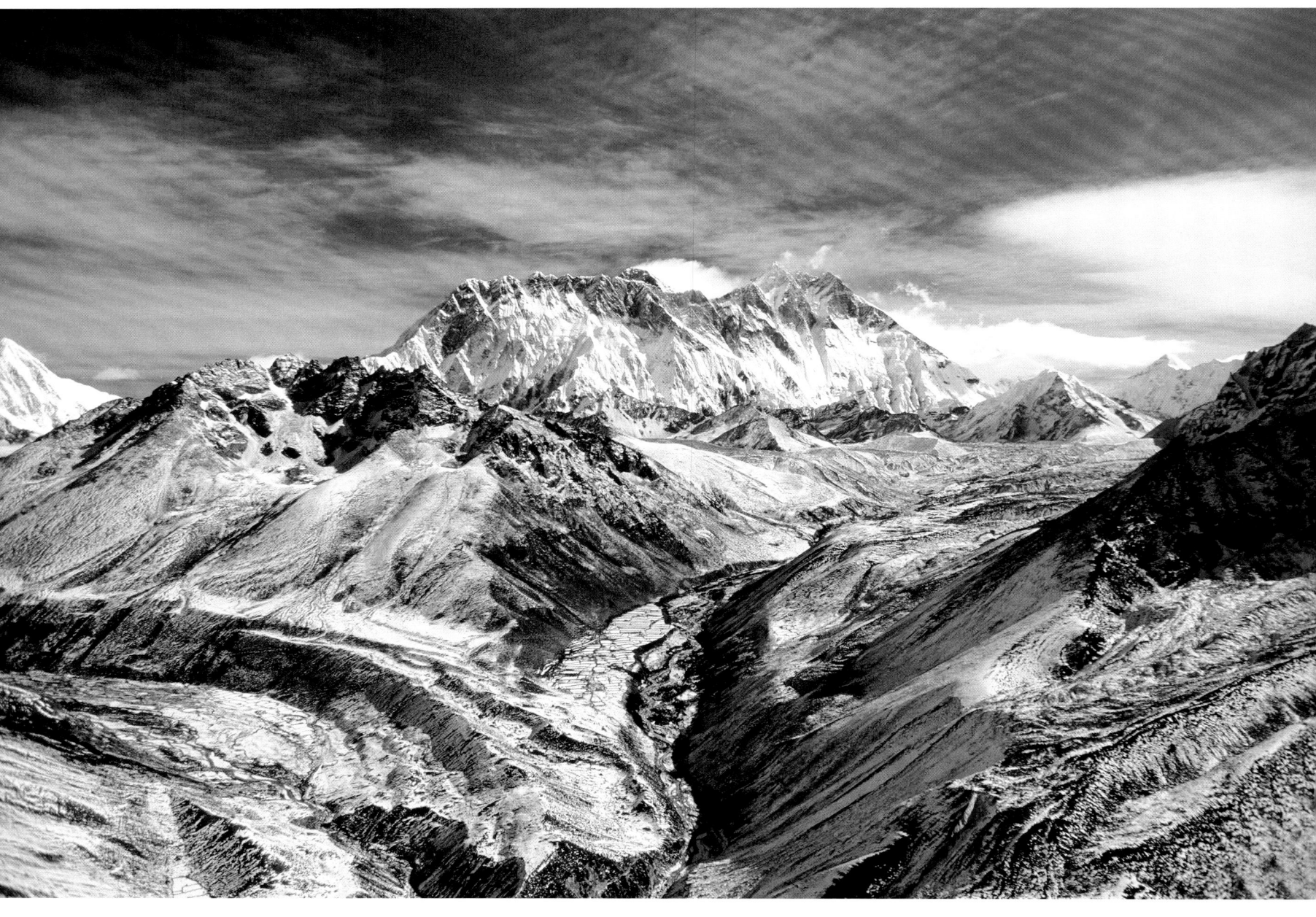

ABOVE
NEPAL
MOUNT EVEREST
27°59' N – 86°56' E

Despite their forbidding peaks, the Himalayas are not uninhabited. Nepal has 30 million inhabitants for a surface area one-third the size of France. Of course, most of the population resides in the lower areas, around Kathmandu and the valleys of the Terai. Tourism is the leading source of income in Nepal and agriculture is the principal activity.

Glacial areas are on the front lines of climate change. And the Himalayas are considered a particularly sensitive region. The rapid melting of glaciers is causing large glacial lakes to form: Enormous quantities of water are accumulating, held by moraines, the rocky debris from a glacier. In Nepal, a total of 2,323 glacial lakes and 3,252 glaciers have been identified. The vast majority of the lakes are small and considered to pose little threat. However, some twenty glacial lakes are potentially dangerous: The moraines holding them back could break under pressure, unleashing massive flooding.

On average, this kind of large-scale flooding, known as a glacial lake outburst flood (GLOF), takes place every seven years and can kill hundreds of people. In the Bhote Kosi River valley, thirty-five smaller floods caused by the bursting of glacial lakes were recorded from 2003 to 2007.

Predicting and anticipating these events is clearly essential. Satellite observation is particularly well suited for monitoring glacial areas, given the difficulties accessing these locations by land. Since August 2008, ICIMOD (International Centre for Integrated Mountain Development, an intergovernmental agency) has focused on monitoring seven glaciers.

The images provided by Planet Action contribute to mapping glacial lakes, notably by supplying data on the various zones' reflection and absorption capacity. The images obtained make it possible to distinguish vegetation from ice, snow, and water. ICIMOD has mapped the risk factors for various lakes and estimated the quantities of water involved and their traveling time, to calculate how long it would take a flood to reach vulnerable areas where populations and infrastructures are concentrated. By educating local populations about the floods, this plan could reduce the risk of disastrous consequences from future glacial lake outburst floods.

PAGES 70-71

UNITED STATES, FLORIDA
FLORIDA KEYS
24°43' N – 81°38' W

Consisting of about one thousand islands and coral islets, the Keys archipelago is an extension of Florida's southern tip. Some islets remain uninhabited and are covered in mangroves. These are sometimes completely submerged by storm waves. This phenomenon is likely to happen more frequently, given that global warming is accompanied by an increase in the intensity of hurricanes. The fragile ecosystem of the Keys, which are home to more than 250 bird species, is legally protected.

PAGES 72-73

MALDIVE ISLANDS
MALOSMADULU ATOLLS
5°20' N – 72°50' E

Located in the Indian Ocean, the Maldives peak at 13 feet (4 m) above sea level, but 80 percent of their component 1,199 islands are no higher than 3 feet (1 m). The Maldives' 399,000 inhabitants currently live off fishing and tourism. But they are in danger of having to leave their country if sea levels continue to rise: According to experts, the sea could rise by more than 3 feet (1 m) by the end of the century.

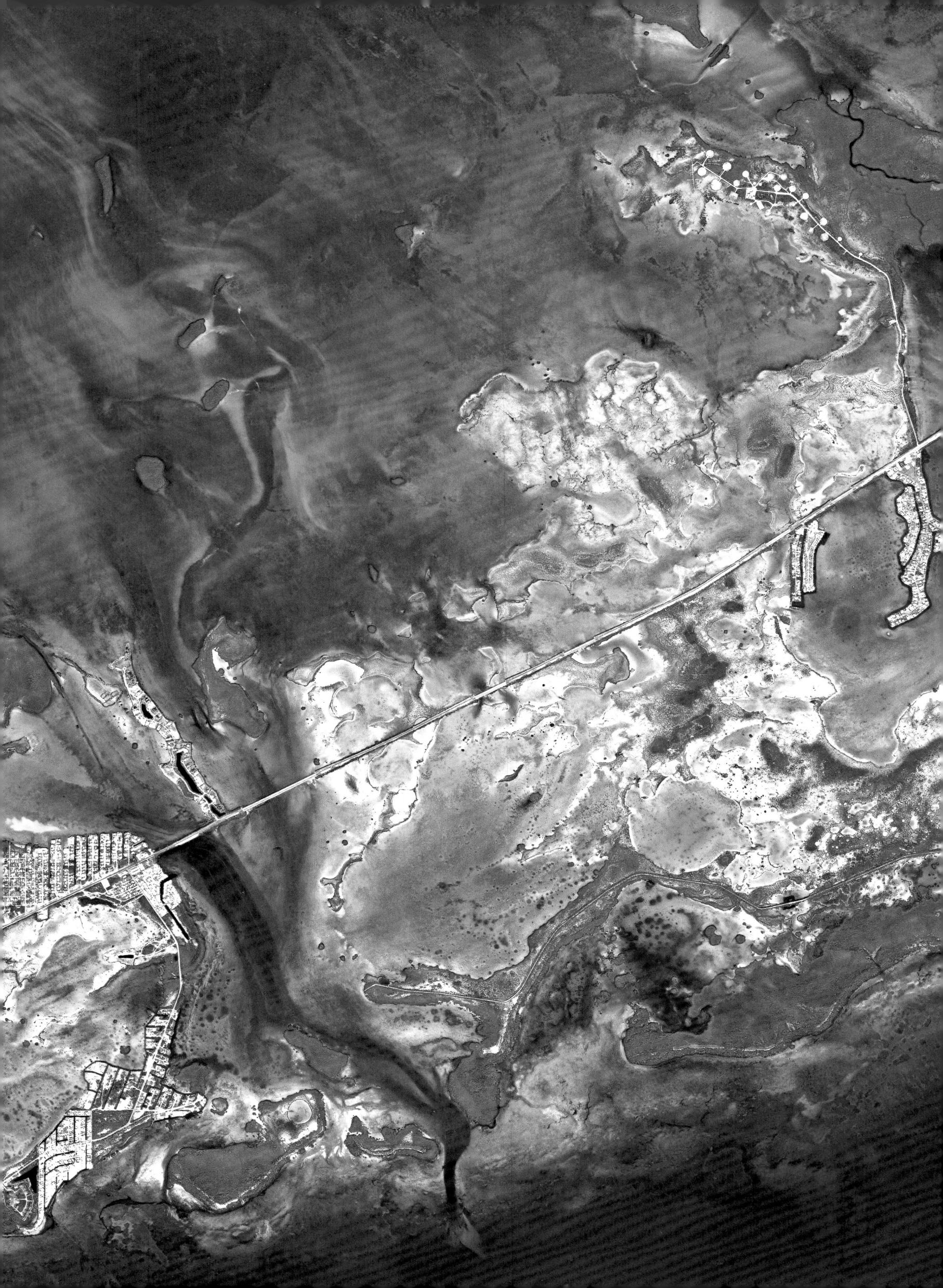

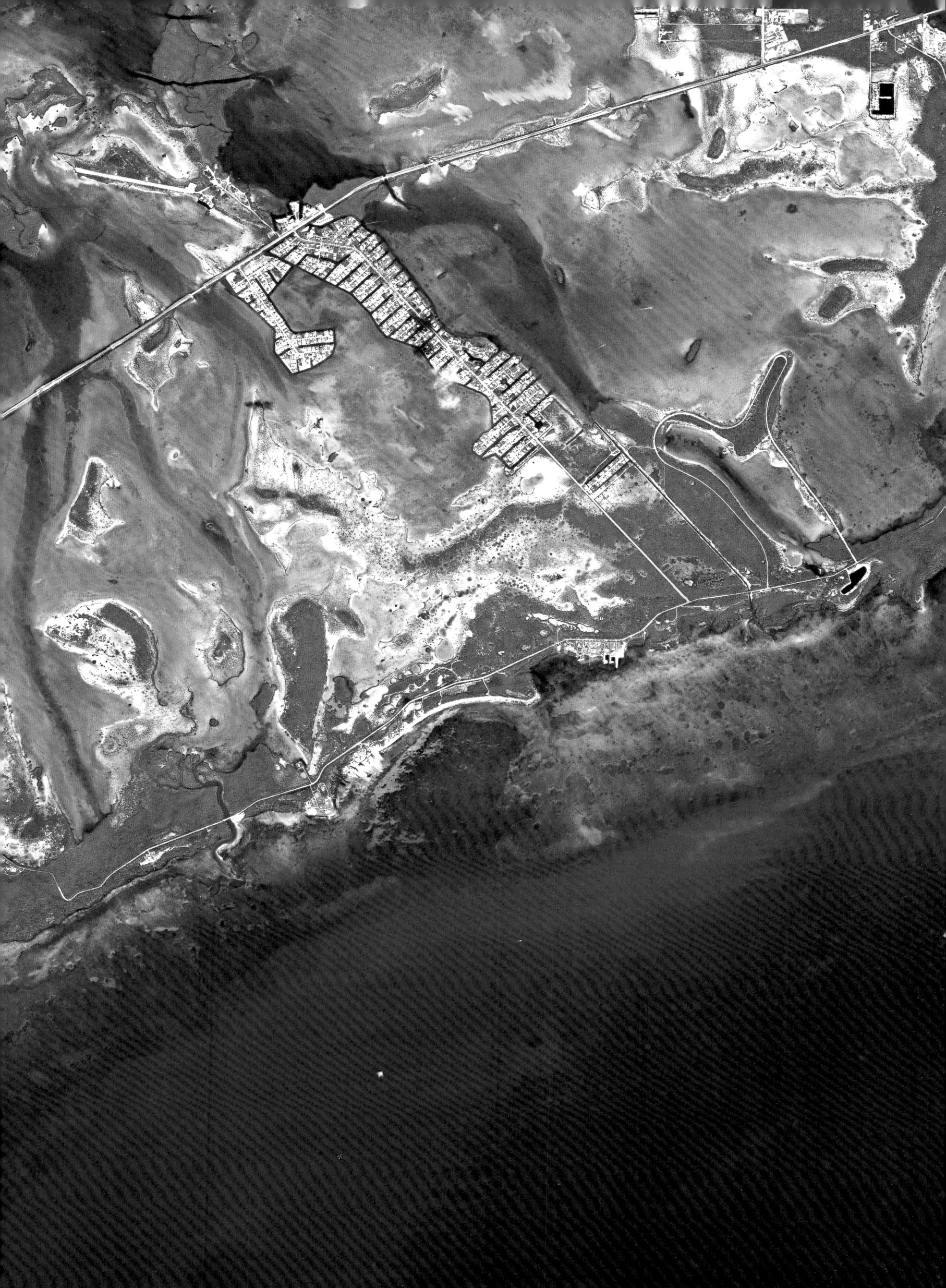

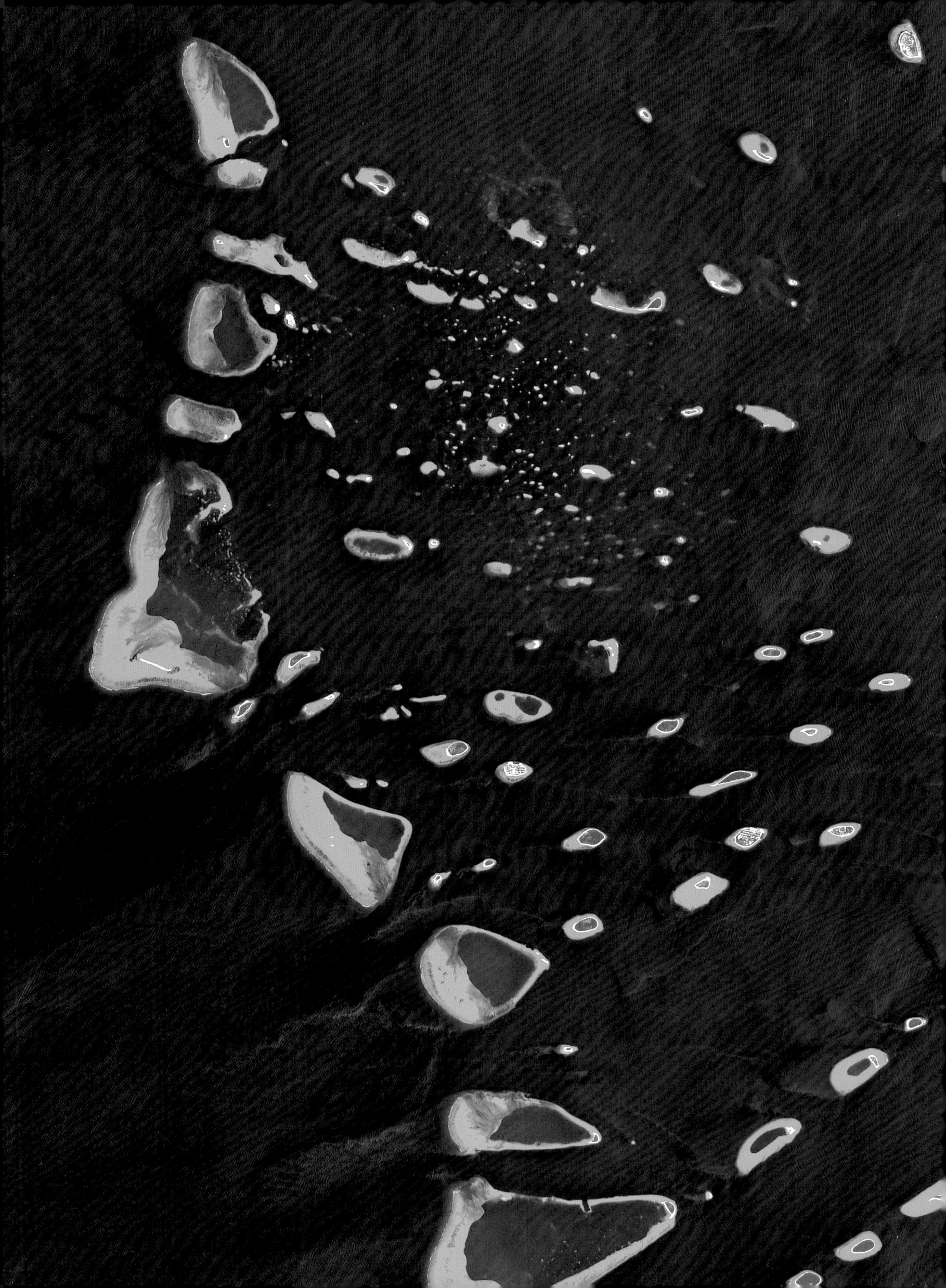

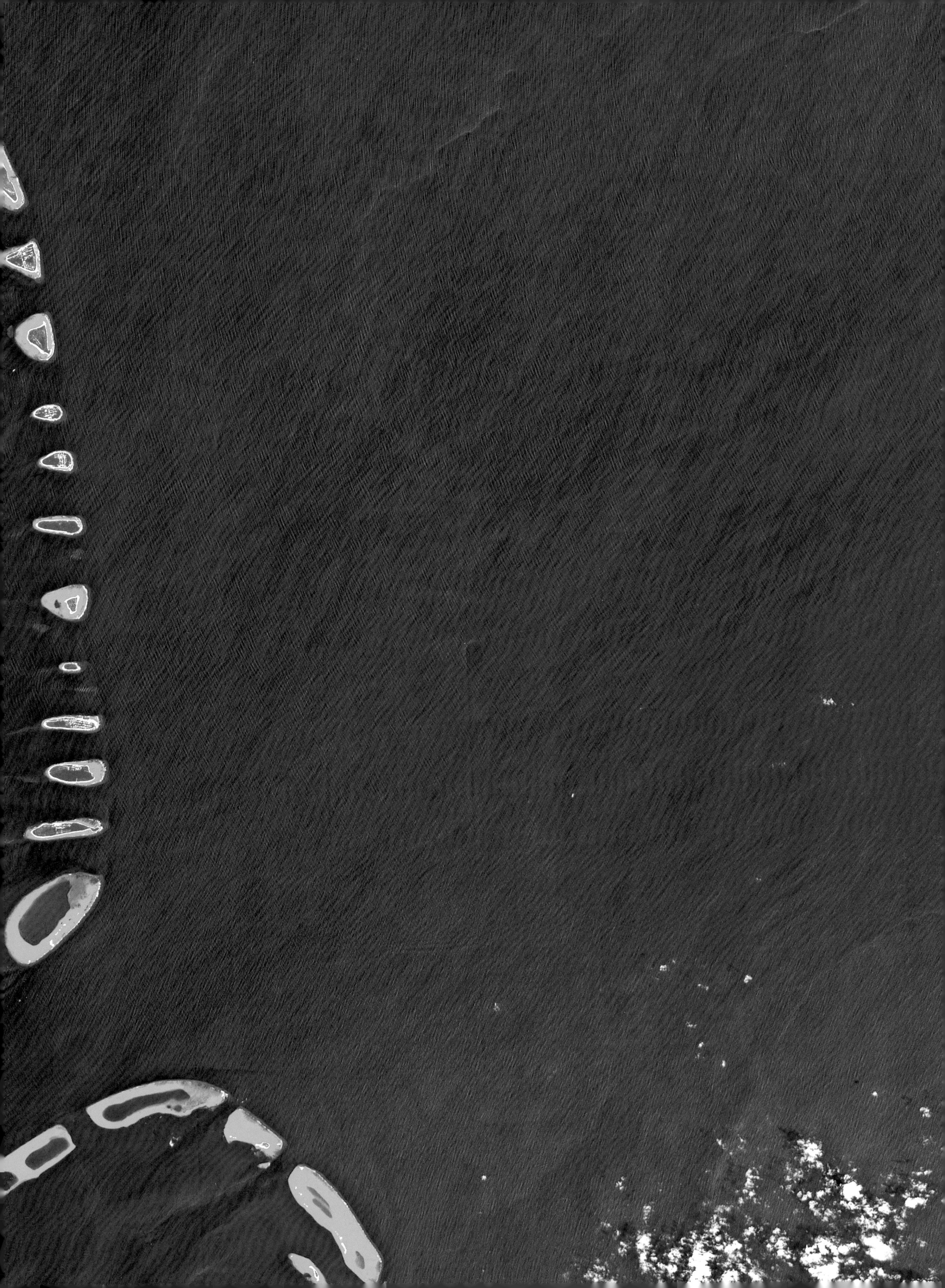

BELOW

ANTARCTICA
MOUNT EREBUS
77°32' S – 167°10' E

Mount Erebus is the southernmost volcano in the world. But even though it stands 12,447 feet (3,794 m) tall, it is not the highest point in Antarctica, which peaks at 16,049 feet (4,892 m) tall. A continent of extremes, Antarctica covers 5.4 million square miles (14 million sq km) and registers temperatures as low as –94°F (–70°C), with wind speeds up to 186 miles (300 km) per hour. Studying the continent's ice is particularly useful in order to understand the origins of contemporary climate change. We have been able to study the ice at a depth of 10,728 feet (3,270 m), which is the equivalent of eight hundred thousand years of climate archives.

OPPOSITE

CANADA, NUNAVUT
AKPATOK ISLAND
60°25' N – 68°07' W

Standing up to 797 feet (243 m) above sea level, Akpatok Island is part of the Canadian Arctic archipelago located north of Quebec, in Nunavut. The island is home to many birds, including the Brünnich's guillemot, or *akpat* in the Inuit language, from which the island gets its name. It also attracts walruses, seals, and whales. As for polar bears, their future is uncertain: Only a few more than twenty thousand are still in the wild. In the United States, the animal has been classified as an endangered species.

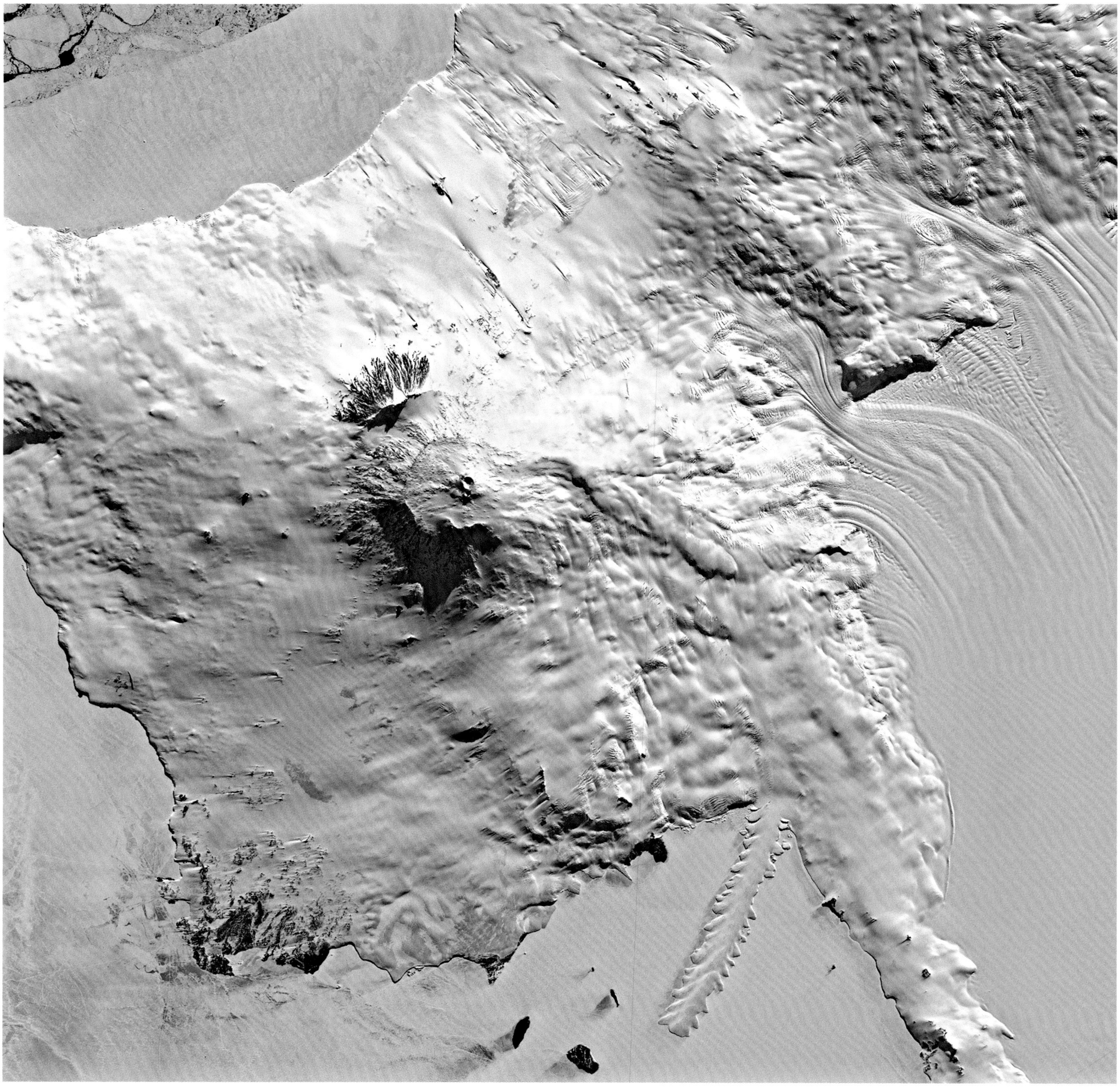

PAGES 76-77

NEPAL
MOUNT EVEREST

27°59' N – 86°55' E

Mount Everest, the tallest point on the planet, stands 29,028 feet (8,848 m) tall. Its peak was first conquered in 1953. Since then, its perilous ascension remains limited to elite mountain climbers. Yet even those who set foot on the world's tallest peaks have noted that the volume of ice is shrinking.

PAGES 78-79

CHILE, PATAGONIA
LAGUNA SAN RAFAEL NATIONAL PARK

46°41' S – 73°50' W

San Rafael Glacier is the only glacier located so far away from the poles as to reach the sea. It is characterized by an exceptionally blue ice wall forming a cliff more than 300 feet (100 m) tall and approximately 1.2 mile (2 km) wide. The calving (formation of icebergs) in the massive ice field in the depths of a fjord in Southern Patagonia has become a tourist attraction. The glacier has shrunk by about 0.6 miles (1 km) over the last decade.

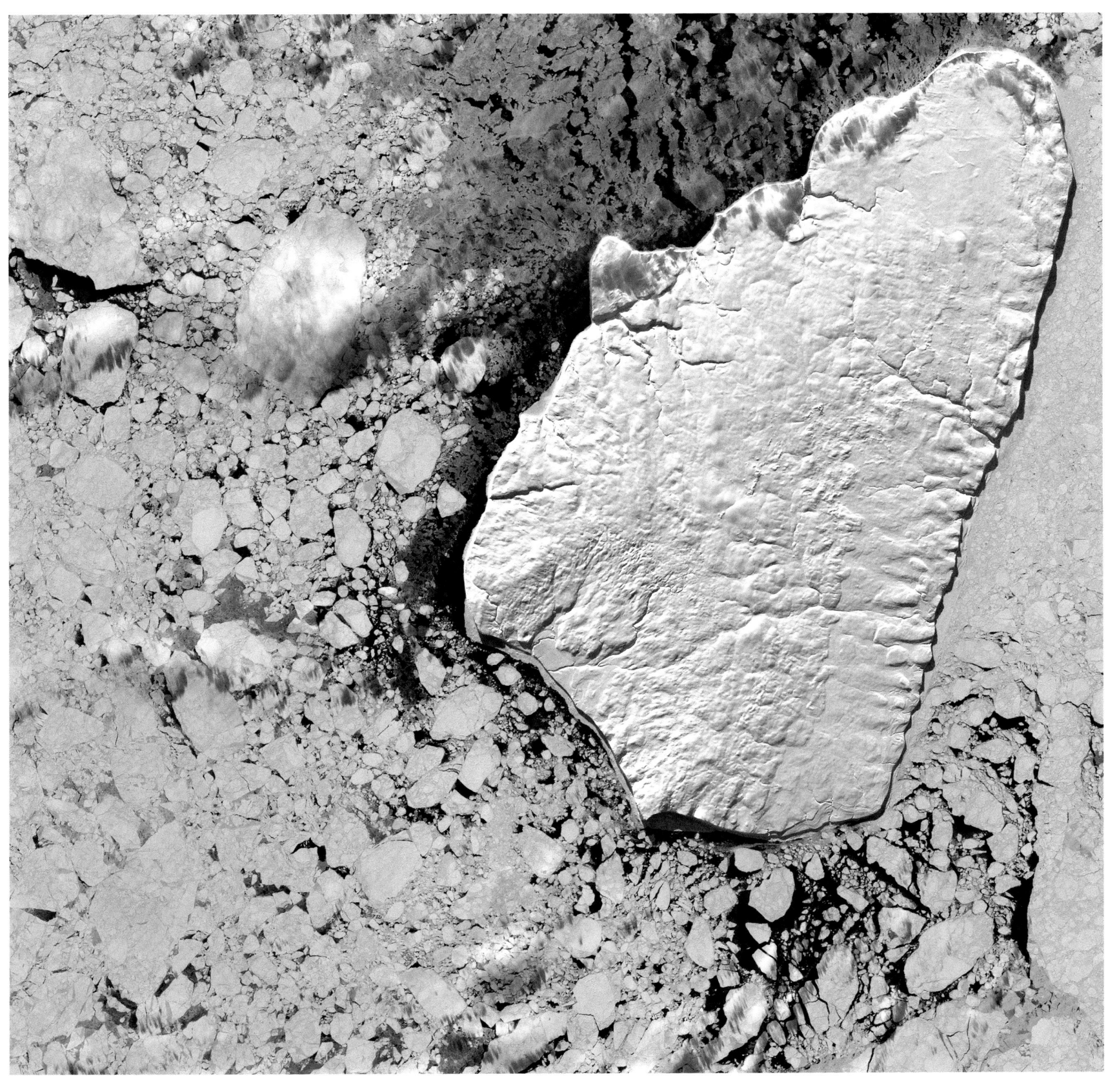

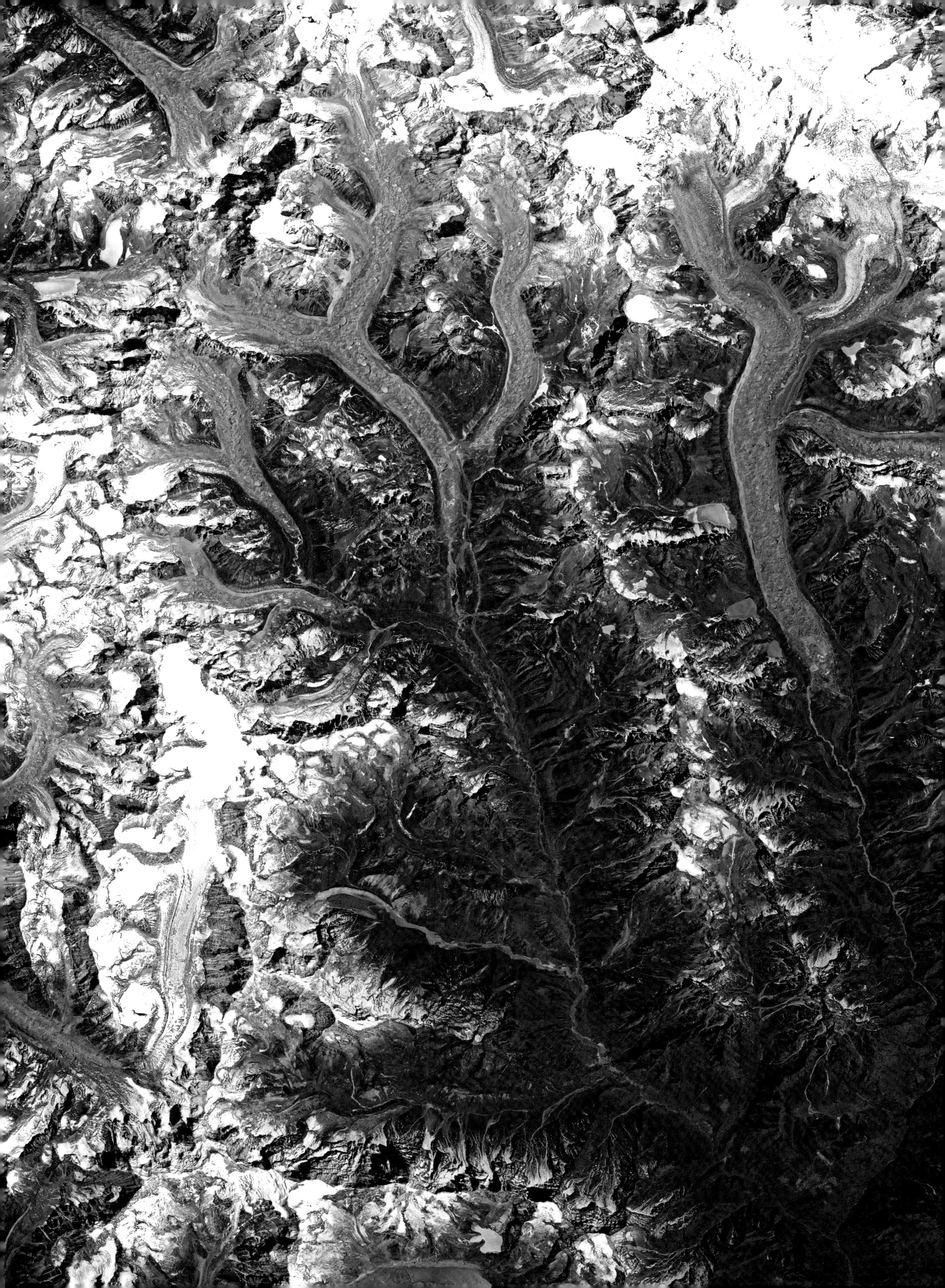

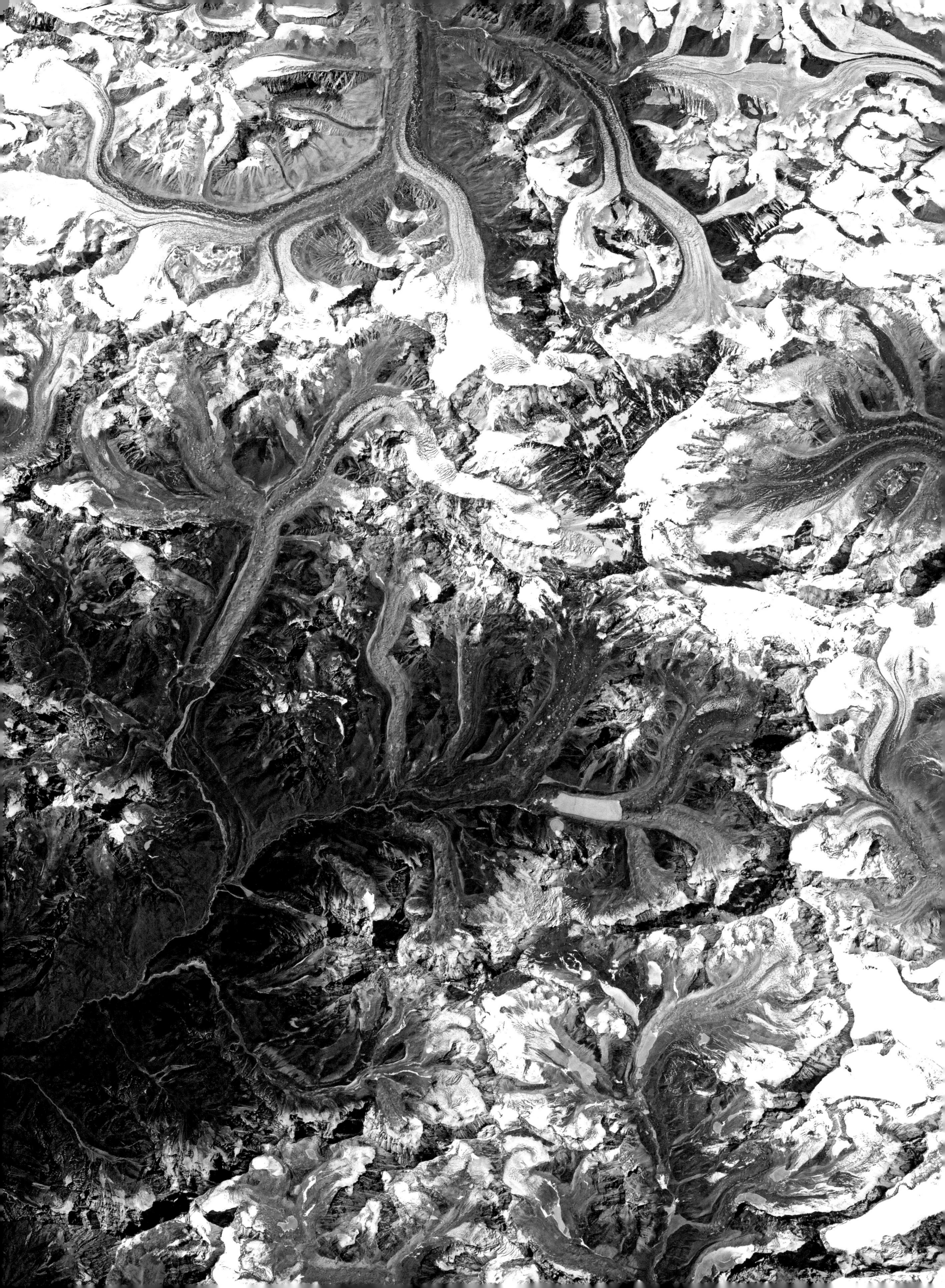

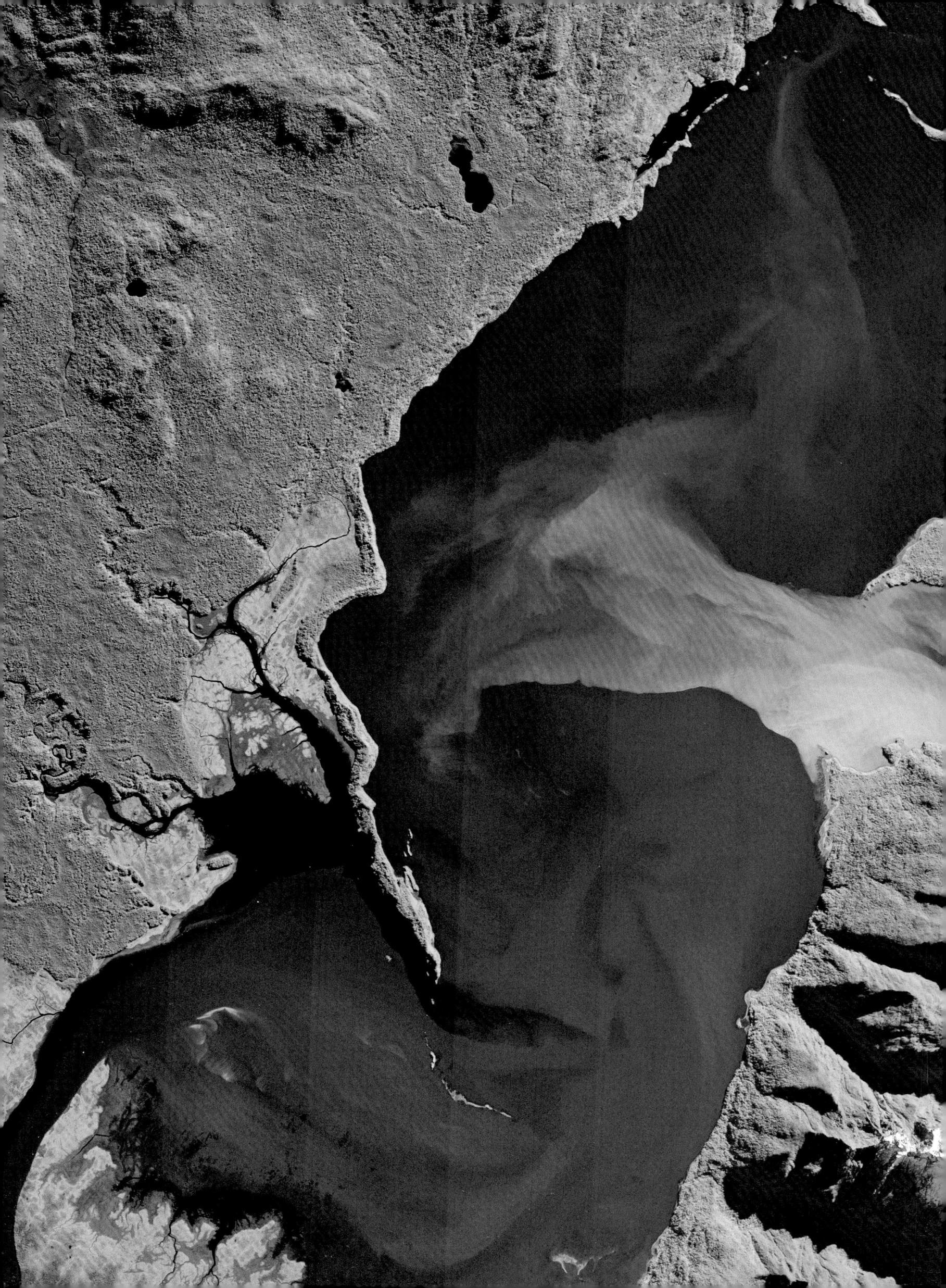

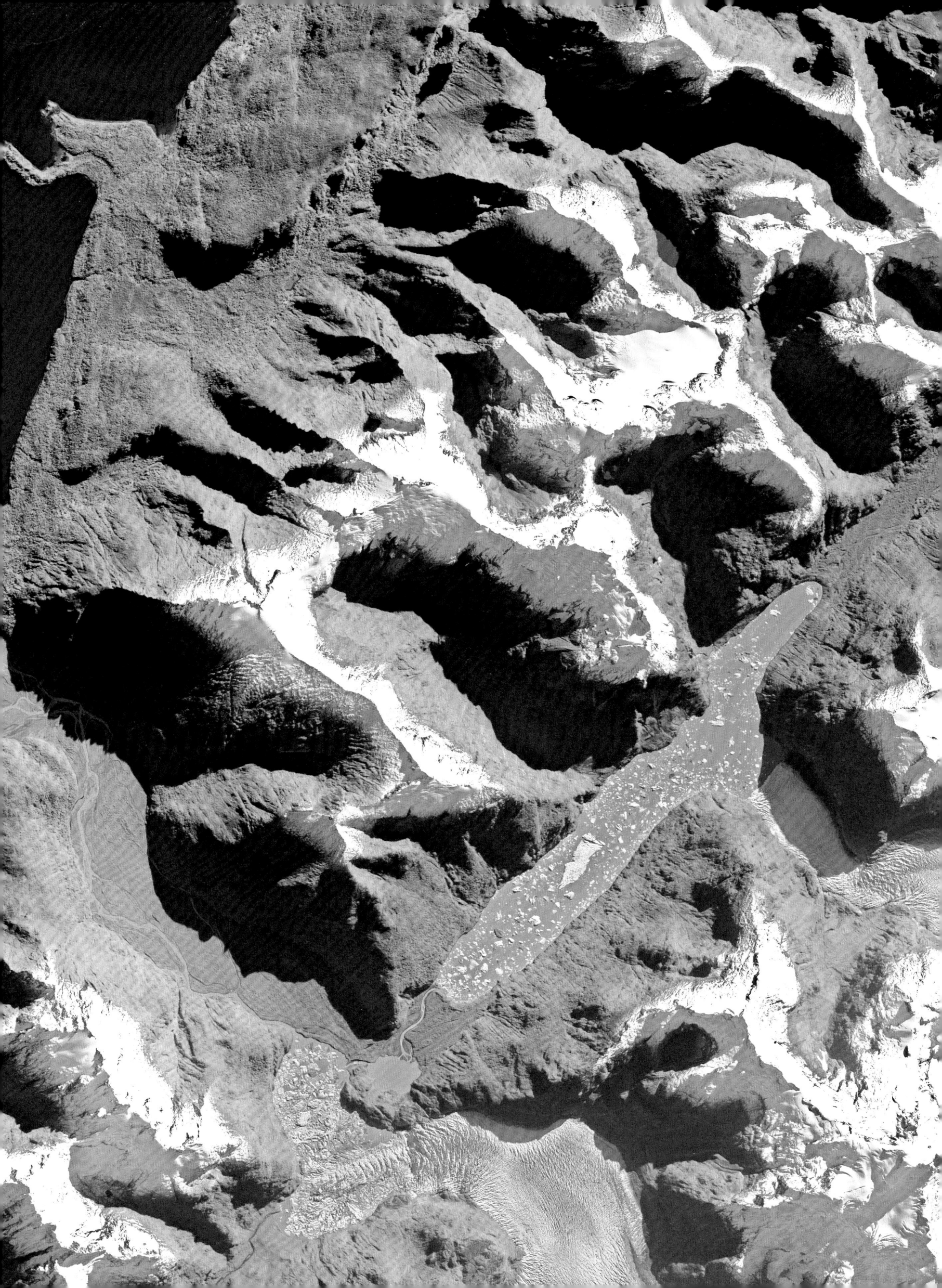

The fight against deforestation is now being managed from space. Thanks to satellite images, we can identify areas being illegally cleared of trees in very close to real time and dispatch helicopters carrying government agents to the scene. And the results are spectacular, at least in Brazil.

___

PROTECTING FORESTS

___

Each year, 32 million acres (13 million hectares) of forest are destroyed around the world due to deforestation—the equivalent of one soccer field every two seconds. The Brazilian Amazon has lost 16.3 percent of its forested surface since the 1970s, or 252 million square miles (653 million sq km), which is equivalent to France and Portugal put together.

The causes are diverse but nearly all are attributable to man: Deforestation is due to the expansion of agricultural zones, the sprawl of urban areas, and the demand for wood for construction, heat, and other basic needs. Even so-called "natural" causes are often due to human activity: More than 80 percent of fires in nature are started by man.

Deforestation spreads at the same rate as man's advance into the wilderness. New roads open up land for agriculture and logging. The most eloquent example came in Brazil in the 1970s, with the construction of the Trans-Amazonian Highway, a 3,100-mile (5,000-km) road that bisected the Amazon forest. Satellite images clearly reveal that the spread of deforestation branched off perpendicularly from the main road as smaller access roads were created.

Deforestation is a major issue in the protection of our planet. It is responsible for 20 to 25 percent of global emissions of carbon dioxide (CO_2), the principal greenhouse gas. But this is not all. Forests hold exceptional biodiversity (close to 50 percent of all the species on the planet) and provide many services by regulating watercourses and floods, purifying the hydrographic network, influencing precipitation, stabilizing soil, and producing oxygen. It is simply essential to preserve these processes.

THE BIGGEST FOREST IN THE WORLD

The biggest equatorial forest in the world is the Amazon, covering almost two million square miles (more than five million sq km). In the Brazilian Amazon, deforestation peaked in 2003–2004, with the clearing of 10,722 square miles (27,772 sq km). The state of Mato Grosso was particularly hard-hit: Since 1988, when statistics began being recorded in this field, the state has experienced 35 percent of the deforestation in Brazil. Soybean cultivation and cattle farming are the state's two major agricultural activities . . . and the leading causes of deforestation. Due to this deforestation, Brazil is among the countries with the highest CO_2 emissions on the planet: In 2004, it produced more than 355 million tons of CO_2, 1.3 percent of the global total.

CONTROLLING DEFORESTATION

To fight deforestation, the phenomenon must be accurately evaluated. How does one follow activities taking place over a vast territory in the heart of the jungle, sometimes hours and hours away from the closest city by road?

In the late eighties, Brazil established a satellite surveillance system to measure the evolution of forest cover—at that time, deforestation in the Brazilian Amazon was affecting about 5,800 square miles (15,000 sq km) of forest a year. In 1988, the Brazilian National Institute for Space Research (INPE) put in place its first monitoring system, PRODES, which covers the 1.5 million square miles (4 million sq km)

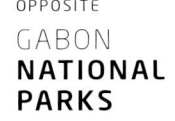

OPPOSITE

GABON
**NATIONAL
PARKS**
1°02' S – 14°04' E

During Earth Summit 2002 in Johannesburg, Gabon announced the creation of thirteen national parks covering 11.25 percent of its territory. Six of these parks are in forest zones: Ivindo, Minkebe, Moukalaba-Doudou,

Mwagne, Bateke Plateau, and Lope. With a small population of only 1.5 million inhabitants and tourism still limited, wild fauna and flora in Gabon currently remain relatively protected from human activities.

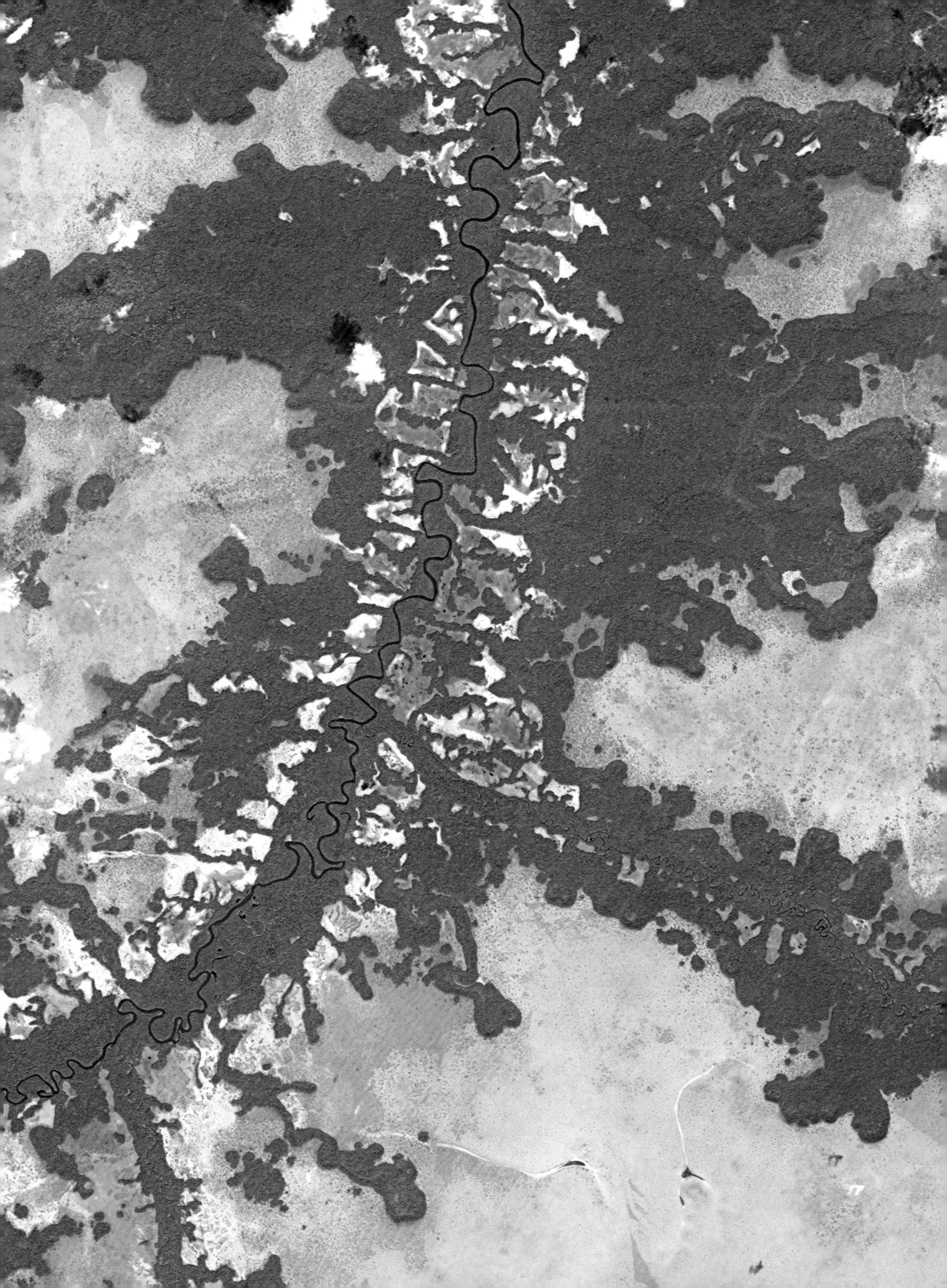

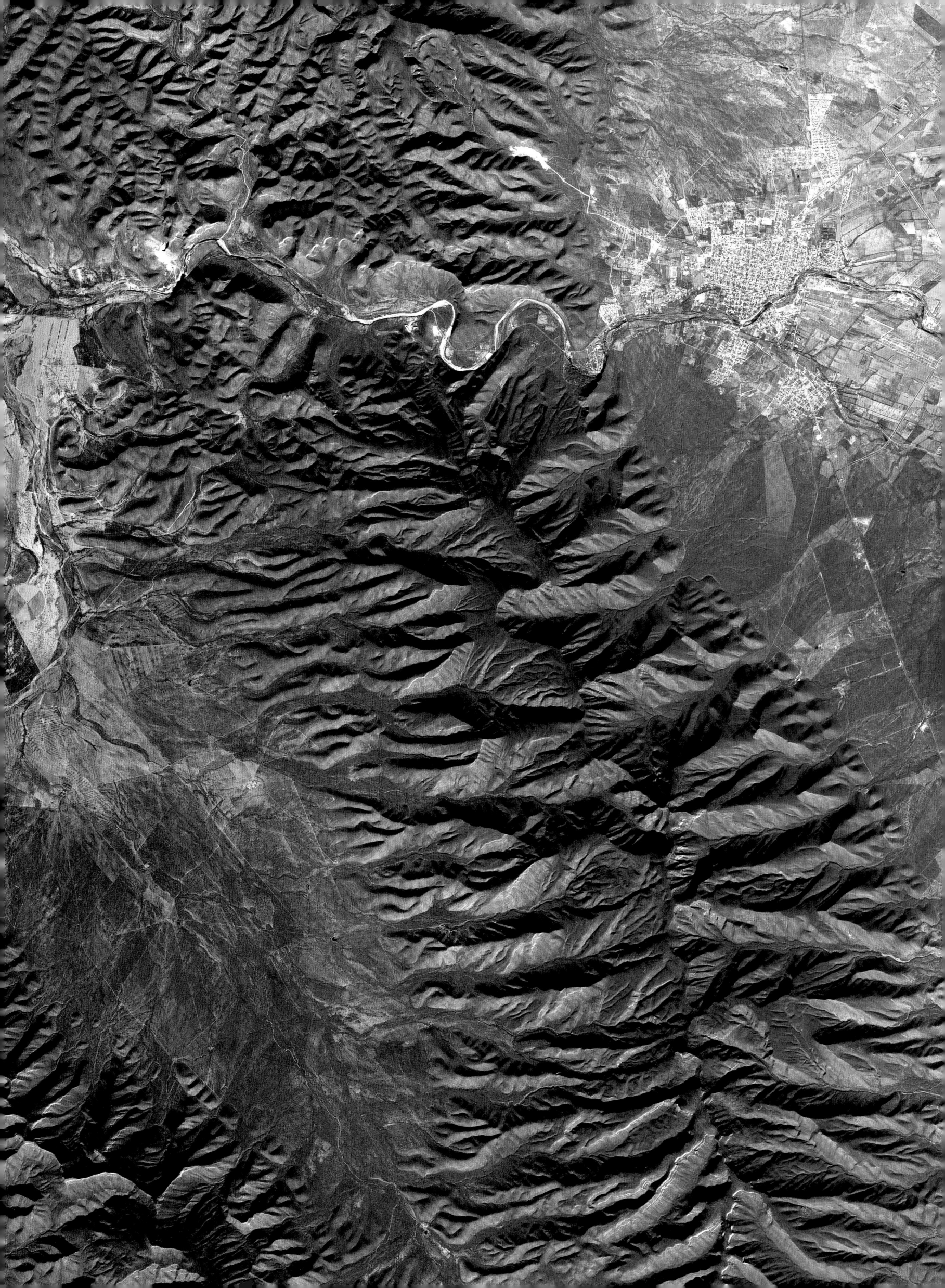

of the Brazilian Amazon. Every August, it provides an annual estimate of the amount of deforestation by clear-cutting, or the clearing of the entire forest floor. But PRODES has its limits: Only surfaces of more than 16 acres (6.5 hectares) can be observed, and the satellite does not detect partial deforestation due to fire or selective cutting. In 2003, INPE launched a new satellite system for detecting deforestation in real time: DETER, which distinguishes between clear- and selec-

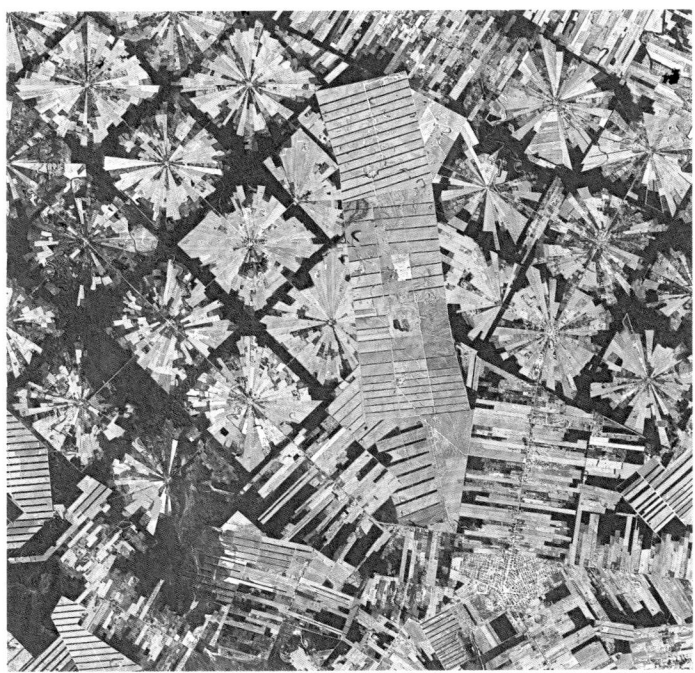

tive cutting. Every two weeks, this system sends detailed information about cleared areas and areas being deforested, data which is analyzed by the Brazilian Institute of Environment and Renewable Natural Resources (IBAMA), part of the Ministry of the Environment. DETER's primary mission is to provide the necessary monitoring tools: Once well informed,

the relevant authorities can choose to send the police to fight deforestation as it occurs.

The system allows IBAMA's agents to be particularly responsive: They are guided to the zones identified by GPS, speeding to apprehend the culprits by helicopter or 4×4. In 2011, IBAMA arrested numerous criminals and seized 650 trucks, 60 bulldozers, and 200 chain saws.

The two systems' approaches are complementary: DETER has lesser spatial resolution, but it provides data on a daily basis. PRODES is slower, but more precise; it is used to estimate the annual deforestation rate.

WHAT SATELLITES TELL US

The data provided by these satellites was made public in 2003. From that point, the rate of deforestation in Brazil dropped rapidly, with only 4,498 square miles (11,651 sq km) of forest cleared between 2006 and 2007. These figures, which were immediately heralded as proof of the success of the Lula administration's anti-deforestation measures, have come to have major political stakes. The disappointing 2008 figures led to major controversy in Brazil: The nation had set itself a goal of reducing the amount of forest surface area cleared in 2007–2008 to less than 4,247 square miles (11,000 sq km), but the first figures published that year showed that close to 2,700 square miles (7,000 sq km) had been cleared in the first five months. Yet the announcement was not a real surprise: There had been more forest fires and the ESA had suggested bad news was on the way. Nonetheless, the figures were met with a mixture of alarmism and denial from certain Brazilian political and media circles. Several governors put tremendous pressure on President Lula, claiming that

ABOVE

BOLIVIA
SANTA CRUZ DE LA SIERRA
17°48' S – 63°10' W

In the mid-1980s, populations from the Andes' high plateau settled in the area of Santa Cruz de la Sierra to escape difficult living conditions and to cultivate soya, whose production was then booming. These almost star-shaped agricultural plots have since totally eroded the forest. A small community was established at the center of each unit, with a bar, a school, and a church, and the soya fields stretching all around it. The large rectangular fields are also soya crops, interspersed with hedges serving as windbreaks.

OPPOSITE

MEXICO, NUEVO LEÓN
SIERRA MADRE ORIENTAL
25°40' N – 100°20' W

The Sierra Madre cuts across Mexico from north to south in two ranges known as the oriental and occidental. The oriental range is watered by clouds from the east and is more humid than the arid neighboring areas. It is home to a pine and oak forest growing at altitudes of up to 11,480 feet (3,500 m).

BRAZIL can claim the MOST SIGNIFICANT
GREENHOUSE GAS PRODUCTION DROP
ever accomplished by a single country.

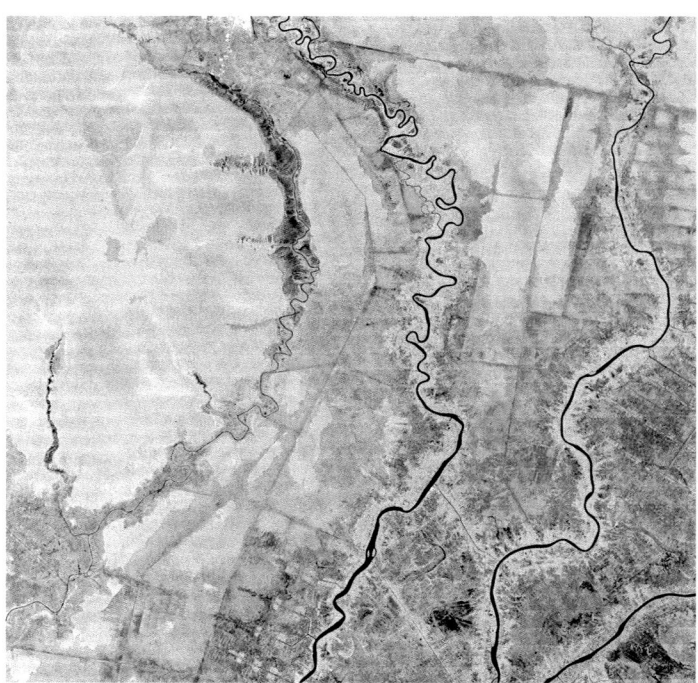

INPE was misleading the public with erroneous and exaggerated data. Eventually, the federal government reiterated its confidence in INPE for its objectives and actions in the fight against deforestation and for the country's economic development. And the figures were confirmed.

EXEMPLARY PROGRESS

At the United Nations climate conference in Copenhagen in 2009, Brazil committed to reducing forest clearing by 80 percent before 2020.

Thanks to satellite surveillance, IBAMA's arrests, political and social initiatives to fight the causes of deforestation, and the government's steadfast support, results were soon apparent. From August 2011 to July 2012, deforestation in the Amazon was at its lowest since statistics began being recorded, with only 1,797 square miles (4,656 sq km) of forest lost, a drop of 27 percent from the previous year and of 83 percent from the 2003–2004 period. Brazil can claim the most significant greenhouse gas production drop ever accomplished by a single country. In CO_2 equivalents, the limitation of Brazilian deforestation from 1996 to 2005 is equivalent to a drop of 870 million tons of CO_2 per year. This is more than the European Union's unrealized objective of 850 million tons.

Precision satellite imagery from the SPOT satellites used in the MT Legal program contributes to establishing an accurate land register, in an area where property borders were traditionally blurry. It also helps to determine the exact rate of plant cover. Brazilian law requires landowners to preserve part of the forests on their property—notably that within 328 feet (100 m) of a watercourse. Based on these images, the authorities identified landowners and enjoined them to preserve their forests or ensure their regeneration if they had been destroyed.

Environmentalists have applauded these Brazilian successes, but remain wary: Agriculture and major infrastructure projects remain threats that could lead to a sweeping recurrence of deforestation in Brazil.

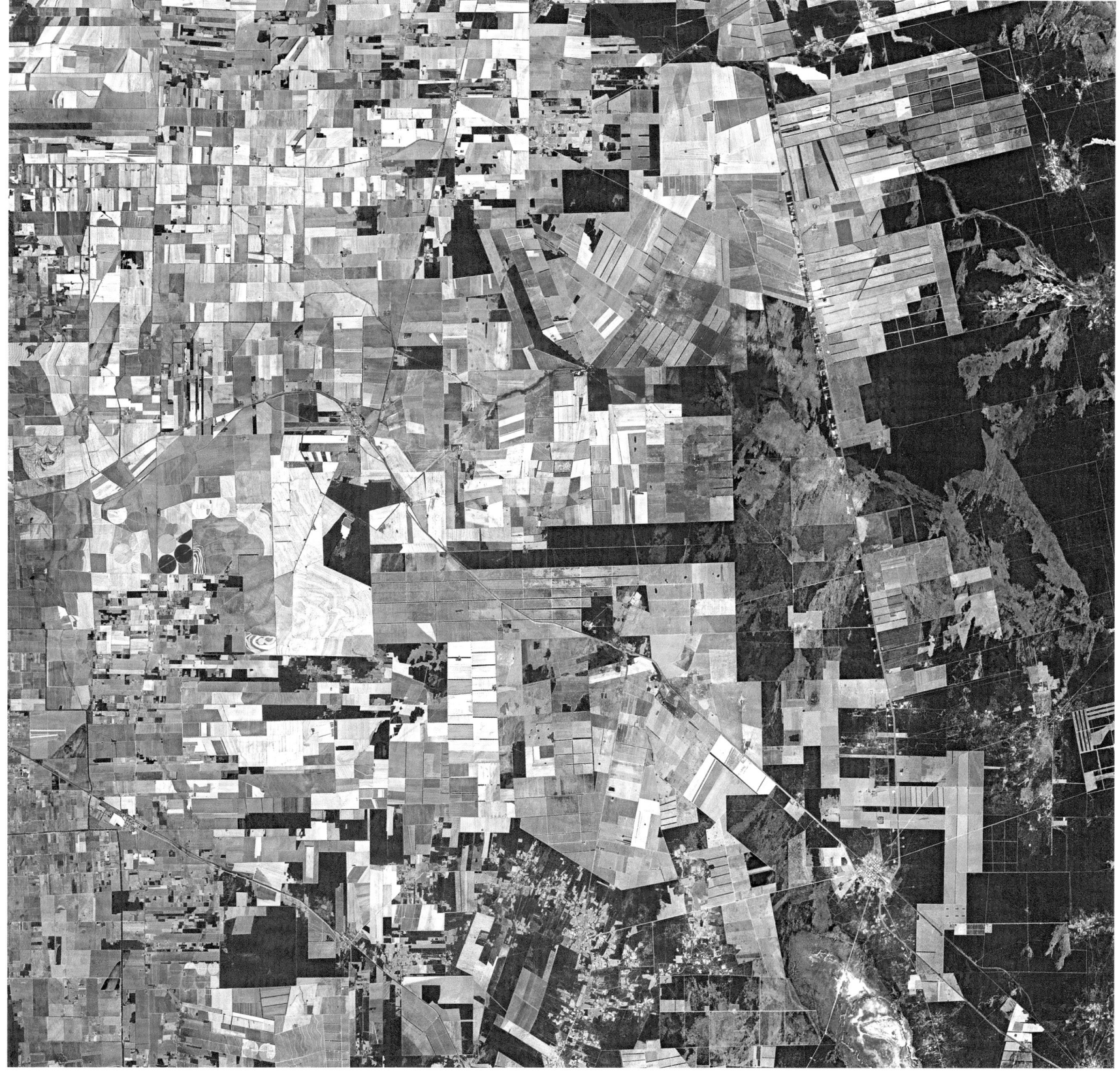

OPPOSITE

INDONESIA, KALIMANTAN (BORNEO)
SEBANGAU NATIONAL PARK
2°33' S – 113°50' E

The peat bog tropical forest of the Sebangau ecosystem in Indonesian Borneo was recently made into a national park, primarily to save Borneo's population of orangutans, which is extremely threatened by deforestation. Satellite data such as this image contributes to monitoring protected forest zones around the world by providing information about their status and helping to identify changes caused by illegal forest exploitation, for example.

ABOVE

ARGENTINA
TUCUMÁN PROVINCE
26°94' S – 65°34' W

By colliding with the Andes, humid Atlantic winds annually generate more than 40 inches (1,000 mm) of rain, whose waters come down on Tucumán Province in the northwest of Argentina. Abundant precipitation gives rise to lush vegetation and promotes varied agriculture. The province owes its nickname of "garden of the Republic" to its production of sugarcane, lemons, kiwis, chili, corn, rice, tobacco, wood, and flowers.

2010

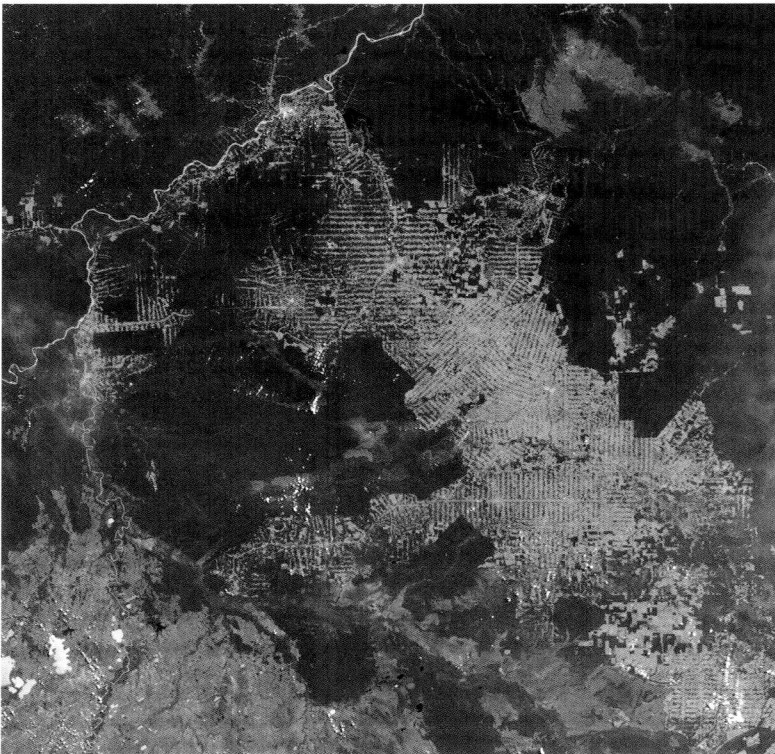

2000

2003

BRAZIL, RONDÔNIA
AMAZON FOREST
10° S – 62°30' W

The Brazilian state of Rondônia, which once held 77,220 square miles (200,000 sq km) of forest—a surface area greater than Kansas—is now one of the most deforested zones in Amazonia. Here, primary forest is shown in green, while deforested zones are in light green (farmland or pastures) or beige (bare land). The evolution of deforestation is predictable: The first trees to be cut are usually found along roads, then with time the areas of deforestation expand.

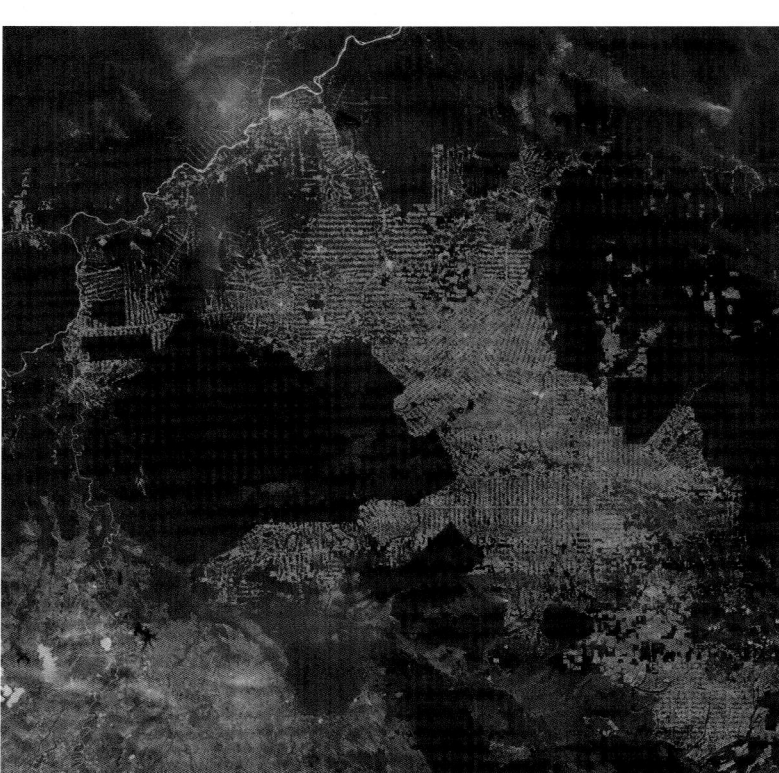

2007

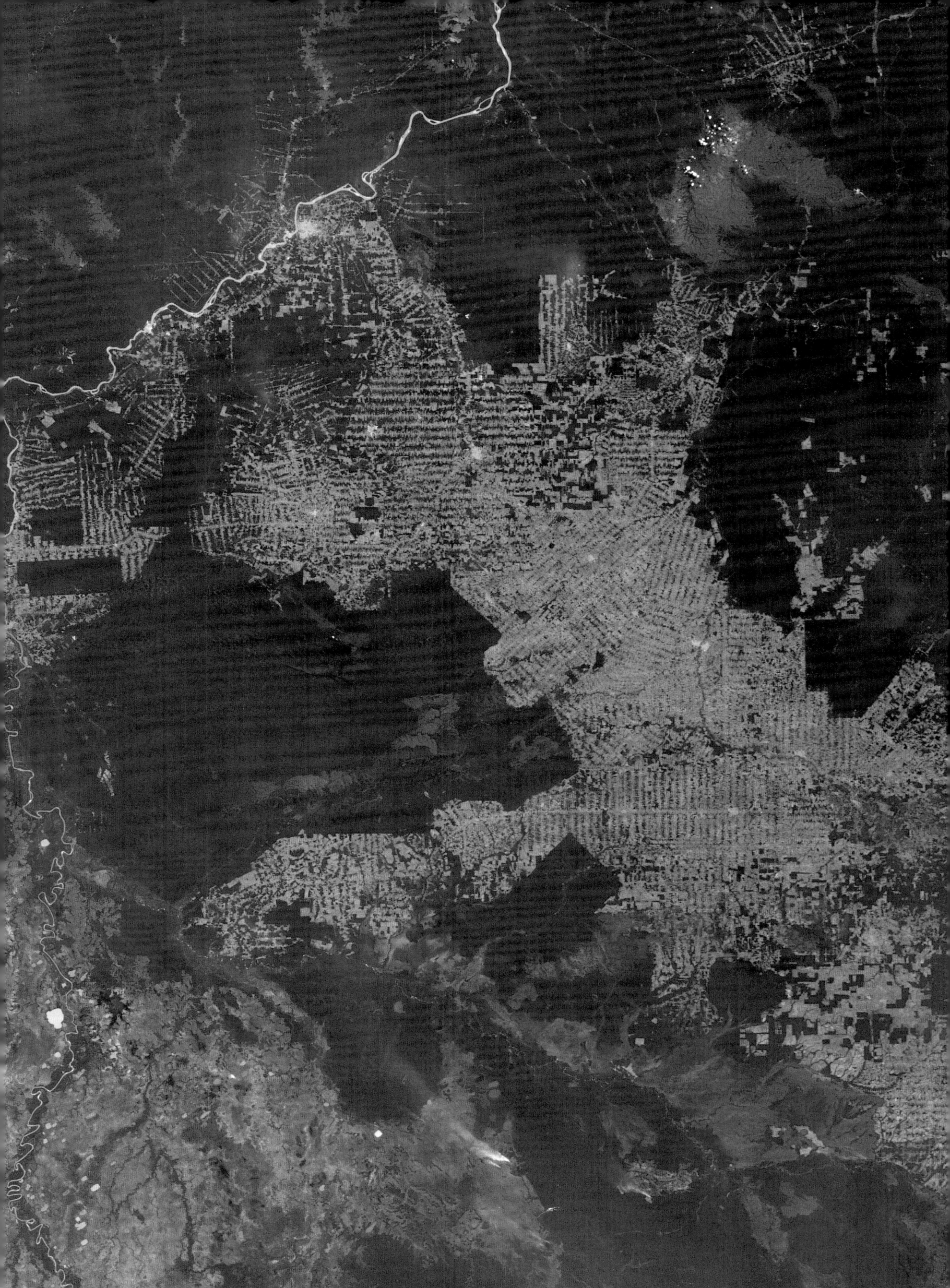

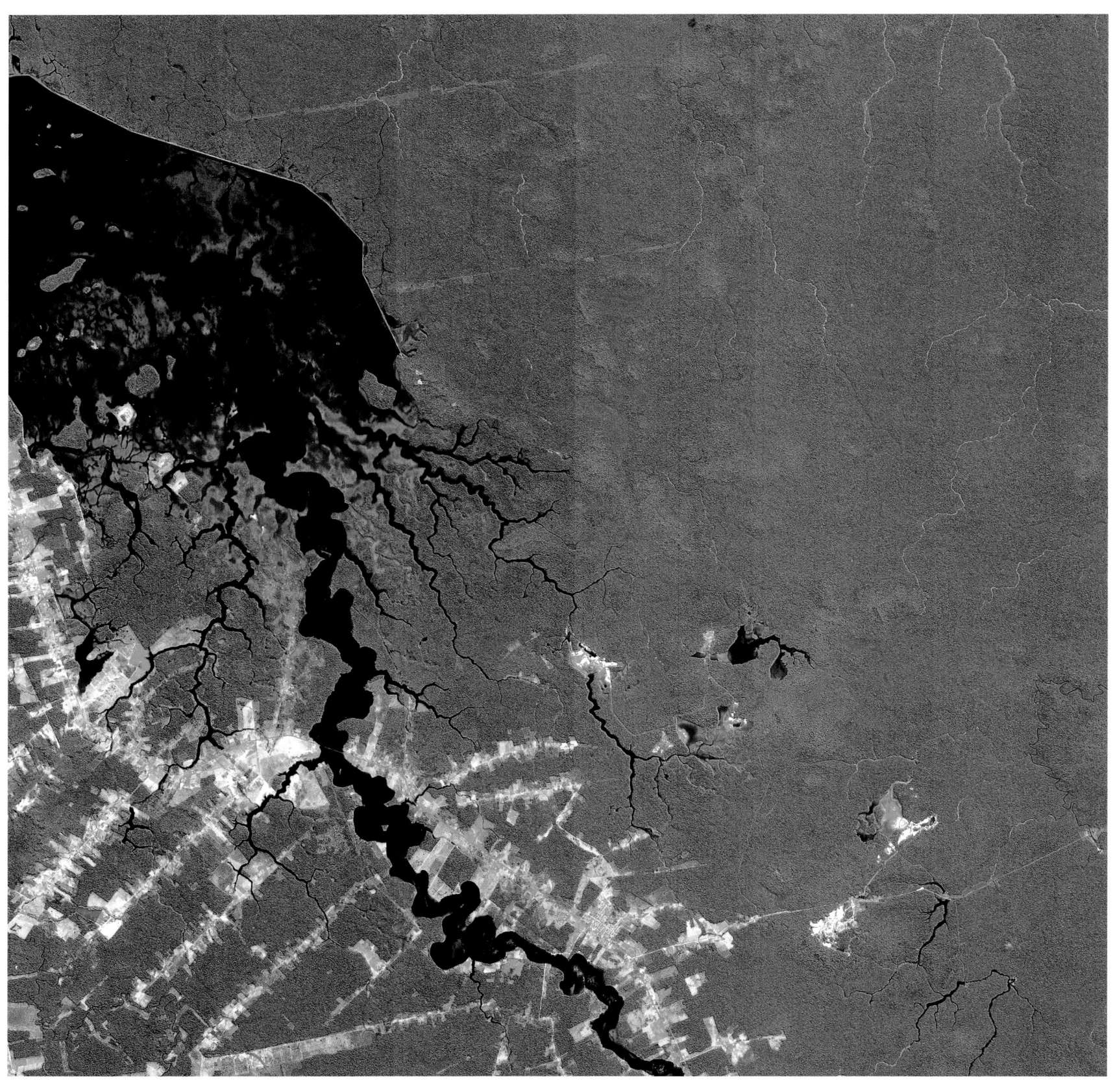

ABOVE

BRAZIL,
RONDÔNIA
SAMUEL DAM
8°46' S – 63°28' W

The construction of hydroelectric factories along the major
Amazonian rivers has required
impressive dam reservoirs to be
built, submerging vast areas of
forest. These factories often incite
the development of agriculture,
which extends in a comblike
network along the flooded zones,
to the forest's detriment.

THE IDEA is a novel one: to FINANCE THE LIMITATION OF DEFORESTATION, or in other words, to PAY PEOPLE TO NOT CUT DOWN TREES.

SATELLITES TO THE FORESTS' RESCUE

Of course, the crucial issue of deforestation does not only concern Brazil. The phenomenon is a global one, concerning many countries.

One of the paths being considered to reduce deforestation on a global scale is to show its connection with global warming, as one of the primary causes of greenhouse gas production.

The United Nations Framework Convention on Climate Change (UNFCCC) developed a strategy known as REDD (reducing emissions from deforestation and forest degradation), which has since been expanded to include sustainable forest management and is now called REDD+. The idea is a novel one: to finance the limitation of deforestation, or in other words, to pay people to not cut down trees. In late 2012, the REDD programs coordinated by the United Nations included forty-six countries and had $117 million in financing.

REDD makes use of satellite images to observe deforestation and measure its impact. They have created a reference scenario, which compares actual deforestation with projected deforestation without the implementation of the REDD program, followed by an evaluation of the corresponding quantities of CO_2.

Other projects have also used satellite images to estimate that the forests of the Congo Basin are storing between twenty-five and thirty billion tons of CO_2. Combined with archival images, recent images will serve as a foundation for studying the area. These images will also be used as the basis for forest maps. Finally, new data should make it possible to update information over time, and thereby follow the evolution of the forest cover. These are all essential instruments to ensure the sustainable management of forests through the REDD program.

Aside from image satellites, GPS satellites also play an important part in the fight against deforestation. Indeed, they allow populations to accurately delimit the land that belongs to them and which must be protected. This is particularly useful in Brazil, where in recent years Amerindians have begun registering their land in order to enjoy legal protection.

Of course, satellites will not save forests alone: Human beings must protect them on the ground. Every preservation operation depends on the participation of the populations concerned.

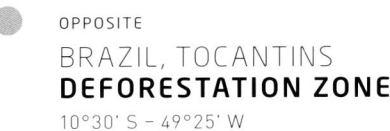

OPPOSITE
BRAZIL, TOCANTINS
DEFORESTATION ZONE
10°30' S – 49°25' W

In just a few years, Brazil reduced its rate of deforestation by nearly 80 percent. This was mainly due to its pioneering satellite monitoring program.

Talking with Gilberto Câmara

"Transparency" Is the Watchword of Our Fight

After eight years as the head of INPE, what is your assessment of the situation?

From 2004 to 2012, deforestation in the Brazilian Amazon was reduced fivefold: It dropped from 10,425 square miles (27,000 sq km) to 1,853 square miles (4,800 sq km). This success is due on the one hand to the implementation of systematic satellite monitoring of deforestation activity in the area and on the other to data transparency. The data is accessible to all, which sets Brazil apart from many other countries facing deforestation. It is the keystone of our advances.

What was the decisive factor that allowed this reduction in deforestation?

Besides technology and the Institute itself, it seems to me that the most important step in the fight against deforestation in our country was President Lula's speech at the United Nations climate conference in Copenhagen in 2009. He announced that Brazil was committing to reduce its deforestation by 80 percent by 2020. By doing so, Brazil set itself apart from other countries blamed for deforestation, such as Indonesia and China, as well as those that are waiting for international aid to act. What appeared to be a utopian idea in 2009 has become a reality: The goal has already nearly been reached and it is only 2013.

How do you explain this success?

In Brazil, "transparency" is the watchword of our program to fight deforestation. Of course, both the methods developed and the action on the ground are indispensable, but these are only the visible tip of the iceberg. Without transparency and an institution responsible for the results and figures, we would not have had the same success.

This is also what allowed us to survive our most significant crisis. In 2008, we used satellite data to establish a list of critical deforestation zones. In these areas, the government prohibited banks from giving loans to agriculture companies and farmers who wanted to expand their activities. While this method appears unfair toward honest farmers, most of them are not and destroy the forest without giving any thought to the law. Elected officials from these areas tried to discredit us. But thanks to the transparency of our activities, our data could be verified and the press, national and international scientific communities, and even the president supported us. Our legitimacy has not been called into question since. To quote a leading daily newspaper, "It's not by switching thermometers that you cure the patient."

Gilberto Câmara was the director of the Brazilian National Institute for Space Research (INPE) from 2005 to 2012 and implemented the national program to combat deforestation by satellite.

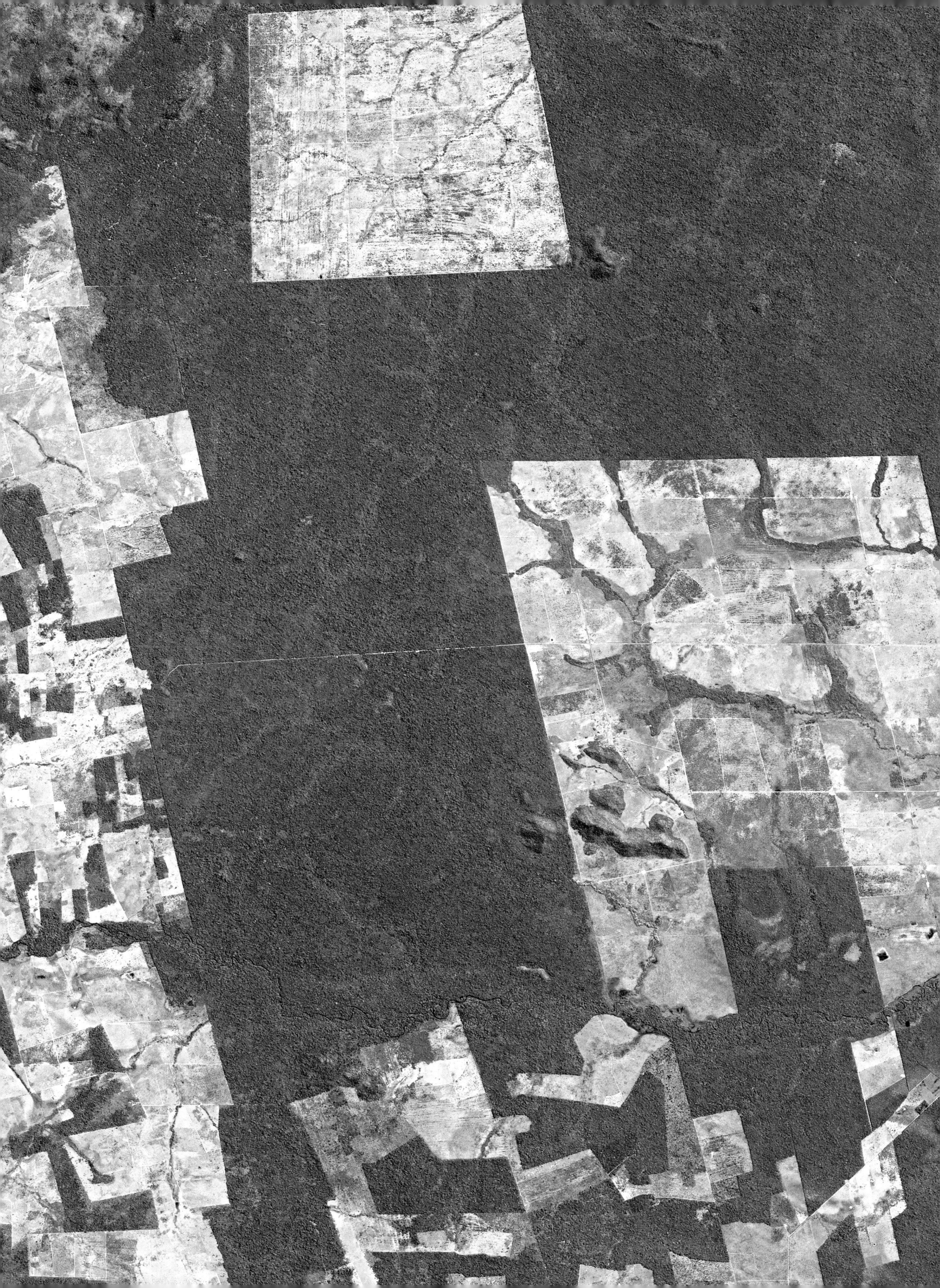

THE FORESTS OF MADAGASCAR

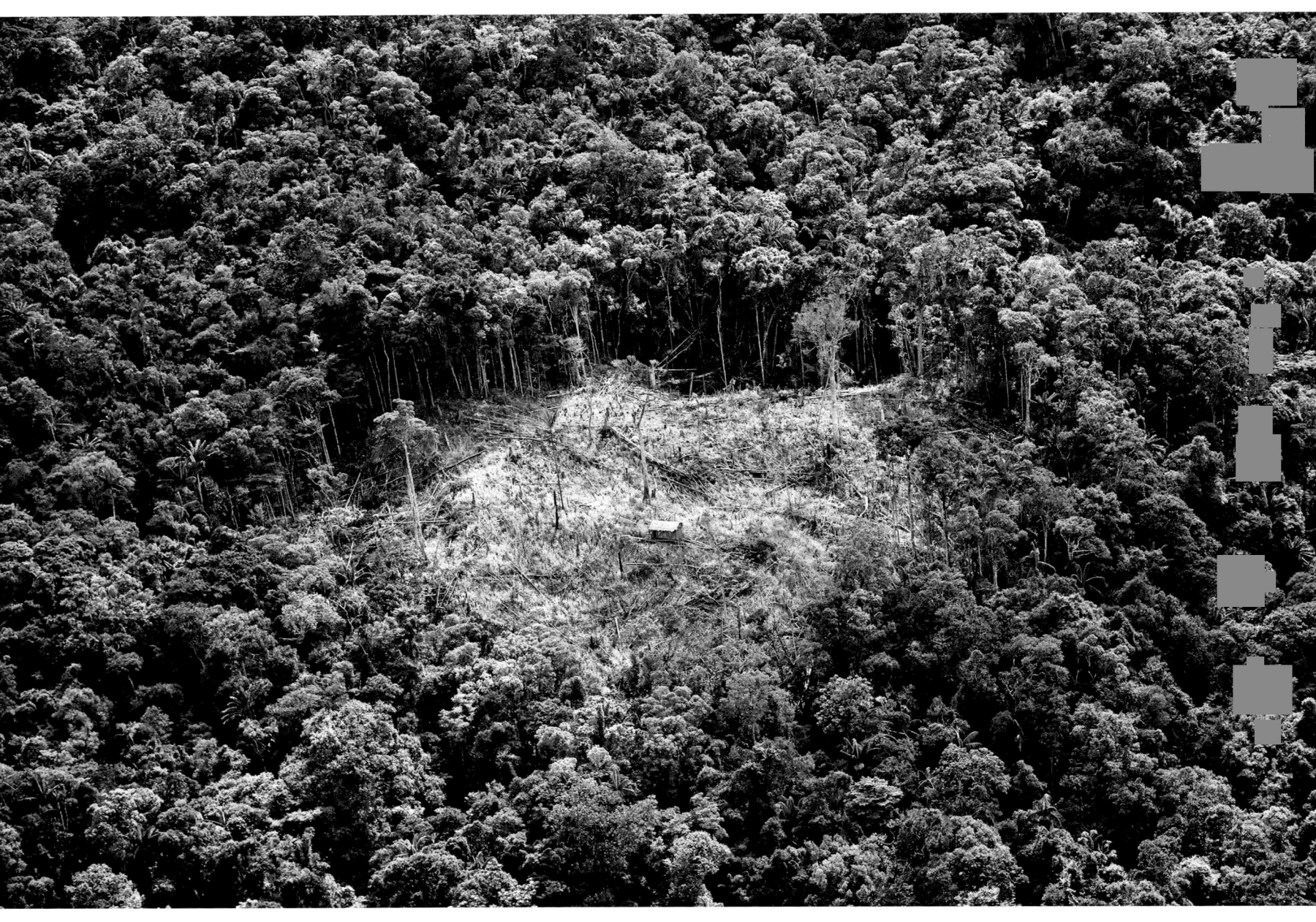

ABOVE

MADAGASCAR
**DEFORESTATION
BETWEEN
MAROANTSETRA
AND TOAMASINA**
17°05' S – 49°10' E

Deforestation in Madagascar has multiple causes. In the east, on hills enjoying heavy rain, rice is primarily grown by slashing and burning the forest. Elsewhere, the expansion of pastures and the production of wood coal for cooking are decisive factors in forest clearing. Combating deforestation on the island therefore requires a specific and multi-headed approach. Combined with studies in the field, satellite images enable close monitoring of the forest's health, particularly in a country like Madagascar where roads are very limited.

Measuring 228,000 square miles (590,000 sq km) and with a population of twenty million, Madagascar is the fifth-largest island on the planet. This Noah's Ark of the Indian Ocean is home to a unique range of biodiversity. But its natural capital is threatened by large-scale deforestation affecting 124,000 acres (50,000 hectares) of natural forest a year.

In 2008, the GoodPlanet Foundation partnered with WWF and Air France to launch a forest conservation program in Madagascar covering more than 1,236,000 acres (500,000 hectares). In recent years, the project has evolved, leading to the creation of the Etc Terra association, which draws on a wide range of strategies: protection, reforestation, restoration of fragmented forests, and support of "green" agricultural practices, as well as the promotion of initiatives aimed at efficiently reducing greenhouse gas emissions. Because the destruction of forests releases vast quantities of CO_2, protecting forests is an effective tool in the fight against global warming. However, a close evaluation of the quantities of carbon dioxide stored in trees, plants, and the soil (and thus preserved by the project) is a challenge, for it requires an analysis of three types of information.

In order to quantify the benefits of preserving more than 1,000 acres of forest, first, land use has to be analyzed in detail. Planet Action provided very high-resolution satellite images, which researchers used to study the distribution of grassy savannas, agricultural land, bare soil, infrastructures, and different types of forests: dense, coniferous, second-growth, exotic, and the like.

Once this data was mapped, engineers used a laser light aboard an airplane flying over the forests to measure—within 20 inches (50 cm)—the height of the canopy and the trees. This technology is known as LIDAR (Light Detection and Ranging). Then the team and its volunteers carried out land inventories to measure the diameter and height of the trees and identify the species present in various sample sites.

Analysis of this data has allowed specialists to estimate the carbon storage potential of the forests in the area protected. Based on these surveys, if everything goes according to plan, Etc Terra believes the project will prevent the emission of 39 tons of CO_2 by 2030.

———

PAGES 94-95

UNITED STATES, ARIZONA
DESERT AND APACHE-SITGREAVES NATIONAL FOREST
33°38' N – 109°17' W

In the American Southwest, the transition from one ecosystem to another can be sudden. This image is a perfect illustration of that fact: The desert lands of

Arizona, seen here in purple, collide with the Apache-Sitgreaves National Forest, in green. Black zones in the desert indicate an ancient volcanic chain.

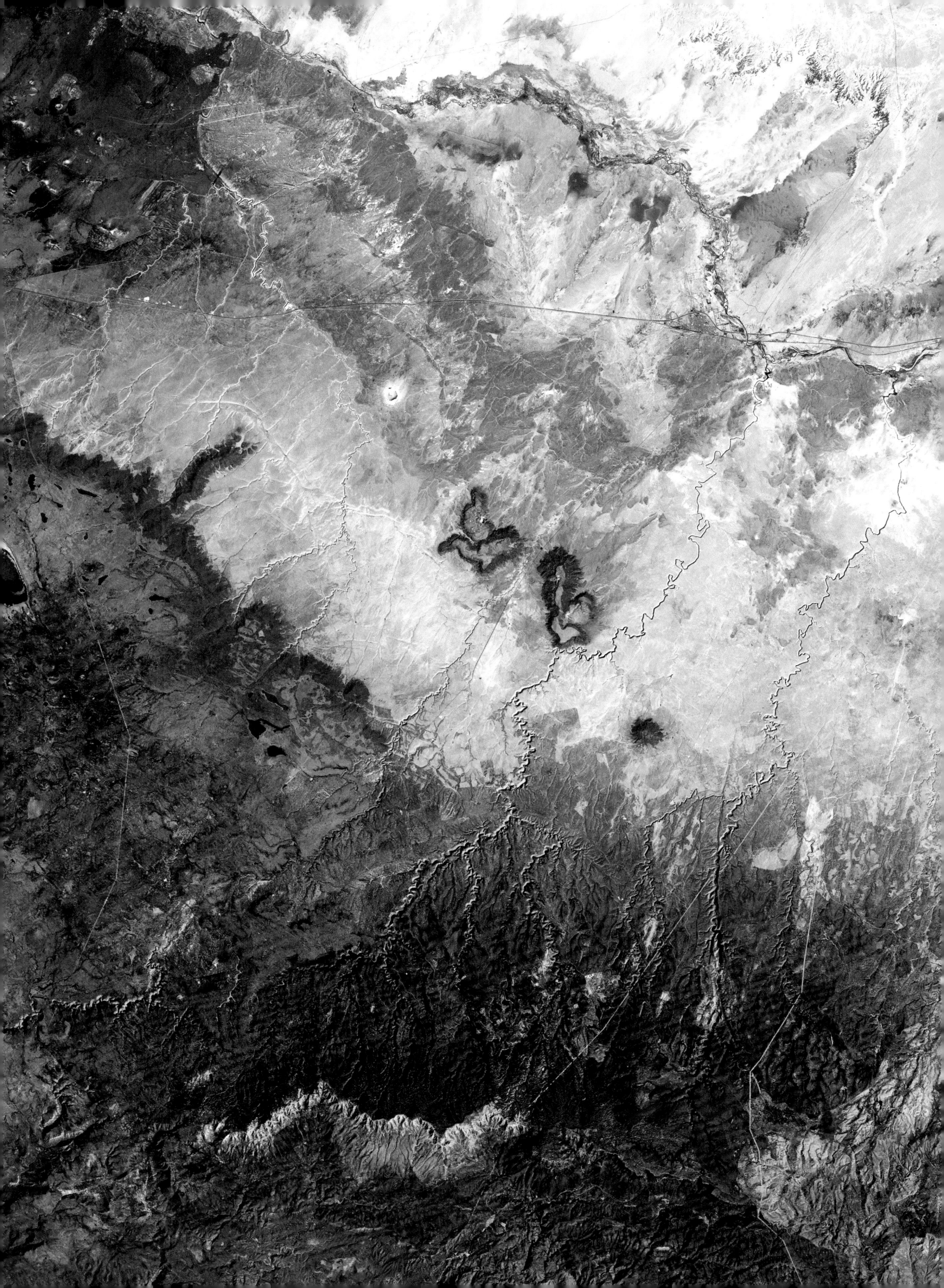

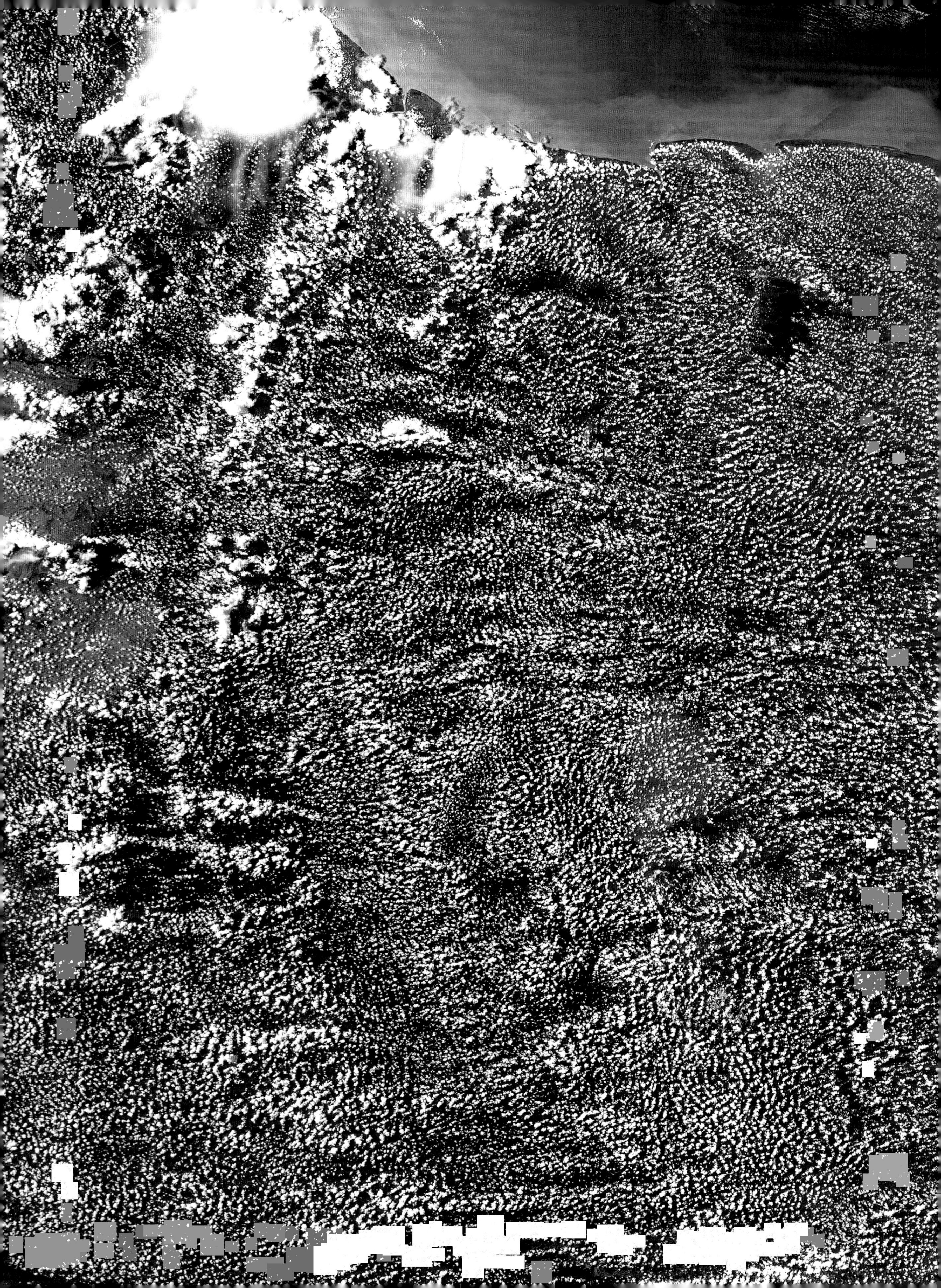

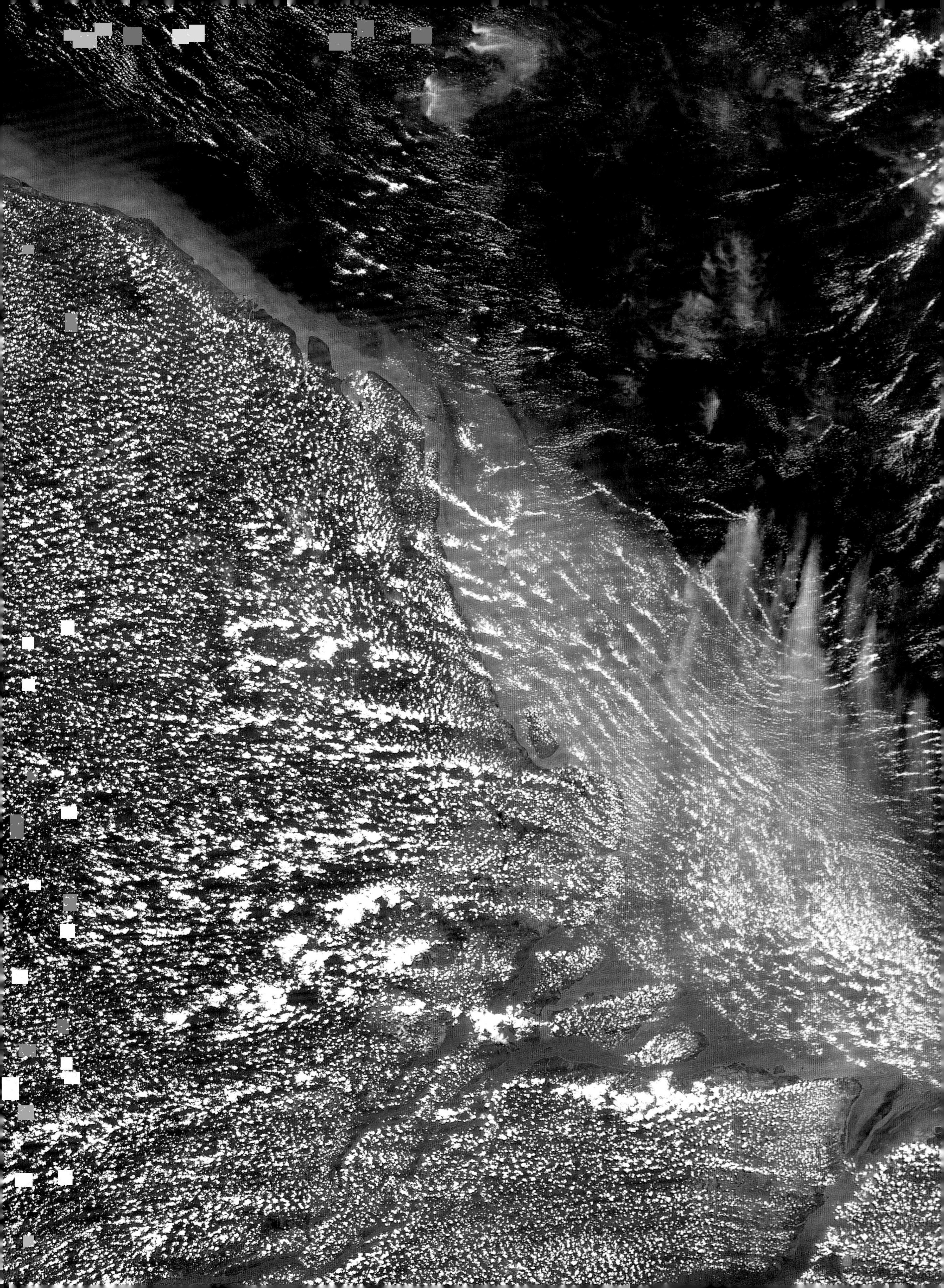

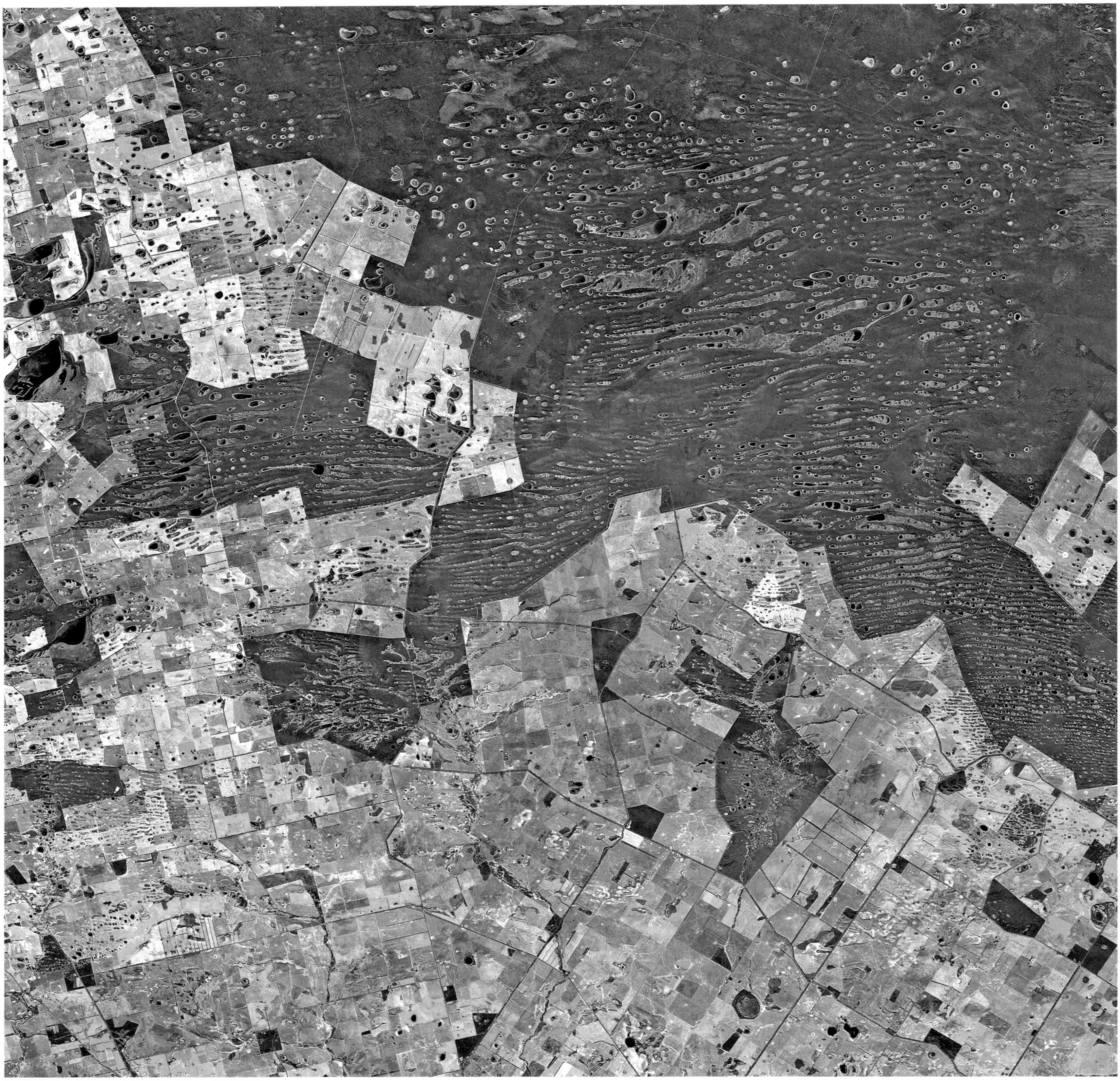

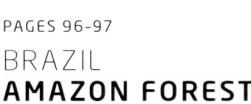

PAGES 96–97
BRAZIL
AMAZON FOREST
5° S – 40° W

Forests play a major role in regulating climate and
precipitation. The clouds covering the Amazon here were
formed by water vapor released by plants and trees.
Vegetation sweats when it absorbs carbon dioxide; a
tree can produce half a ton of water a day. In the heat of
daytime, humid air rises and condenses into clouds.

ABOVE
AUSTRALIA
ESPERANCE
33°17' S – 121°58' E

This image shows the expansion of agricultural land
into the forest north of the city of Esperance. From 1990
to 2010, Australia annually lost close to 642,470 acres
(260,000 hectares) of forest, or 0.17 percent of its forest
cover, due to agricultural and urban growth.

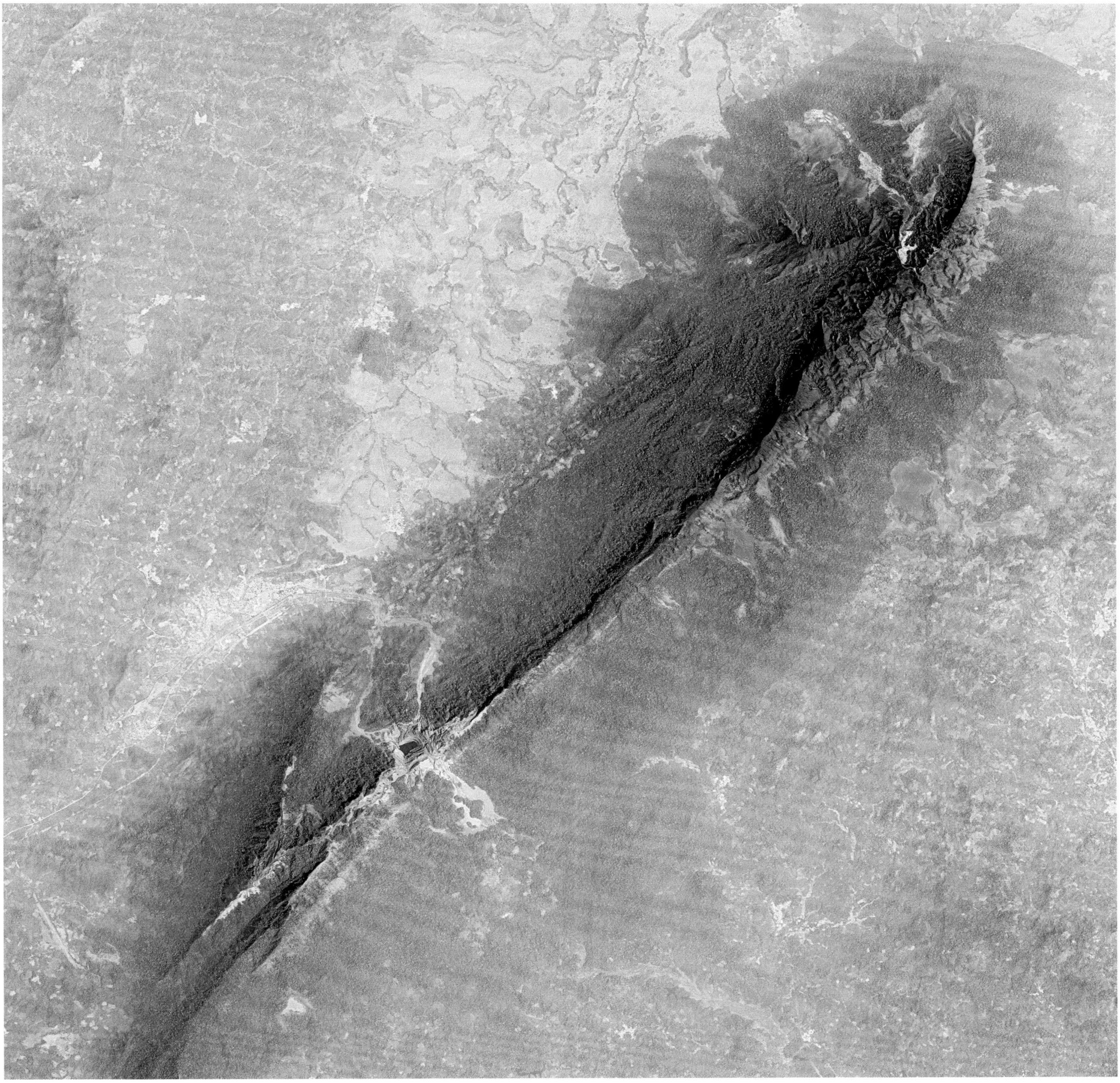

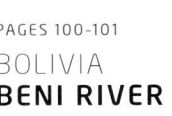

PAGES 100-101

BOLIVIA
BENI RIVER
14°06' S – 67°33' W

The Beni River basin stretches from Bolivia to Peru over
more than 109,200 square miles (283,000 sq km), a surface
area equal to that of the United Kingdom. The river springs
from the eastern slopes of the Andes, south of La Paz, and
flows into the Madeira, the Amazon's principal tributary.
The forests of the Peruvian Amazon are known as South
America's richest area for biodiversity.

ABOVE
IVORY COAST
NIMBA RANGE
7°37' N – 8°24' W

The Nimba Range is West Africa's water tower and
a UNESCO World Heritage site. It is home to dense
forests and grassy mountair pastures, as well as
protected animals such as the chimpanzee.

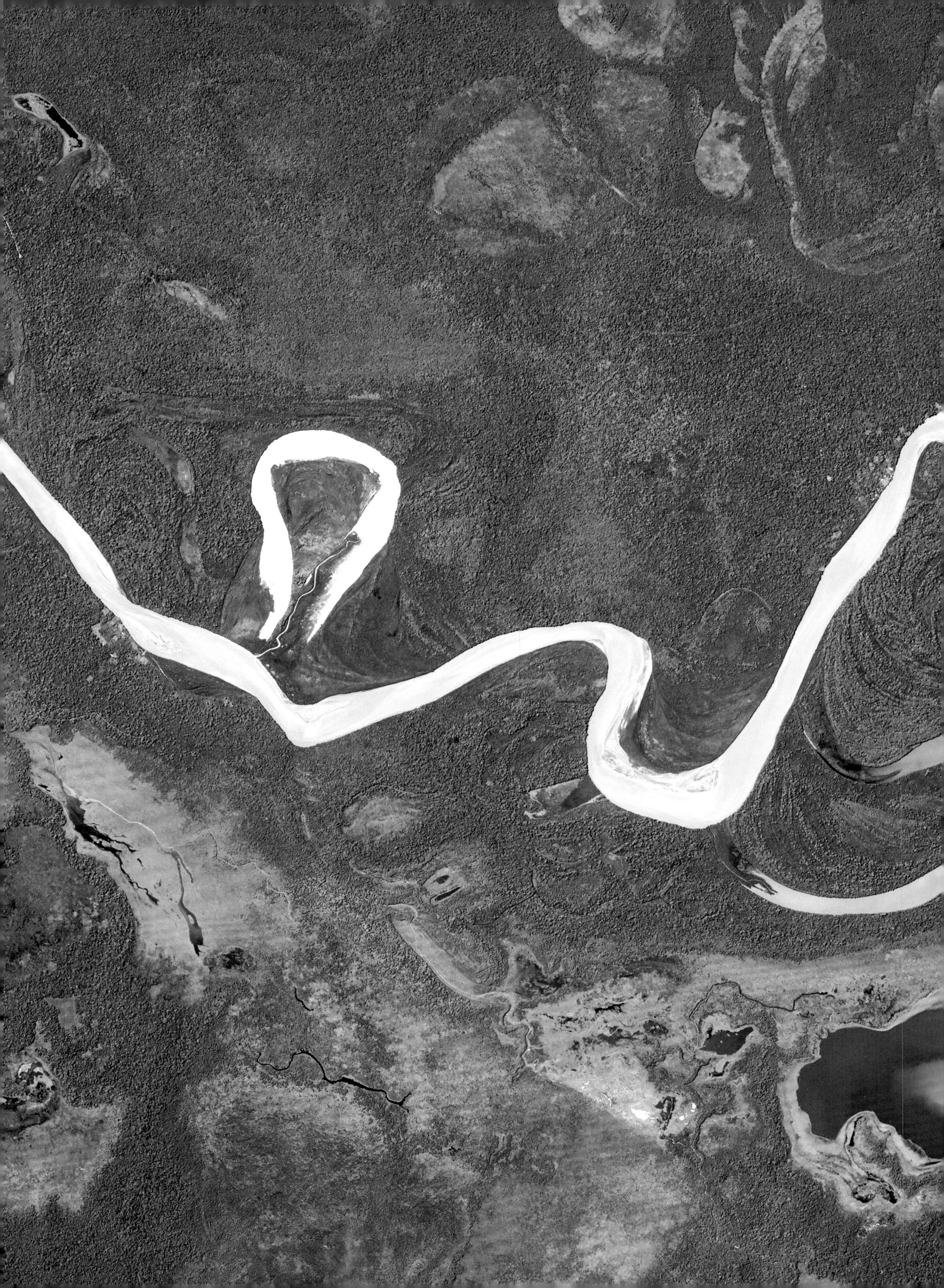

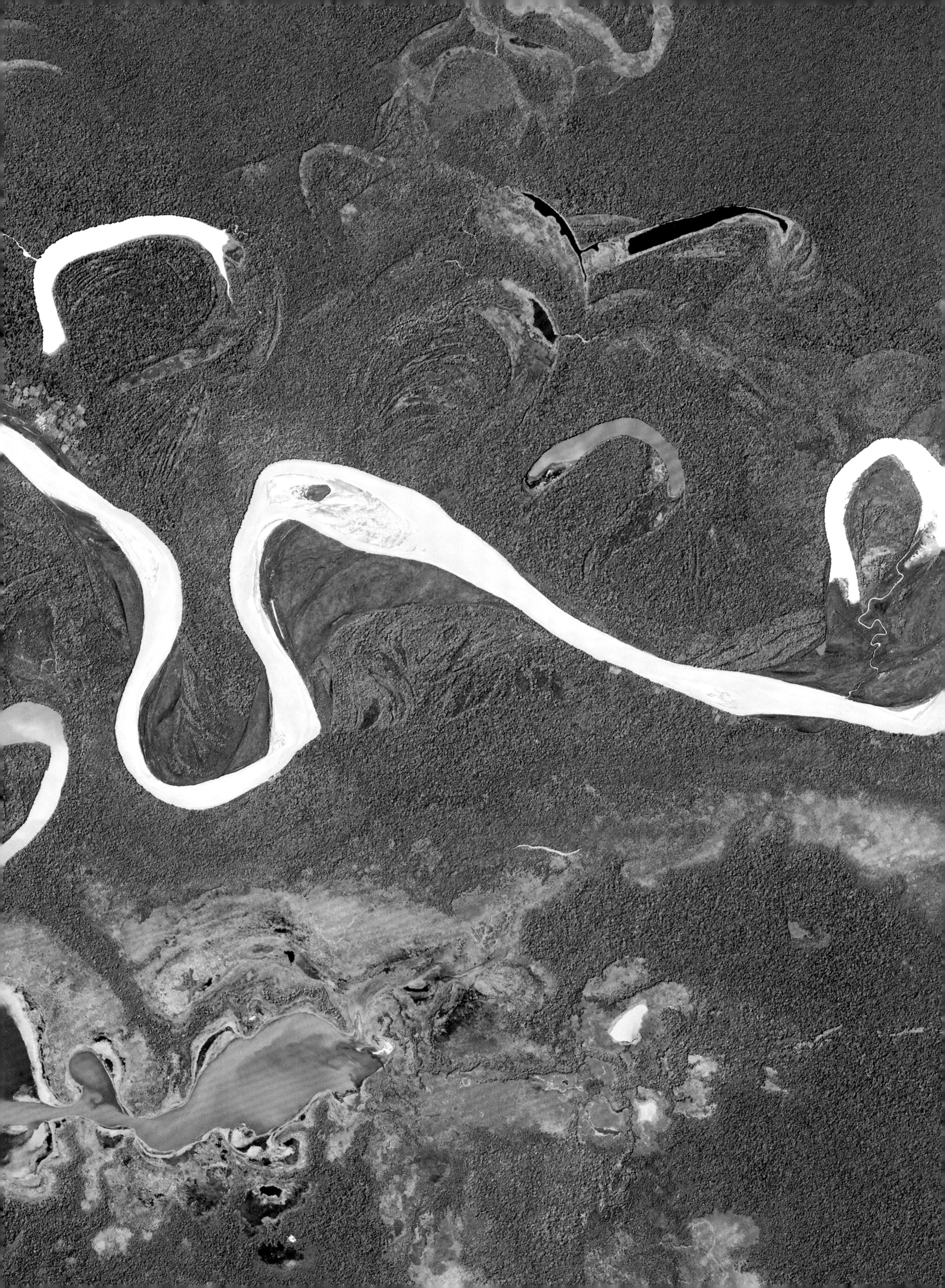

The story of the hole in the ozone layer is that of both an exceptional exploit in satellite imagery and a historical success for protectors of the planet. In the last twenty-five years, it has provided the most striking example of the international community's mobilization to save the planet.

___

POLLUTION FROM ABOVE

___

It all began in the early 1980s, when a team of British researchers led by Joe Farman discovered what would come to be known as the "hole" in the ozone layer over Antarctica. They could not believe their eyes—or rather their instruments, which they promptly sent back to England to be checked. But it was no mistake. The ozone layer had truly shrunk by 50 percent over Antarctica. Similarly, NASA reacted to disturbing data from its Nimbus-7 satellite by assuming its sensors were poorly calibrated. But by reexamining the satellite data in the light of Farman's findings, experts uncovered a gigantic "hole" covering some 11.5 million square miles (30 million sq km).

Scientists set off the alarm. Nothing less than the survival of the human race was at stake. Located at altitudes of 15 to 18 miles (25 to 30 km), this thin layer of gas—ozone—protects Earth from ultraviolet light. Without ozone, UV light could destroy all life on Earth.

The causes of stratospheric ozone's disappearance are well known: chlorofluorocarbons (CFCs) and hydrochlorofluorocarbons (HCFCs), gases used in industry as refrigerants and propellants. They were identified in 1974 by two chemists, Mario Molina and Frank Sherwood Rowland, who were awarded the 1995 Nobel Prize in Chemistry.

International decision makers gathered in 1987 to discuss the issue and decided to prohibit these products through what is known as the Montreal Protocol. The ban was easy to apply given that substitutes existed and were available on the market.

In just a few years, the problem of the hole in the ozone layer was diagnosed and a solution was found, adopted by the international community and implemented. Since then the quantities of CFCs and HCFCs in the atmosphere have slowly diminished. Yet because these gases have a long half-life, remission is slow. Experts who follow the annual evolution of the "hole," which also exists over the Arctic, predict it will take several decades to disappear.

MEASURING AIR POLLUTION

While the problem of stratospheric ozone depletion has nearly been resolved, the same is not true of many other types of atmospheric pollution. According to the World Health Organization (WHO), air pollution kills 3.2 million people a year, making it a major public health issue.

In most Western cities, air quality has considerably improved in recent years due to the adoption of stricter norms and deindustrialization. Yet elsewhere the situation is getting worse. The combustion of fossil fuels (oil, gas, and especially coal) for transportation and energy production is the main culprit. But cooking and heating in poorly ventilated rooms inside houses and buildings are even more lethal than industrial activities.

An increasing number of observation satellites are being put in place to track these pollutants in real time. For instance, researchers can follow what is known as the Asian brown cloud: First detected from space in the early 2000s, this "cloud" refers to a vast zone of pollution, covering several hundred thousand square miles (several million square kilometers) and being several hundred thousand square miles (several thousand meters) thick, which hangs over India and China several months a year and sometimes reaches as far as the western shores of the United States. The cloud can expand when there are gigantic forest fires in these areas.

OPPOSITE

UNITED STATES, GULF OF MEXICO
DEEPWATER HORIZON OIL SPILL IN 2010
28°12' N – 88°48' W

The biggest oil spill in American history took place in April 2010, when the Deepwater Horizon oil rig exploded 50 miles (80 km) off the Louisiana coast and killed eleven employees. It took BP eighty-seven days to stop the oil leak, which was at a depth of 13,123 feet (4,000 m), while dispersants and floating dams were deployed to prevent the oil from reaching the coasts. Meanwhile, close to 210 million gallons (800 million liters) of crude petroleum spread in the sea.

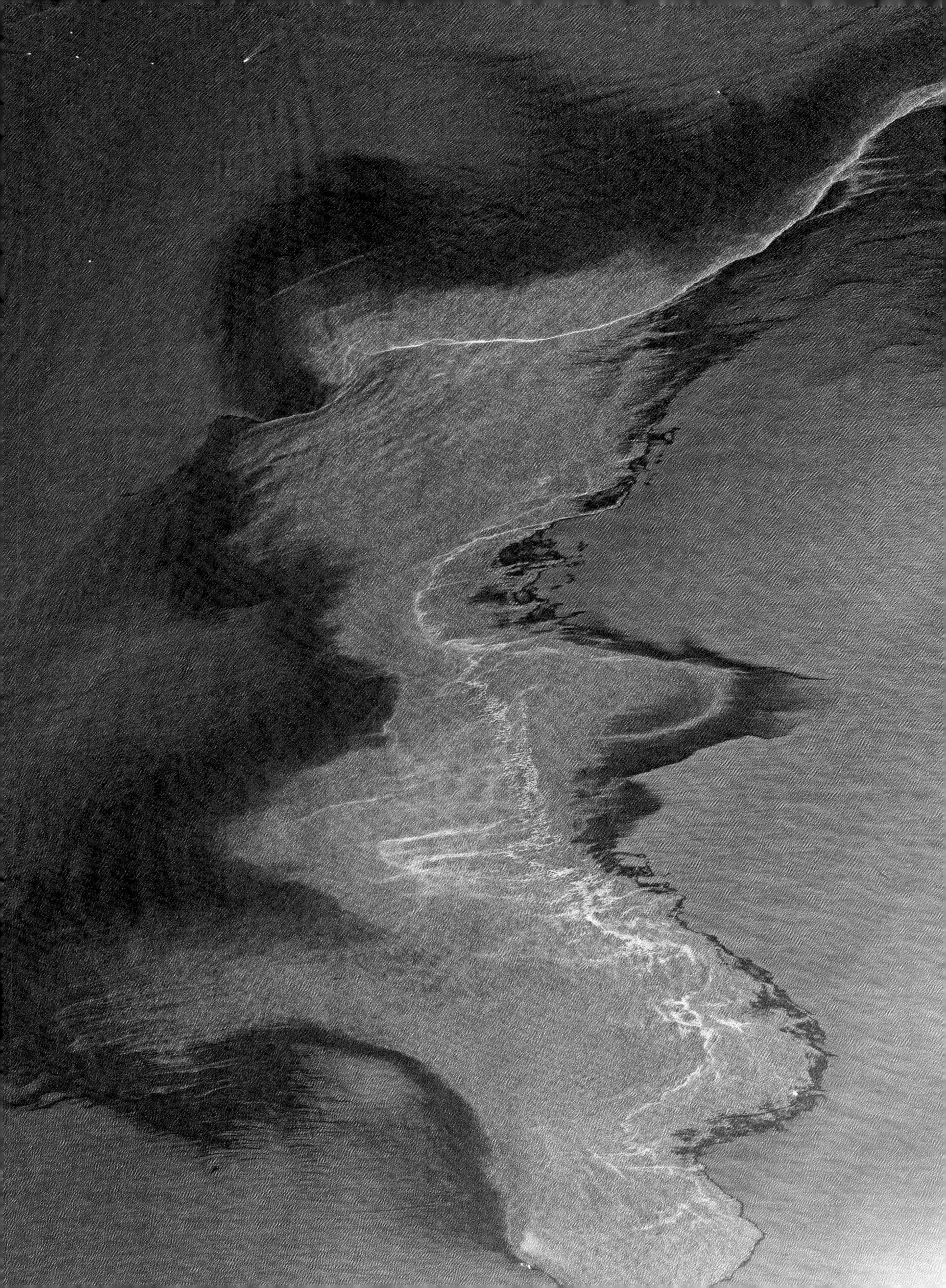

AUSTRALIA
HUTT LAGOON
28°10' S – 114°15' E

Located in an arid region, the Hutt Lagoon salt lake stretches 8.6 miles (14 km) by 1.2 miles (2 km) along Australia's west coast. Its red color is due to the presence of a microalgae known as *Dunaliella salina*. While it is green in low-salinity waters, the algae turns pink and red when salinity increases. It is cultivated and harvested for its carotenes, pigments used to create food coloring.

SOUTH AFRICA
THE BUSHVELD IGNEOUS COMPLEX
25°40' S – 27°22' E

The Bushveld Igneous Complex near Pretoria, South Africa, holds 90 percent of the world's reserves of platinum group metals (composed of six metals including palladium and iridium). Approximately 80 percent of the platinum annually extracted comes from these deposits. The blue polygons in this image are water reservoirs used for mining operations. Vegetation is shown in green and rocks in red and brown. Mining operations often require part of the ground to be destroyed and demand large quantities of water.

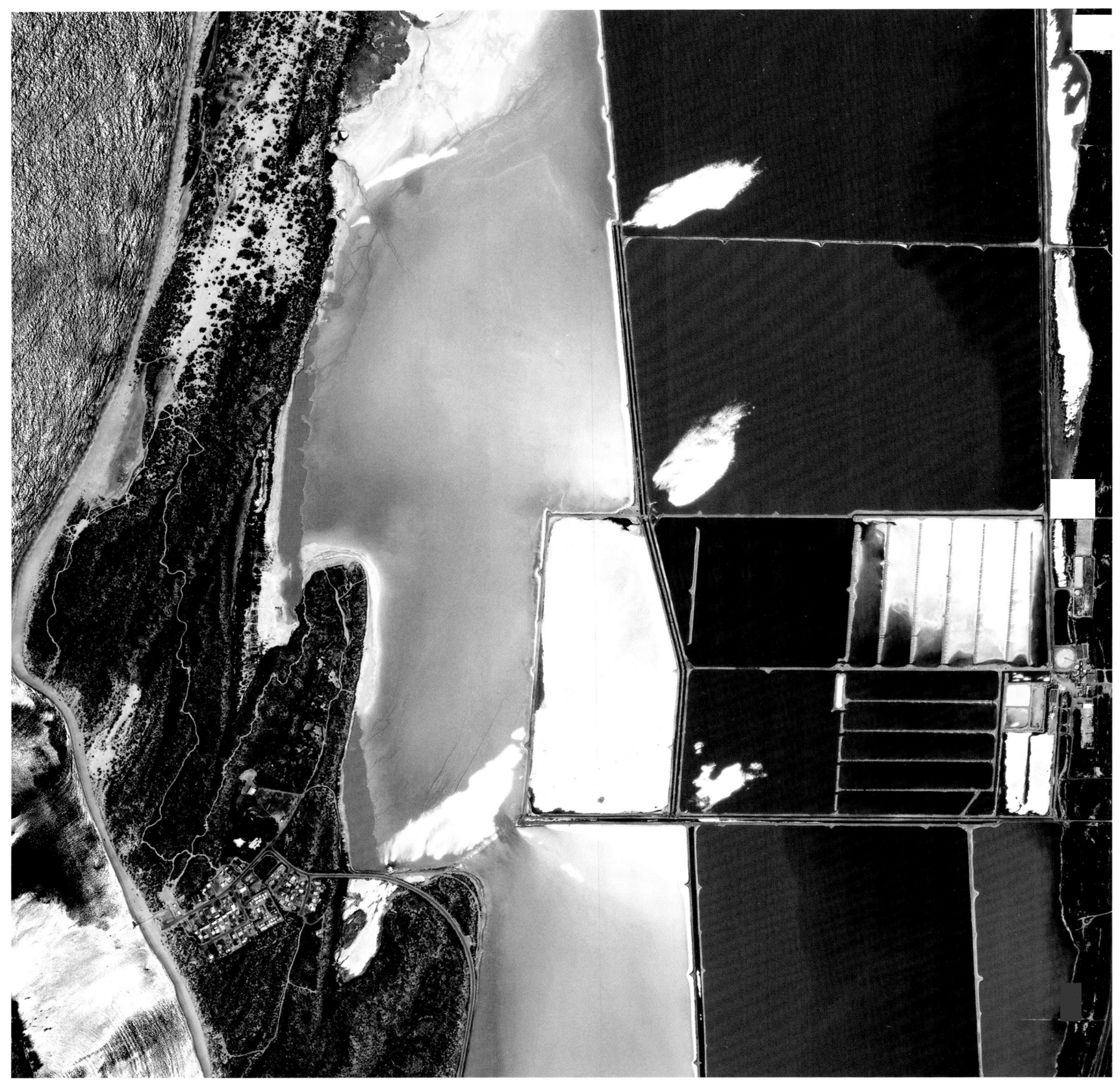

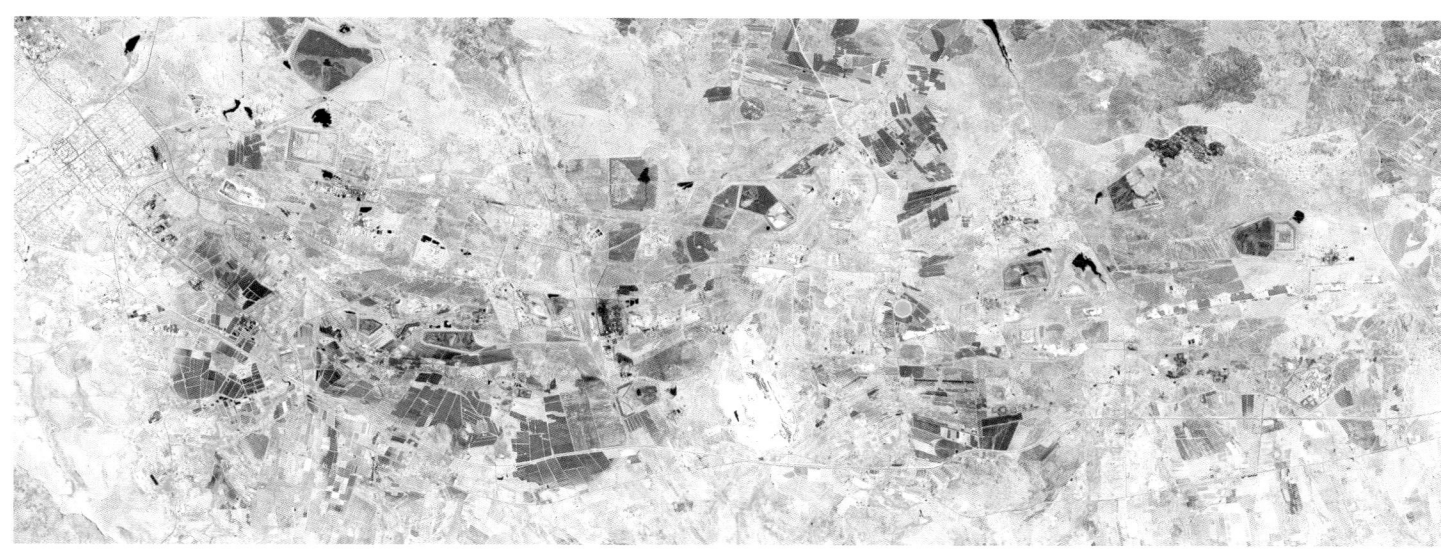

MONITORING WASTED GAS

Another significant though relatively little-known form of pollution is gas flaring, which consists of burning unused gases on gas or oil production sites. This is the cause of the flames visible on oil rigs. Flaring is both a waste and a source of pollution. Thanks to their satellite thermal sensors, experts spot flare stacks in the night and deduce the volume of gas wasted. From 2005 to 2011, the volume decreased from 5,438 to 4,944 billion cubic feet (or 154 to 140 billion cu m), which is equivalent to 360 million tons of greenhouse gas. Russia is the leading flarer in the world, followed by Nigeria, Iran, Iraq, and the United States, where this practice is becoming more widespread with increasing use of schist oil and gas. Satellite data makes it possible to evaluate the actual extent of the phenomenon, which is often underreported by companies and governments.

THE EARTH'S SCARS

Flaring makes a direct connection between air pollution and land pollution. The latter is principally due to mining. Mining of natural resources has increased by 41 percent in the last twenty years. In 2008, 68 billion tons of natural resources were extracted from the earth, versus 42 billion in 1992. In no particular order, these resources include rocks for construction, minerals used in industry, and coal and oil, which are among the leading energy sources in the world. The resulting dynamited mountains, gigantic man-made holes, deforested areas, and pollution of terrestrial and marine surfaces are among the scars human activity leaves on the planet. And their extent can now be observed with unprecedented accuracy.

The sprawling open-pit copper mines in Zimbabwe, the United States, and Chile can be identified from space by their

In 2008, **68 BILLION TONS** of natural resources were extracted from the earth, **VERSUS 42 BILLION IN 1992.**

Satellite data also sometimes
makes **PRECIOUS RESOURCES** more easily detectable
and is used by GEOLOGISTS SEARCHING FOR NEW MINES.

concentric circles and tailing ponds. They can form gashes up to 2,300 feet (700 m) deep. NASA photographs show the damage done to the forested peaks of the Appalachians by mountaintop removal, a mining technique of blowing up mountaintops to extract coal.

Satellite data also sometimes makes precious resources more easily detectable and is used by geologists searching for new mines, particularly for copper, potash, and platinum. This is the case in Zambia, for instance, which has reserves of copper ore estimated at two billion tons. Copper mining represents 9 to 15 percent of Zambia's GDP and more than 60 percent of its exports: The country used satellite images to assess its geologic potential and mineral resources.

TRACKING OIL SLICKS AND GREEN TIDES

Oceans cover 71 percent of the Earth's surface and are exposed to various kinds of pollution: excessive amounts of agricultural inputs (which can lead to the proliferation of algae), trash, and fuel spills.

From space, green tides and oil slicks can practically be spotted in real time. This happened when the oil tanker *Prestige* sank in 2002 and with the oil slick caused by the explosion of the Deepwater Horizon oil rig in 2010. By combing this data with their knowledge of tides and weather conditions, experts were able to predict some of the oil's movement on the surface. Yet such major disasters account for only 7 percent of this pollution, with the lion's share coming from small shipwrecks, voluntary degassing at sea, and various leaks. Degassing can also be observed by satellite. Some organizations, notably in the European Union, have already implemented measures to keep a close watch on maritime spaces. In the future, this might dissuade those responsible for the reckless practices that cause this pollution.

ALWAYS FASTER, HIGHER, FARTHER

For a time people imagined sending rockets full of trash toward the sun in order to deal with the accumulation of garbage on Earth. Yet today both Earth and space are already polluted. More than seven hundred thousand objects at least 0.3 inch (1 cm) long are floating in the Earth's direct orbit, of which twenty-one thousand are more than 3.9 inches (10 cm)

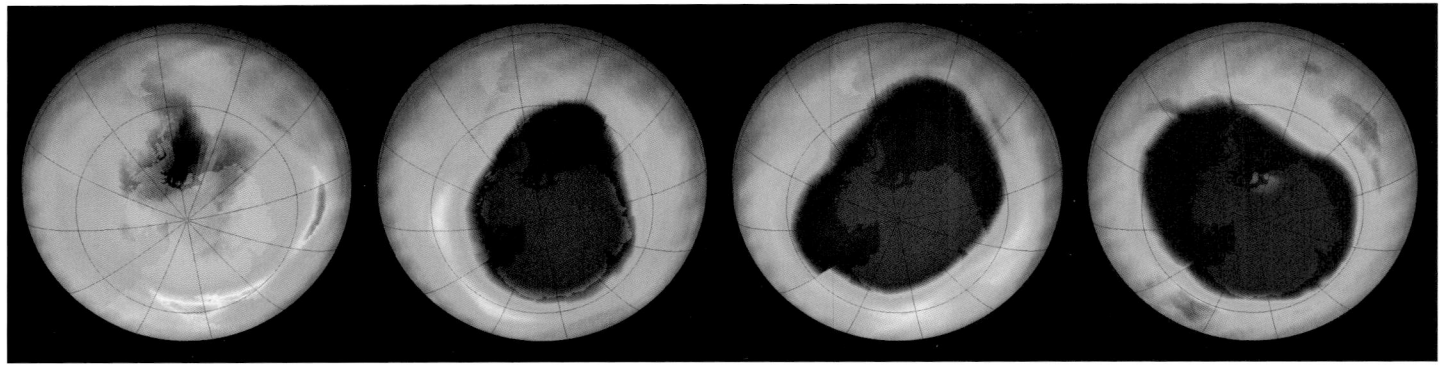

OPPOSITE

ANTARCTIC POLAR CIRCLE
HOLE IN THE OZONE LAYER
1979, 1990, 2000, 2011

The colors on these world maps show the progression of the hole in the ozone layer from 1979 to 2011. Dark blue and purple indicate the areas where the stratospheric ozone layer is less than 220 Dobson units; the lowest ozone level recorded was of 73 Dobson units in 1994 and the most extensive hole was measured in 2006. Thanks to international efforts initiated by the Montreal Protocol signed in 1985, and despite fluctuations, the hole is starting to shrink.

BELOW

UNITED STATES, LOUISIANA
MISSISSIPPI DELTA IN 2010
29°38' N – 89°54' W

Two months after the explosion of the Deepwater Horizon oil rig, NASA released this image of the oil spill on the Mississippi delta. Red indicates plant cover, which contrasts with the shades of white and blue representing the water. Oil slicks increase reflection on the surface and therefore appear as a brighter white. The image reveals the extent to which hydrocarbon pollution penetrated the delta's marshlands, which are so rich in biodiversity.

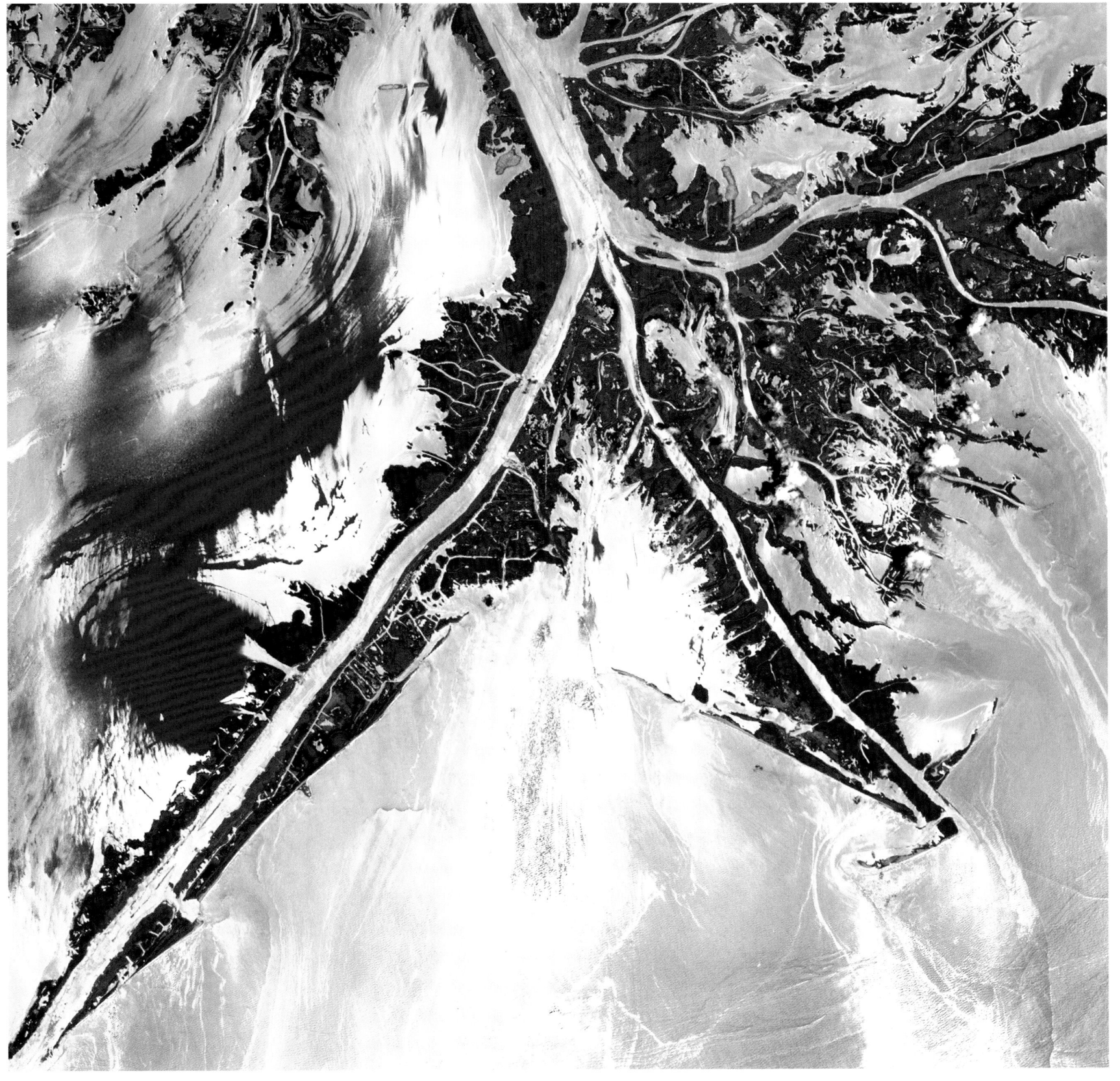

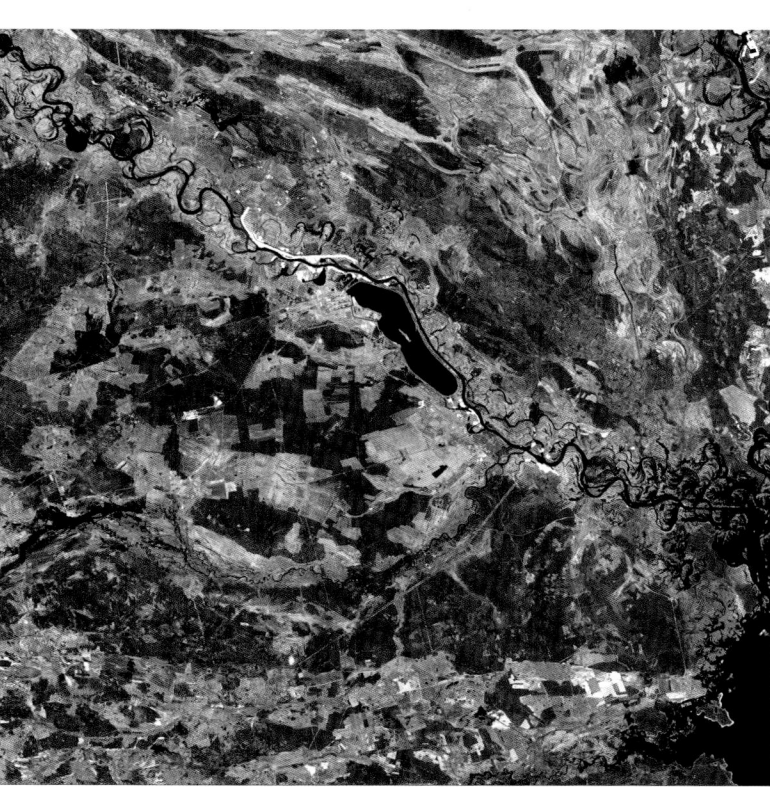

2012

1986

1988

1996

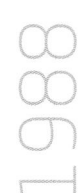

UKRAINE
CHERNOBYL
51°25' N – 30°06' E

On April 26, 1986, reactor number 4 of the Ukrainian nuclear power station Chernobyl exploded, sending a significant quantity of radioactivity into the atmosphere. More than one hundred thousand people were evacuated; the number of casualties, ranging from several thousand to several hundred thousand, continues to be debated.

Over the years, with agricultural activity having come to a stop, vegetation has gradually taken over. Plots of farmland disappeared and the city of Pripyat was deserted. The Pripyat River was canalized to secure navigation near the power station and to avoid any change to the riverbed following the thaw. Traces of this project are clearly visible in the 1996 image, where rubble and mounds of soil form two white strips along the riverbanks. The north embankment stretches 7 miles (11.3 km).

Because the sarcophagus built to prevent radioactive leakage is showing signs of weakness, a new sarcophagus must be installed in the coming years at an estimated cost of 1.5 billion euros. To this day, an 18-mile (30-km) zone around the power station, known as the exclusion zone, remains closed to the public.

The photograph on the right was taken by the Pléiades satellite. Launched in 2012, it is equipped to produce images in natural colors.

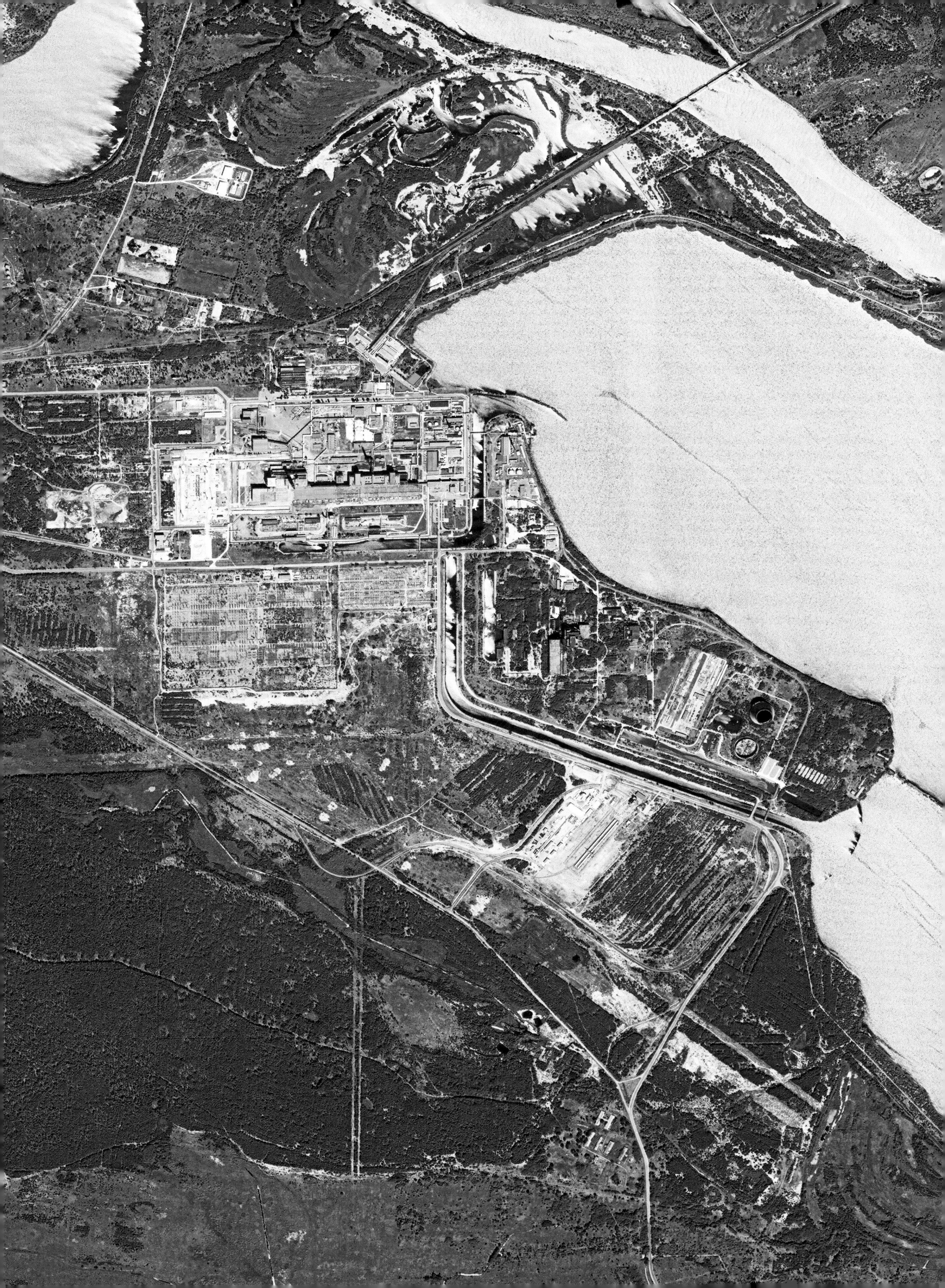

BELOW

VIETNAM BAC LIEU PROVINCE
AQUACULTURE
9°10' N – 105°50' E

There are many aquaculture facilities in the area of the Mekong delta—notably shrimp farms. But fish farming causes problems similar to those raised by factory farming, particularly in terms of pollution: high concentrations of nitrogen, vulnerability to disease, and the use of antibiotics to fight infections, as well as other hormones and harmful additives.

OPPOSITE

CHINA
SHANGHAI AND ITS SUBURBS
31°13' N – 121°28' E

With 23 million inhabitants in 2010, Shanghai is one of the most highly populated cities in the world. From 2000 to 2010, the population of this city along the Huangpu and near the mouth of the Yangzi increased by 35 percent. With its modern buildings and a port which has been the leading commercial port in the world since 2010, Shanghai is a symbol of China's contemporary success.

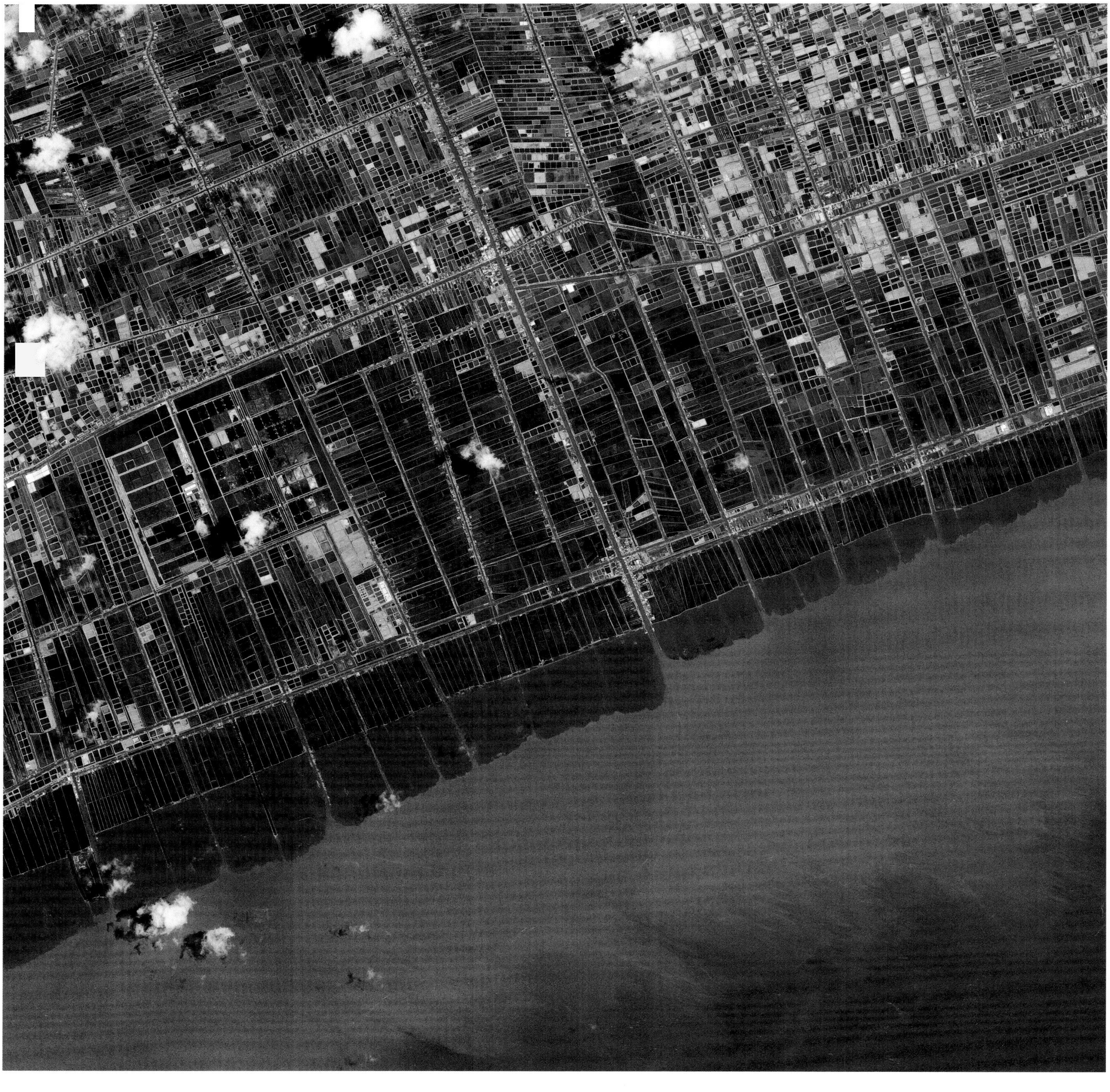

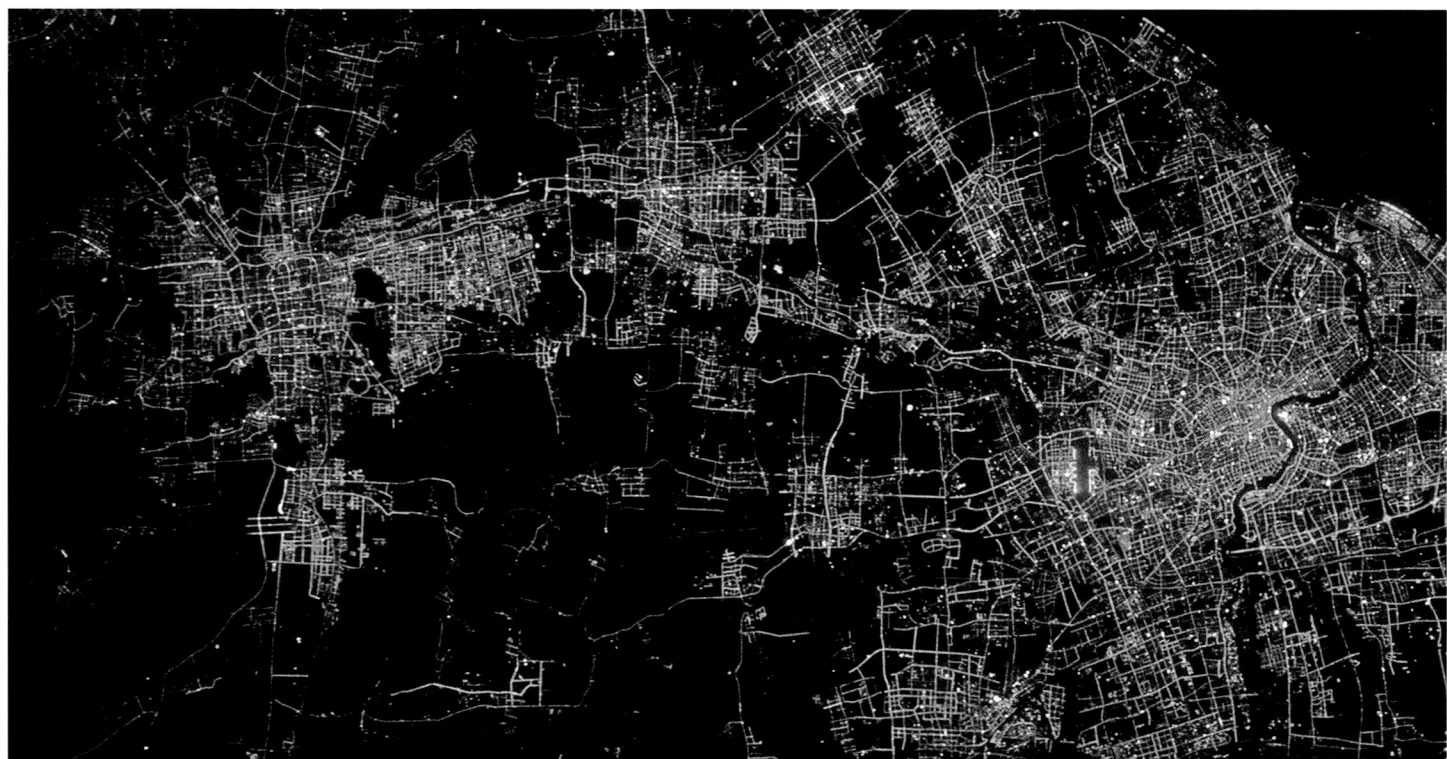

launches that have occurred since 1957. Some orbits are now so polluted that it is recommended satellites not be sent into them.

Space debris—fragments of devices sent into space, tools left by astronauts, even decommissioned satellites—is a threat to orbital satellites, space vessels, and space stations. With the debris moving at 4.9 miles (8 km) per second, any impact causes major damage. The largest objects might even be a threat to Earth's inhabitants if they do not disintegrate upon entering the atmosphere. But this danger remains low. To date, no man-made object launched into space has fallen back to Earth and cost human lives, mostly because the Earth is primarily covered in oceans.

Even so, space agencies and manufacturers are trying to make launched objects deteriorate and return into the atmosphere more easily. The international space community has coordinated its efforts to actively study, measure, model, and reduce the risks associated with space debris.

Yet whether in space, on land, or in the oceans, pollution sadly seems to be a corollary of all human activity.

OPPOSITE

UNITED STATES
**NIGHT VIEW OF NORTH
DAKOTA'S CITIES AND
OIL FLARES**
47° N – 100° W

Tele-epidemiology monitors and anticipates mosquitoes'
movements in order to prevent the transmission
of diseases such as malaria and dengue.

Talking with Murielle Lafaye

The Mosquito Weather Report

Why use satellites in the health field?
Space telecommunication technologies allow us to
develop telemedicine, notably long-distance diag-
nostics. Satellite observation data is also particularly
useful for monitoring the transmission factors of an
infectious disease and the spread of a disease.

What is tele-epidemiology?
It's a method aimed at mapping the risk of the spread
of an infectious disease. We proceed in three stages:
First we have to identify the key factors in the field,
then we ensure they can be monitored with satellite
sensors. Finally, we establish the risk maps. The satellite provides constant monitoring on both a small
and a large scale. In Dakar, for instance, the project
was implemented to fight urban malaria. In this program, the first phase consisted of determining the
conditions favoring the mosquitoes' development: A
study of the area carried out in collaboration with
doctors and entomologists revealed that mosquito
larvae mostly proliferate in temporary little ponds.
Then we turned our attention to how satellites can
help detect these small ecosystems through imaging

and pluviometric data. Once this information was collected, we were able to generate predictive maps of
the risk. Thanks to them, local authorities are better
equipped to make decisions that will prevent the proliferation of larvae in risk zones, notably by sending
teams to spread larvicide in well-determined areas.

**What kind of epidemics can be monitored through
tele-epidemiology?**
This method can be applied to malaria, Rift Valley
fever, and dengue. We worked on malaria in Senegal.
Argentina, Paraguay, and Mexico have also expressed
their interest in our research. In more arid areas, tele-
epidemiology can help develop a risk map for menin-
gococcal meningitis, epidemics of which arise from
droughts and hot winds carrying sand and dust.

This technology does not only apply to human
infections; it is also used to monitor diseases threat-
ening animals.

We are currently trying to develop a kind of mosquito weather report to make the service more useful.

Murielle Lafaye, of the French Centre National d'Études Spatiales
(CNES; National Center for Space Studies), is responsible for
forecasts, strategies, and relations with ministries, and was formerly
responsible for environment, climate, and health.

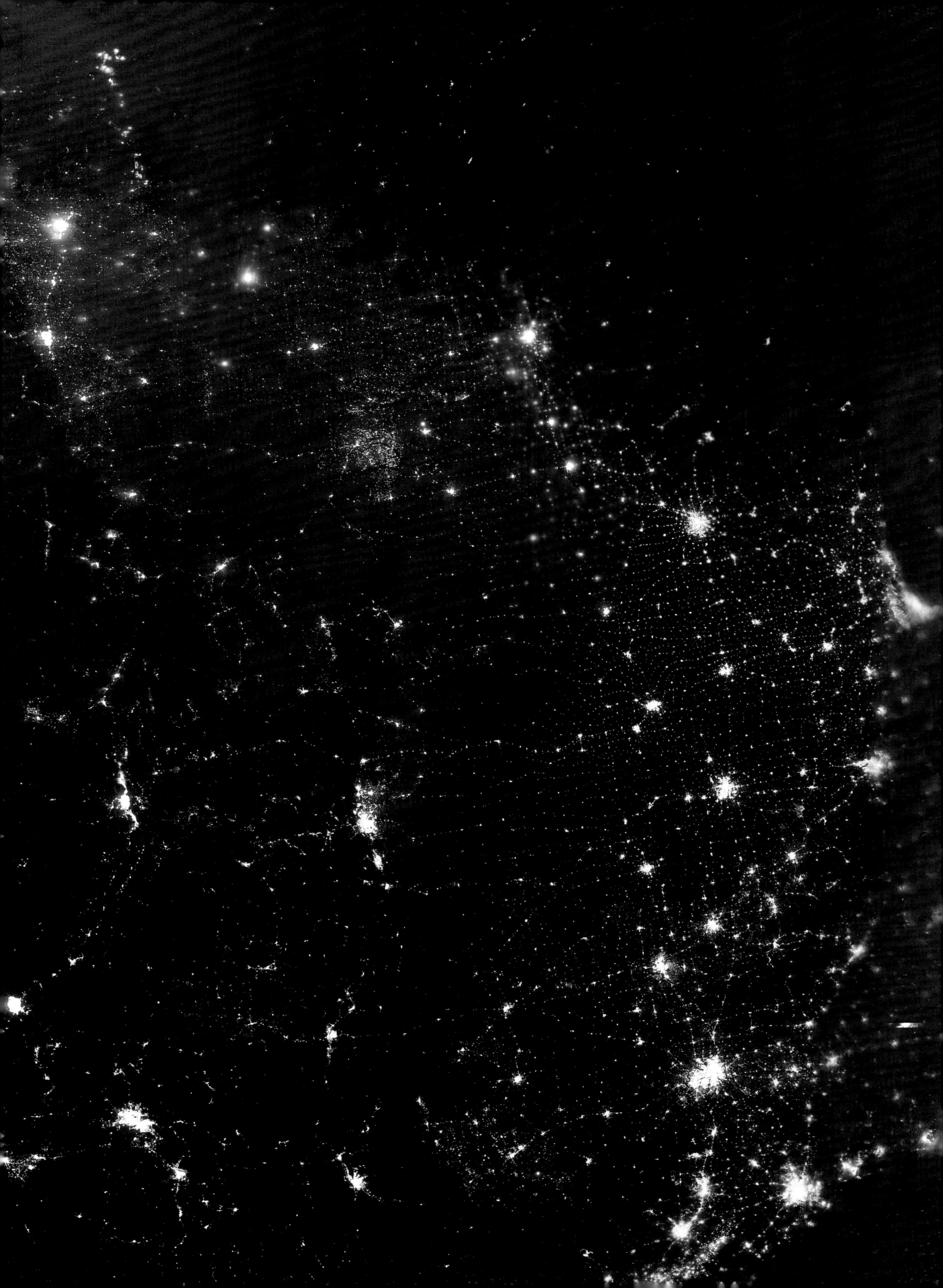

OBSERVING AIR POLLUTION IN DAKAR

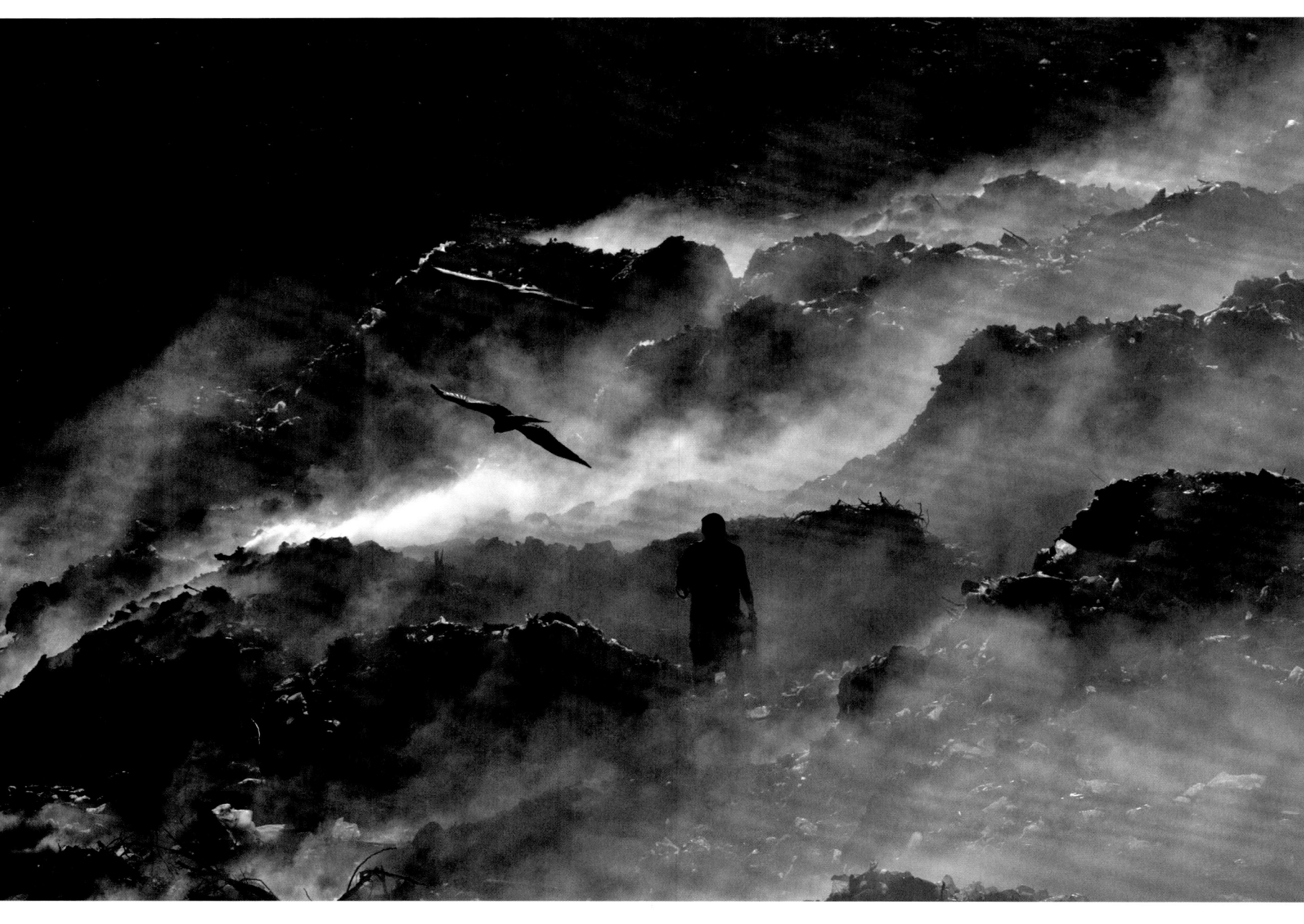

ABOVE

SENEGAL
DAKAR
14°48' N – 17°19' W

The Mbeubeuss garbage dump in Dakar covers approximately 250 acres (100 hectares); some 1,300 tons of garbage are dumped there daily. A thousand men and women scavenge whatever plastics and metals they can find in this mountain of trash. But the task is hard and exhausting, and far from risk-free: Scavengers are exposed to heavy metals such as zinc and lead, as well as pathogenic germs including salmonella.

ncreasingly recognized as a major public health problem, air pollution hits hard in Dakar and other African metropolises, which are facing significant demographic and urban growth.

What if we detected this pollution from outer space, on a city-wide scale, rather than with measurement stations located in different neighborhoods? This was the idea behind a joint study carried out by Dakar's Centre of Air Quality and satellite imagery specialists based in Senegal and France. The study compares the concentrations of particles collected by air quality monitoring stations on the ground with estimations made thanks to Earth observation satellite FORMOSAT-2. Within only a few days of its passage, the satellite provides images of a location in real colors. The level of the air's transparency indicates an "optical layer of aerosol" comparable to results on the ground.

Maps obtained this way make it possible to view fine-particle pollution on the scale of an entire megalopolis—and to determine its causes.

Fine-particle pollution consists of a heterogeneous mix of dust, soot, and molecules defined by their size; they are known as PM10 if they are smaller than 10 microns in diameter or PM2.5 if they are smaller than 2.5 microns. These particles enter our respiratory tracts, particularly the lungs, and can cause various problems and illnesses.

While we know that motor vehicles, industry, and using wood and coal for cooking contribute to pollution, satellite images have revealed desert dust's significant impact on air quality degradation. Additionally, images show strong local differences. In Dakar, for example, the neighborhood of Bel-Air, which is close to the port area, is not true to its name, for it is one of the most polluted in Senegal's capital.

This kind of data has not yet led to the implementation of measures to improve air quality, but the development of this technology could provide a more efficient and less costly alternative to running a monitoring network on the ground.

———

PAGES 116-117

UNITED STATES, ARIZONA
THE DAVIS-MONTHAN AIR FORCE BASE
32°10' N – 110°52' W

The Davis-Monthan Air Force Base in Arizona holds five thousand decommissioned planes waiting to be disassembled. The dry, sunny weather has helped preserve the aircraft, which include the B-52, a famous Cold War symbol. Military equipment often uses pollutants such as asbestos and depleted uranium, which makes it challenging to dismantle.

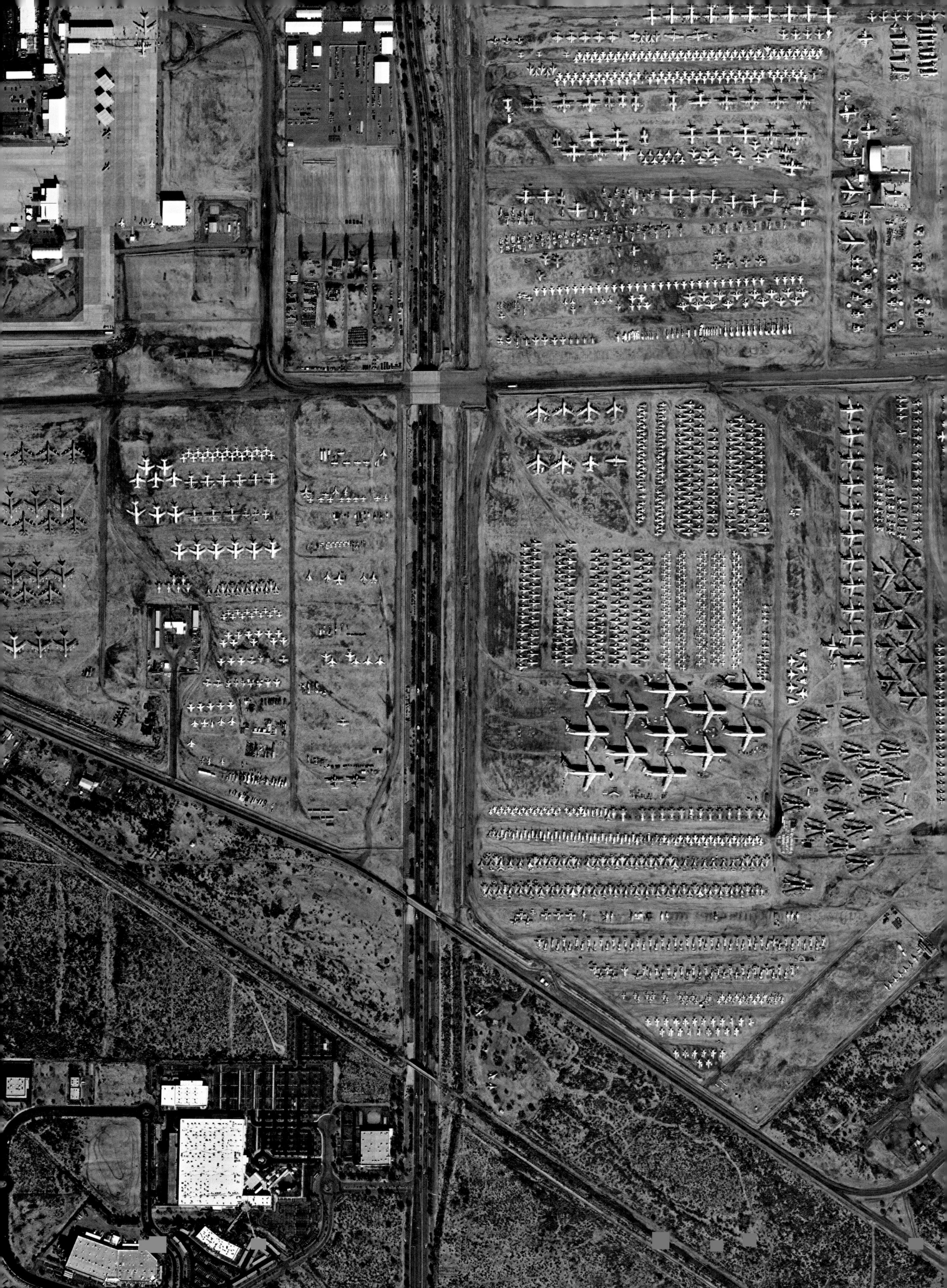

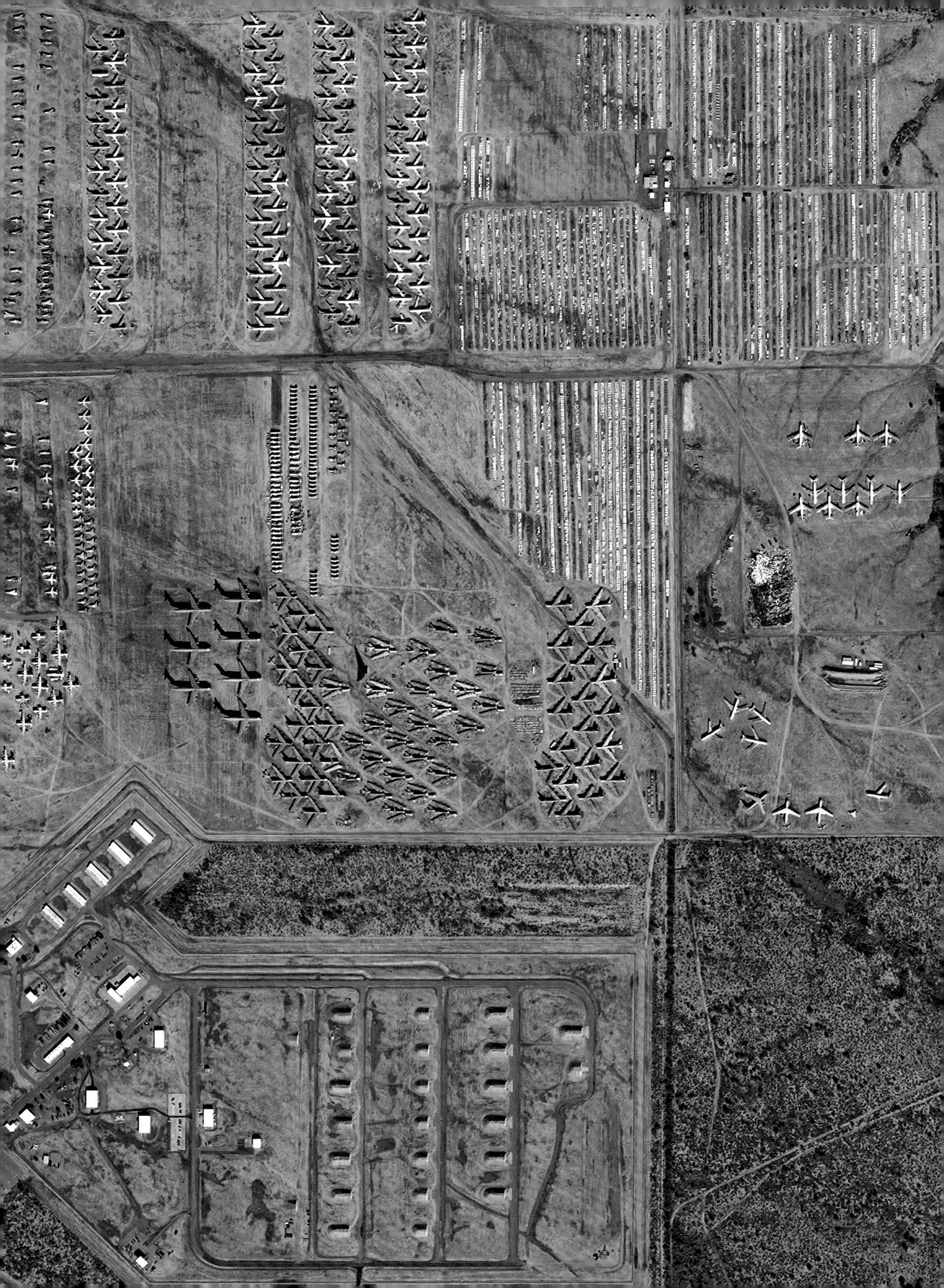

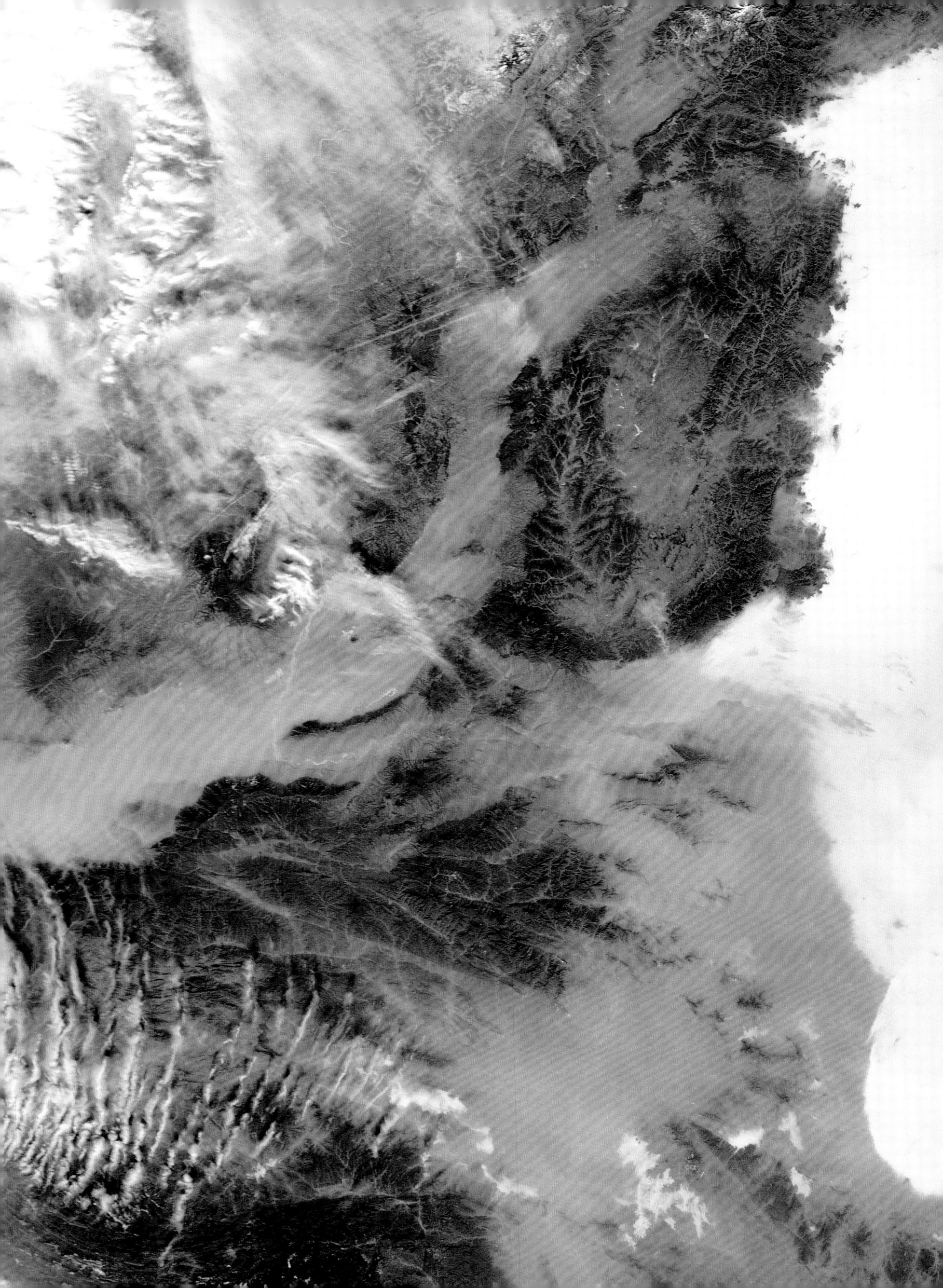

PAGES 118-119
CHINA
SMOG OVER BEIJING
39°54' N – 116°23' E

Smog pollution covers Beijing, where pollution levels are twenty to forty times greater than levels recommended by the World Health Organization. Industry and transportation are the principal culprits—China produces 70 percent of its energy with coal. Beijing's polluted air often makes headlines. Since 2008, the American Embassy has even published its own readings of pollution levels in the Chinese capital.

BELOW
CANADA
LAKE ERIE
41°45' N – 82°30' W

Fed by agricultural fertilizer, microscopic algae multiply in the lake and become visible from space. These algae are not toxic themselves, but are broken down by other microorganisms, which proliferate until they have consumed most of the oxygen available. This creates a dead zone, in which other life-forms—fish, crustaceans, and marine mammals—are asphyxiated and can no longer survive.

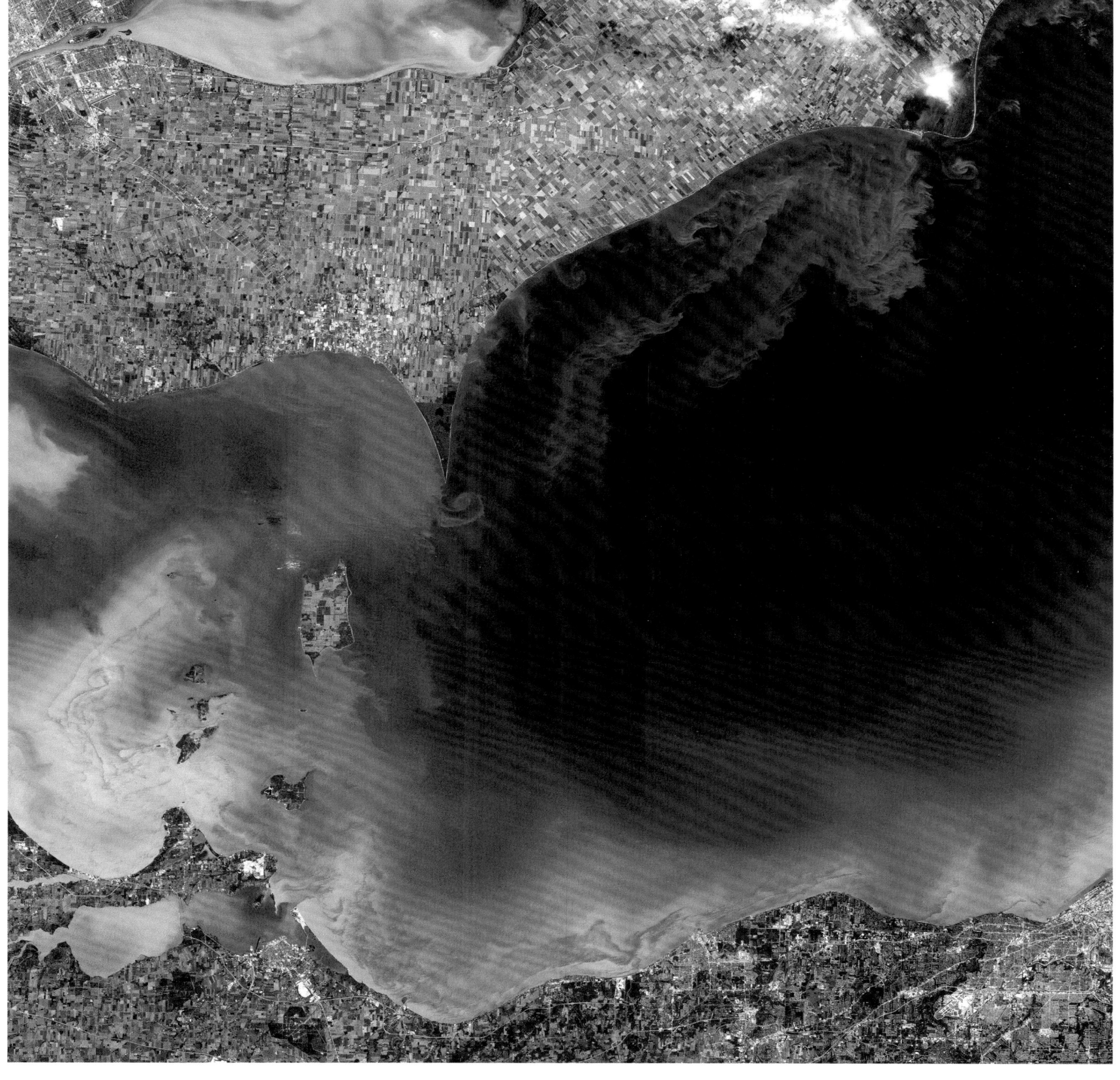

BELOW
KUWAIT
BURNING OIL WELLS IN 1991
29°40' N – 47°58' E

In January and February 1991, Iraqi troops in Kuwait were routed. As they retreated, they set fire to more than six hundred oil wells; ten months and 11,000 people were required to extinguish these giant flares. According to estimates, nearly 5 percent of Kuwaiti territory was covered by a thick black cloud of soot and the equivalent of 1.5 billion barrels of oil was spilled into the environment, leading to one of the largest instances of pollution in history.

PAGES 122-123
CHILE
CHUQUICAMATA MINE
22°19' S – 68°56' W

This open-pit mine—measuring 1.2 miles (2 km) wide, 1.8 miles (3 km) long, and more than 2,600 feet (800 m) deep—is one of the largest ones in the world. Its highly coveted ore is the richest in copper. It also has a high sulfate content, toxic particles of which are found in clouds of dust released by explosions and the endless coming and going of trucks. Workers are not permitted to work in the mine for more than three years and are at risk of contracting lung cancer.

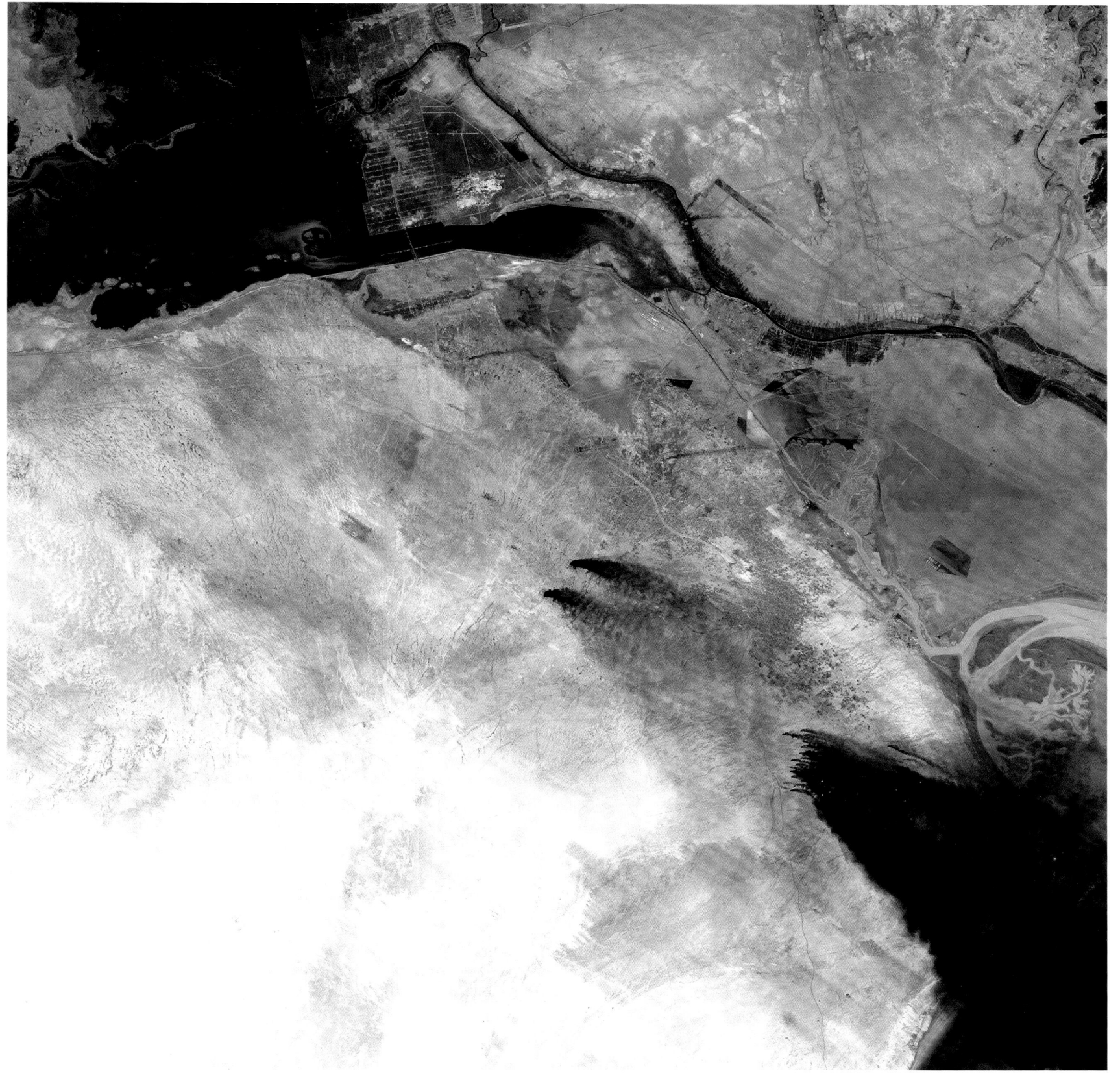

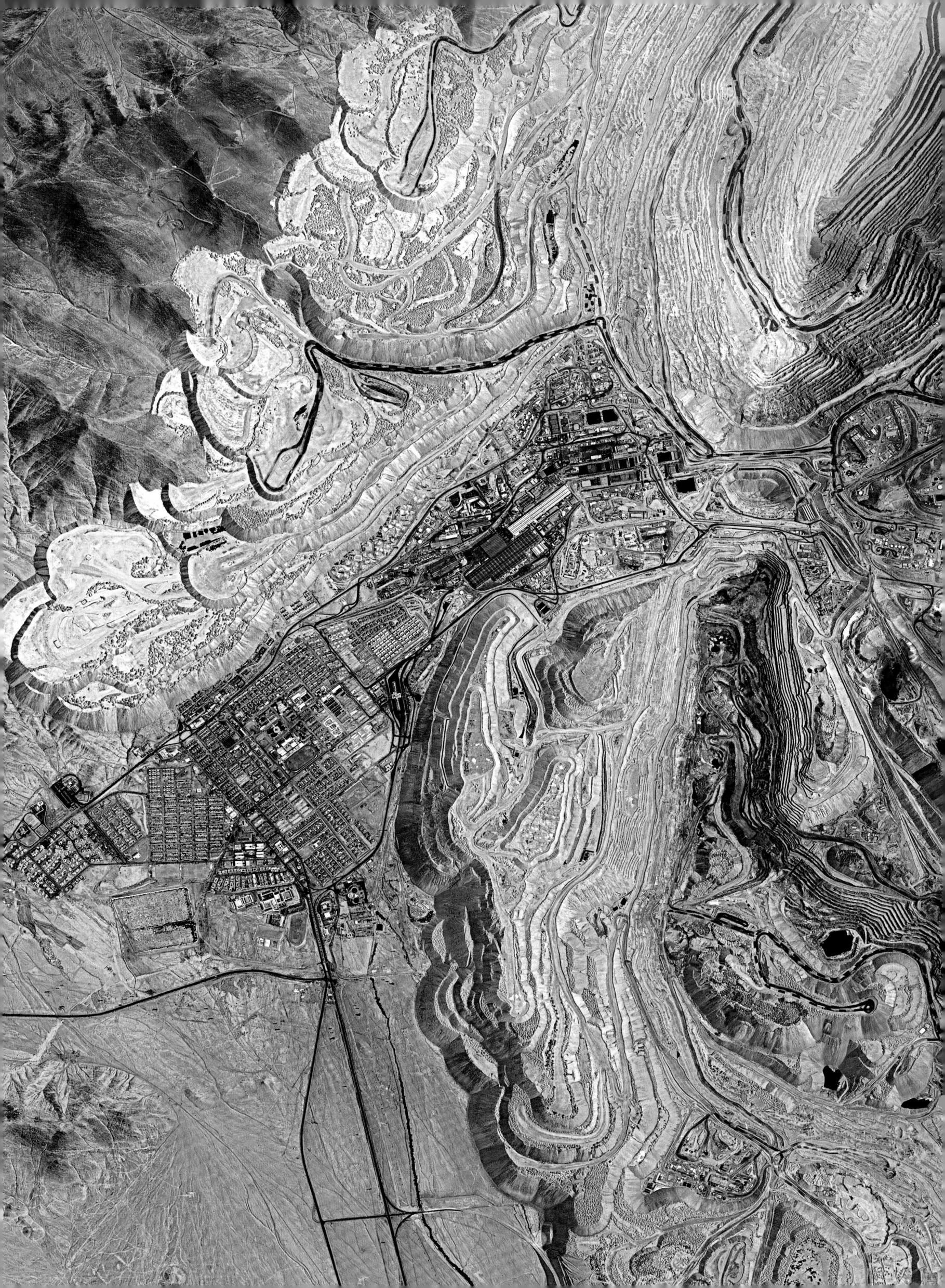

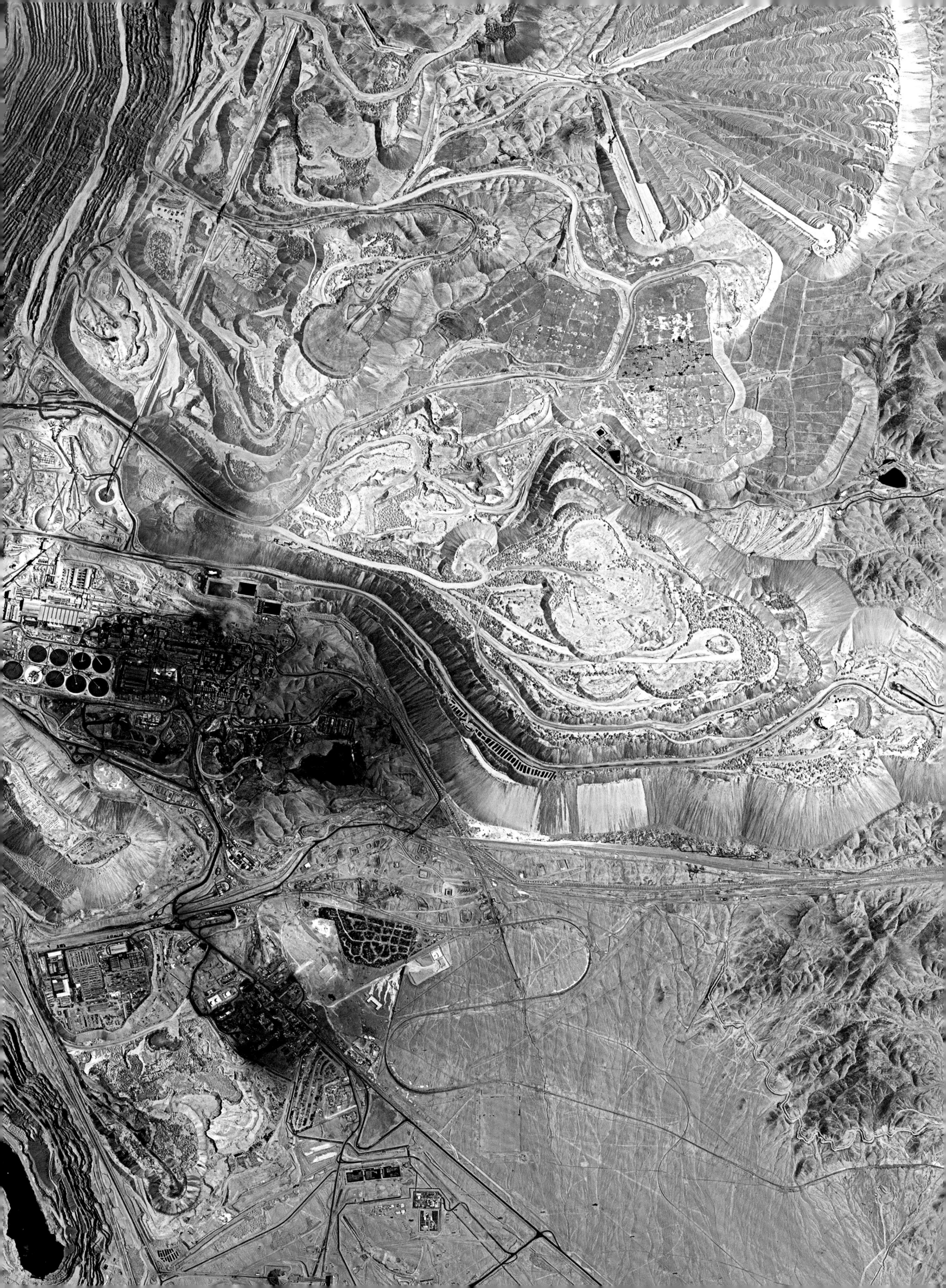

Urban areas are growing, constantly taking a bigger toll on nature, while concentrating wealth and accounting for a growing share of global consumption.

___

URBAN SPRAWL

___

For the first time in human history, since 2007, more than one in two people live in cities. After accelerating all over the world and throughout the twentieth century, urbanization has passed a symbolic milestone: The global population living in cities grew from 13 percent in 1900 to 50 percent in 2007. This phenomenon is likely to continue in the twenty-first century, particularly in developing countries.

Each year, cities spread out and at considerable speed. While the global rate of urbanization, or percentage of people living in cities, grows by 1.85 percent a year, the surface area of cities is increasing at twice that speed, or even more: The area of Mexico City has increased by 7.4 percent a year over the last thirty years, or four times faster than its population; similarly, the surface area of Ouagadougou, Burkina Faso, has quadrupled in the last thirty-five years.

In rich countries, where the majority of the population already lives in cities, urban sprawl predominates. Contrary to common belief, the density of big cities in the developed world is diminishing. Skyscrapers, for instance, are not residential. Located downtown, they are primarily devoted to offices, which are vacant at night. Yet suburbs are endlessly expanding, following the American model of single-family houses with separate yards outside the city center. Individually owned automobiles are ubiquitous and are used for commuting to work, shopping, and traveling to leisure and entertainment centers.

This sprawl has resulted in sweeping land artificialization. Anywhere from 15 to 20 percent of an urban area is covered by roads. On a global scale, approximately 12.3 million acres (5 million hectares) of farmable land are lost annually to urbanization and encroaching infrastructures.

IN THE DEVELOPING WORLD

In the developing world, rural exoduses have led to the densification and growth of metropolises. Each week, close to 1.2 million people move to cities throughout the developing world, with 880,000 of them in Asia and 230,000 in Africa. The metropolises undergoing the strongest urban growth include several Chinese cities—such as the relatively little-known cities of Chongqing, Tianjin, and Shenzhen, each having approximately ten million inhabitants—and Lagos, Nigeria, and Dhaka, Bangladesh, which grew from three hundred thousand inhabitants in 1950 to fourteen million in 2009.

These examples reveal a larger trend, the overall growth in urban populations in the developing world. In 1975, there were only three cities with more than ten million inhabitants—Tokyo, New York, and Mexico City. In 2010, there were twenty-one, more than half of which are in developing countries.

SHANTYTOWNS

In developing countries, population influx on this scale leads to chaotic urban development, characterized by marked social divisions, with beautiful residential neighborhoods for the wealthy surrounded by vast slum neighborhoods.

More than one billion people around the world currently live in shantytowns, favelas, and other poor neighborhoods. But here, too, we need to beware of conventional wisdom. These are also areas of transition and social mobility: While twenty-seven million people settle in these neighborhoods every year, others leave them. In a decade, close to

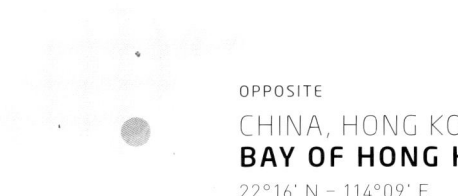

OPPOSITE
CHINA, HONG KONG
BAY OF HONG KONG
22°16′ N – 114°09′ E

Boat traffic is constant in Victoria Harbour, as ships connect the island of Hong Kong to Kowloon, on the continent. Long a British territory, Hong Kong was retroceded to China in 1997 but has conserved a unique status that ensures its prosperity.

A global financial center, the city of Hong Kong is characterized by its skyscrapers and population density, which is among the highest in the world with an average of more than 9,677 inhabitants per square mile (or 6,000 inhabitants per sq km).

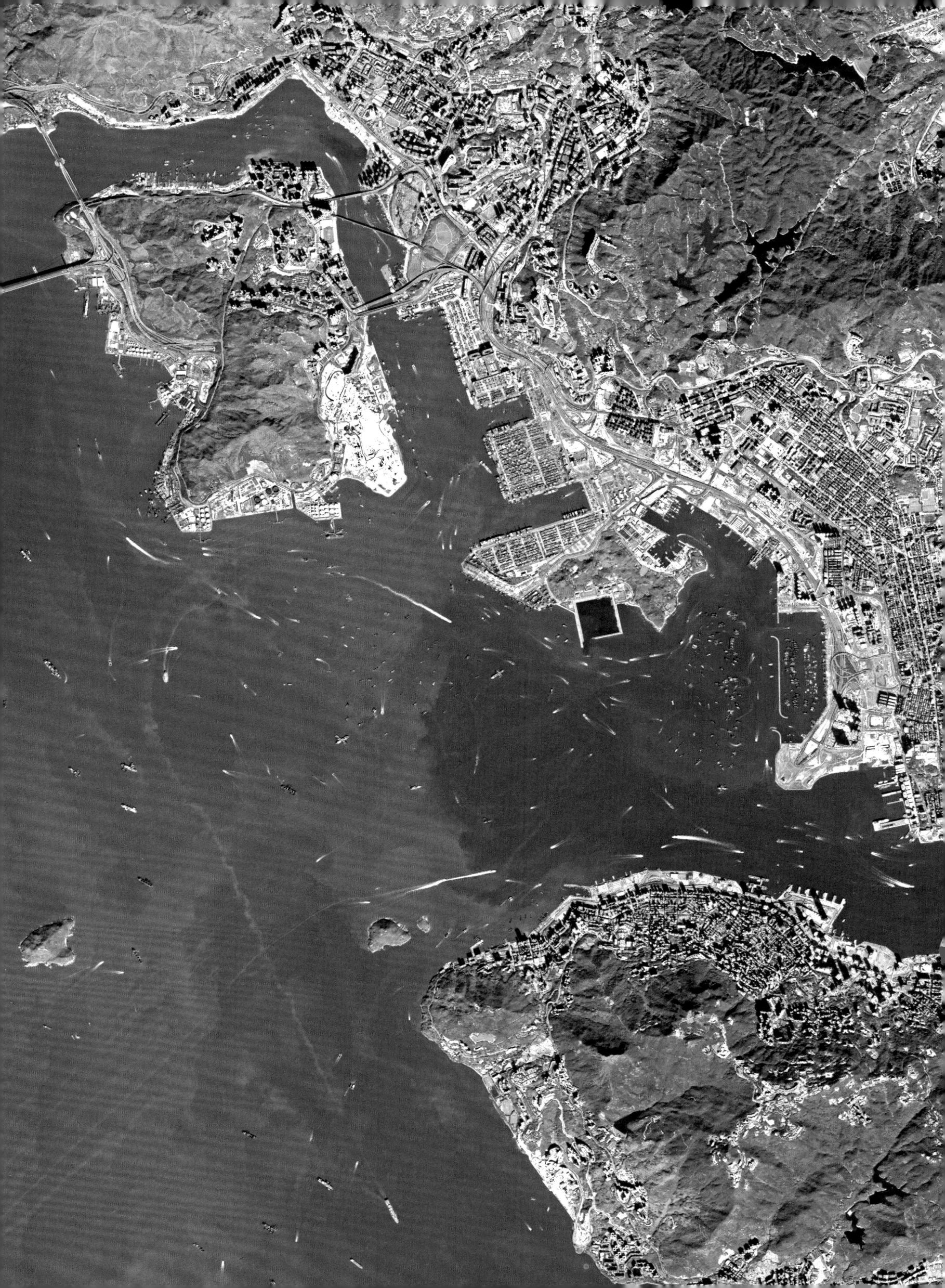

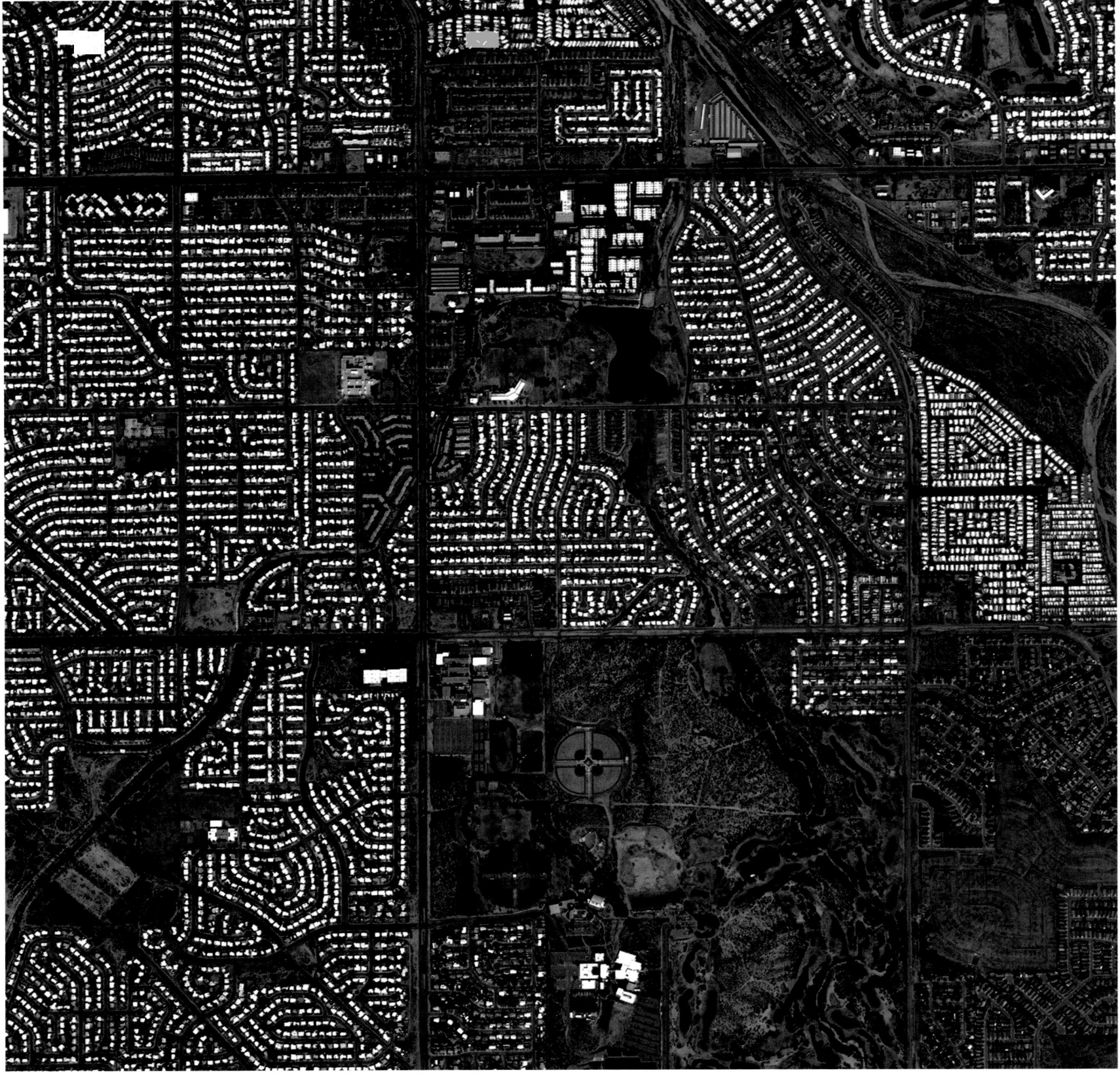

ABOVE

UNITED STATES, ARIZONA
TUCSON
33°13' N – 110°55' W

Tucson, which lies close to the Mexican border, is the second-largest city in Arizona. It has close to one million inhabitants, over a third of whom are of Mexican descent. With its major arteries and blocks of buildings, Tucson follows the American urban model, which promotes residential single-house suburbs lying outside the downtown area. As a consequence, the total number of automobiles in the United States has reached 240 million, for a total population of 313 million inhabitants.

OPPOSITE

PAKISTAN, BALOCHISTAN
GWADAR
25°07' N – 62°19' E

Located on the shores of the Gulf of Oman, Gwadar was once a simple fishing village, but since 2007 it has become one of Pakistan's biggest ports. The new port was financed by China in order to ease the delivery of oil to that country, a project that will be capped with the construction of an oil pipeline over the border between Pakistan and China.

LONDON would require a surface area **THREE HUNDRED TIMES ITS CURRENT SIZE** to meet its own needs and **ELIMINATE ITS WASTE AND POLLUTION SOURCES.**

227 million people have left their slums or seen their living conditions improve.

Occasionally, satellite imagery is used to better understand and track an area's growth, evaluate the population based on the type of buildings, and monitor its evolution, as well as plan ahead for structural developments and the installation of utilities.

Of course, the richest cities also use these images. Satellite imagery is used to monitor zoning: Illegally built swimming pools and villas can be detected from space. In Russia, for instance, Greenpeace used satellite images to protest the illegal construction of villas in protected forests around Moscow.

A GLOBAL IMPACT

In the past, cities relied on the surrounding countryside to supply their provisions. Today, with the globalization of trade, they are no longer exclusively connected to their backcountry but to the entire planet: A fresh food product can come from the other end of the world; a piece of furniture can be made in Asia with wood from Africa. Cities now consume enormous amounts of energy, raw materials, and food. For instance, London would require a surface area three hundred times its current size to meet its own needs and eliminate its waste and pollution sources.

Las Vegas, a city built in the desert and observed by NASA satellites since 1972, continues to expand. Its growth has been accompanied by an exuberant lifestyle and the construction of golf courses and water parks, all of which weighs increasingly heavily on the area's water supply. In fact, data shows that the level of Lake Mead, which serves as a reservoir for the Hoover Dam on the Colorado River and provides water for the entire area, has dropped significantly over the last forty years.

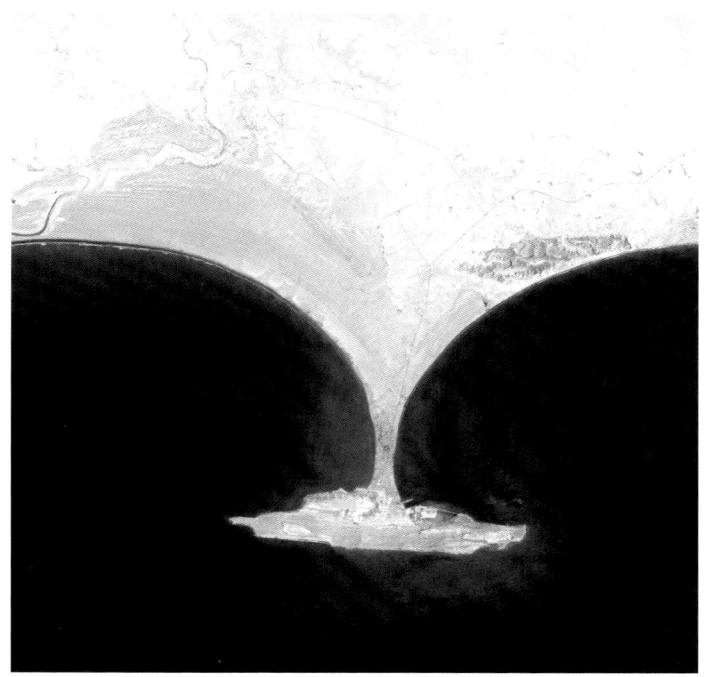

URBAN CONSTELLATIONS

Energy consumption in large urban centers can also be observed from space. At night, a thousand lights shine in cities, visible from a few hundred miles in orbit. These are man's traces on the planet. In the early 2000s, NASA published pictures of the Earth seen from space by night. North America, Europe, and coastlines were by far the most illuminated areas, while the centers of Africa and Australia remained in darkness.

This visible disparity between lighting for the wealthy and darkness for the poor can even be found in the heart of single cities. In Brasilia, a city planned and built in the 1960s, the rich center gleams with artificial lights while the slums around it lie in darkness. Lighting also varies according to location: Lamps are orange in Europe, blue and green in Tokyo. Yet this is only the visible part of energy consumption, for only 3 percent of the energy used in a building is expended on lighting, while 60 percent is used for temperature control. Initial thermal maps from satellite data show as much.

OVERCONSUMPTION

Urbanization leads to a profound transformation in ways of life, particularly with the increasing consumption of manufactured products, which cuts across all cultural and social groups. Though extreme poverty exists beside the greatest wealth, a frequent by-product of the city is the development of a middle class that equips itself with telephones, televisions, computers, washing machines, and refrigerators. The expansion of the number of consumers is largely taking place in developing countries, notably in India and China. But this lifestyle is exhausting the planet: Today the human race consumes the resources of one and a half planets in one year, meaning that it uses 1.5 times the resources the Earth is able to supply in one year.

TOWARD MORE SUSTAINABLE CITIES

Yet cities offer opportunities for environmental preservation. Densification supports economies of scale for energy consumption and transportation optimization. Reducing the energy used by buildings could significantly affect greenhouse gas emissions. A few simple principles could be implemented: Double glazing for windows can decrease the amount of heat required by 15 to 20 percent, while using motion-detection light systems, low-energy lightbulbs, and better managing natural light can lead to a 75 percent drop in energy consumption for domestic lighting, which is responsible for the emission of 1.9 gigatons of CO_2. Intelligent electricity grids, known as "smart grids," are currently being developed to better distribute energy and avoid waste.

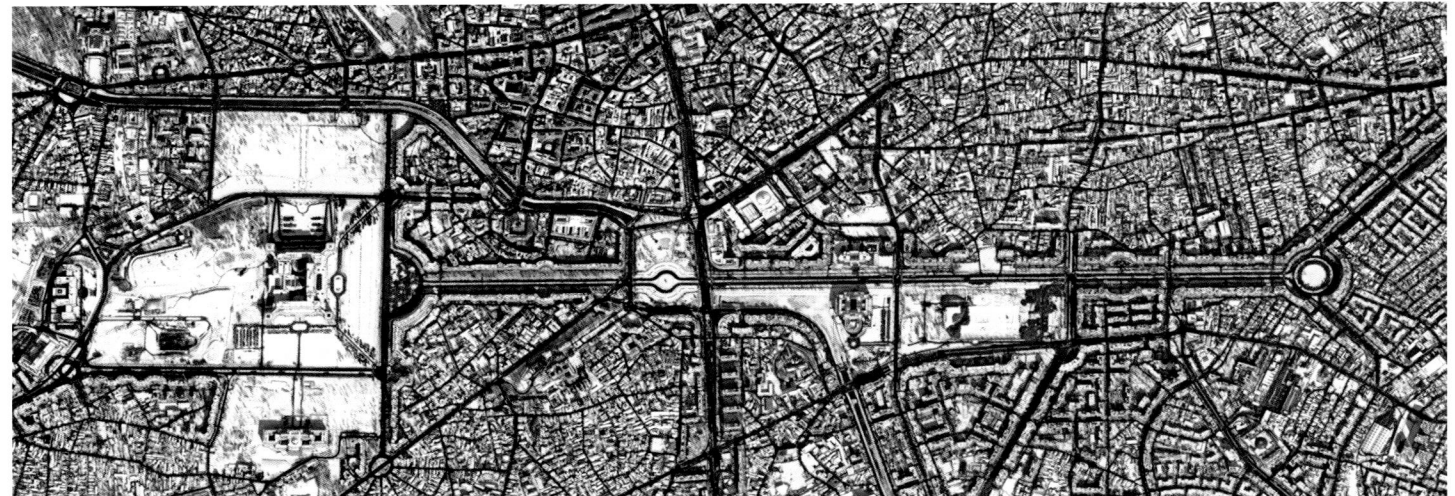

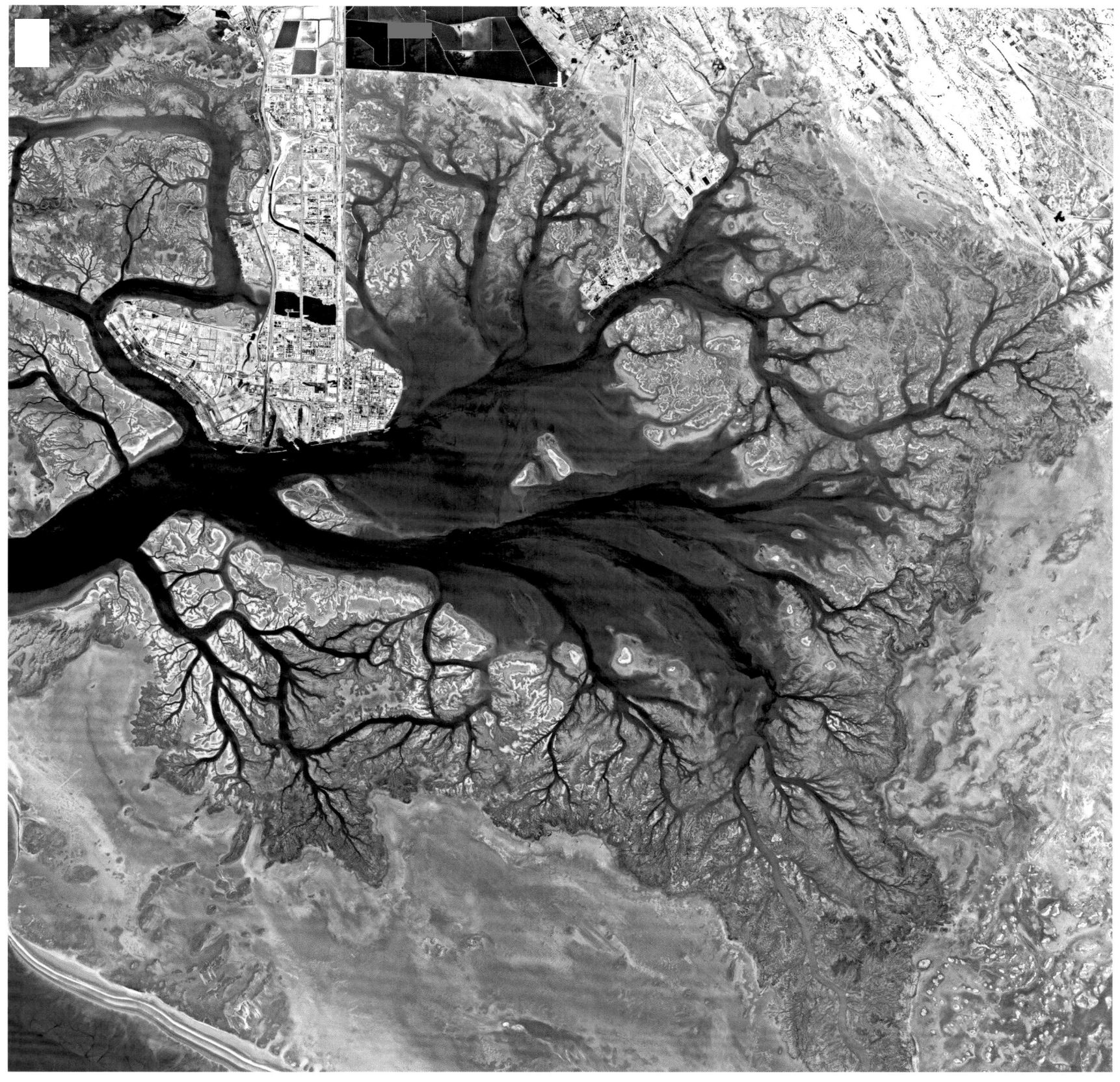

OPPOSITE

ROMANIA
BUCHAREST
44°25' N – 26°06' E

A historic part of Bucharest was razed to make room for the former House of the People commissioned by Nicolae Ceaușescu, who ruled from 1965 to 1989. Representative of the outsize projects implemented by communist regimes, this colossal stone edifice is the second-largest building in the world after the Pentagon. Today it holds the Palace of the Parliament. Romania is one of the last states to have joined the European Union, in 2007.

ABOVE

IRAN
BANDAR-E EMAM KHOMEYNI
30°26' N – 49°05' E

The port city of Bandar-e Emam Khomeyni, population sixty-five thousand, owes its growth to the petro-chemical industry. Iran, which controls part of the Strait of Hormuz, exports some of its oil and exerts its influence on the region. The Strait of Hormuz is an essential axis of the oil business: About one-third of international oil traffic passes through this channel from the Persian Gulf to the Gulf of Oman.

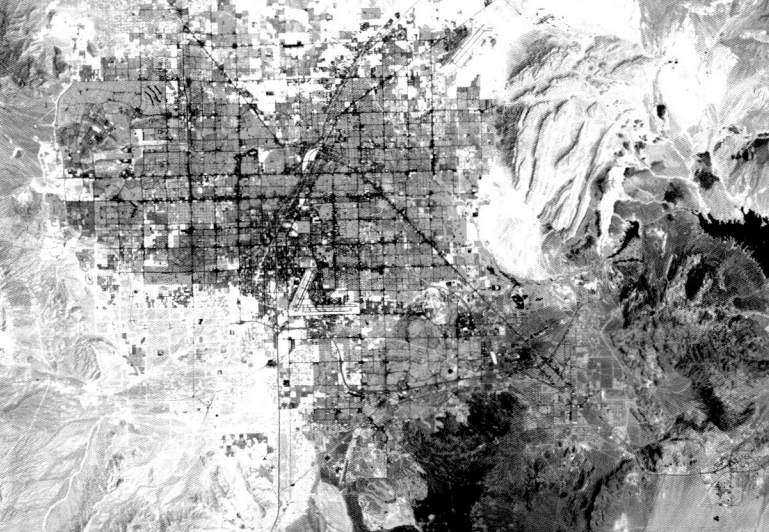

2009

1984

1996

2002

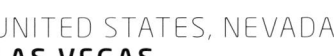

UNITED STATES, NEVADA
LAS VEGAS
36°06' N – 15°09' W

Las Vegas endlessly expands in the heart of the desert, despite its less-than-ideal natural hydrographic situation. From 1984 to the present, the blue, violet, and green have gained ground: These are the roads and buildings that accompany a city's expansion. Las Vegas's economic development primarily rests on the gambling industry and tourism, accompanied by the building of green spaces—gardens, parks, and golf courses—and pools. The urban area's population has tripled in twenty years.

Water consumption has led to a noticeable drop in the level of Lake Mead, the principal reservoir in the region, which has led authorities to multiply initiatives to prevent leaks and encourage locals to use water more responsibly and replace lawns with cactus gardens.

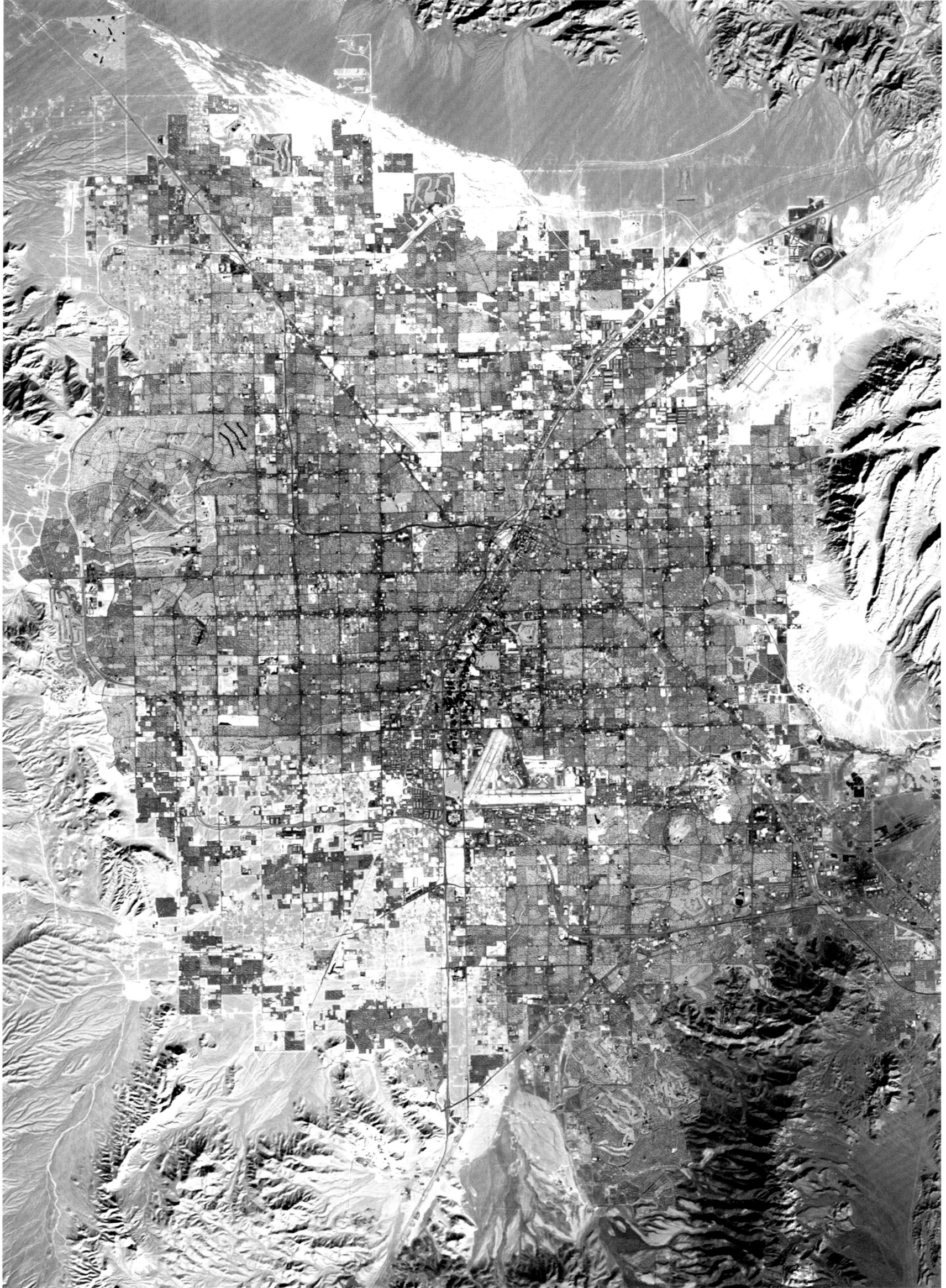

BELOW

PALESTINIAN TERRITORIES
RAMALLAH
31°54' N – 35°12' E

Located about 10 miles (16 km) from Jerusalem, the city of Ramallah is the Palestinian Authority's administrative center. The population is 40,000 within the city itself; the number rises to 200,000 if the surrounding villages are included. This photo shows the wall and Israeli colonies in the hills.

OPPOSITE

UNITED STATES, NEW YORK
NEW YORK CITY
40°42' N – 73°59' W

With more than eight million inhabitants, New York City truly never sleeps. Broadway shines bright into the night. In terms of energy, buildings' energy consumption is one of the principal sources of greenhouse gas emissions, but lighting only has a minimal impact: Currently, 80 percent of the energy consumed in New York City's buildings is for heat and air-conditioning.

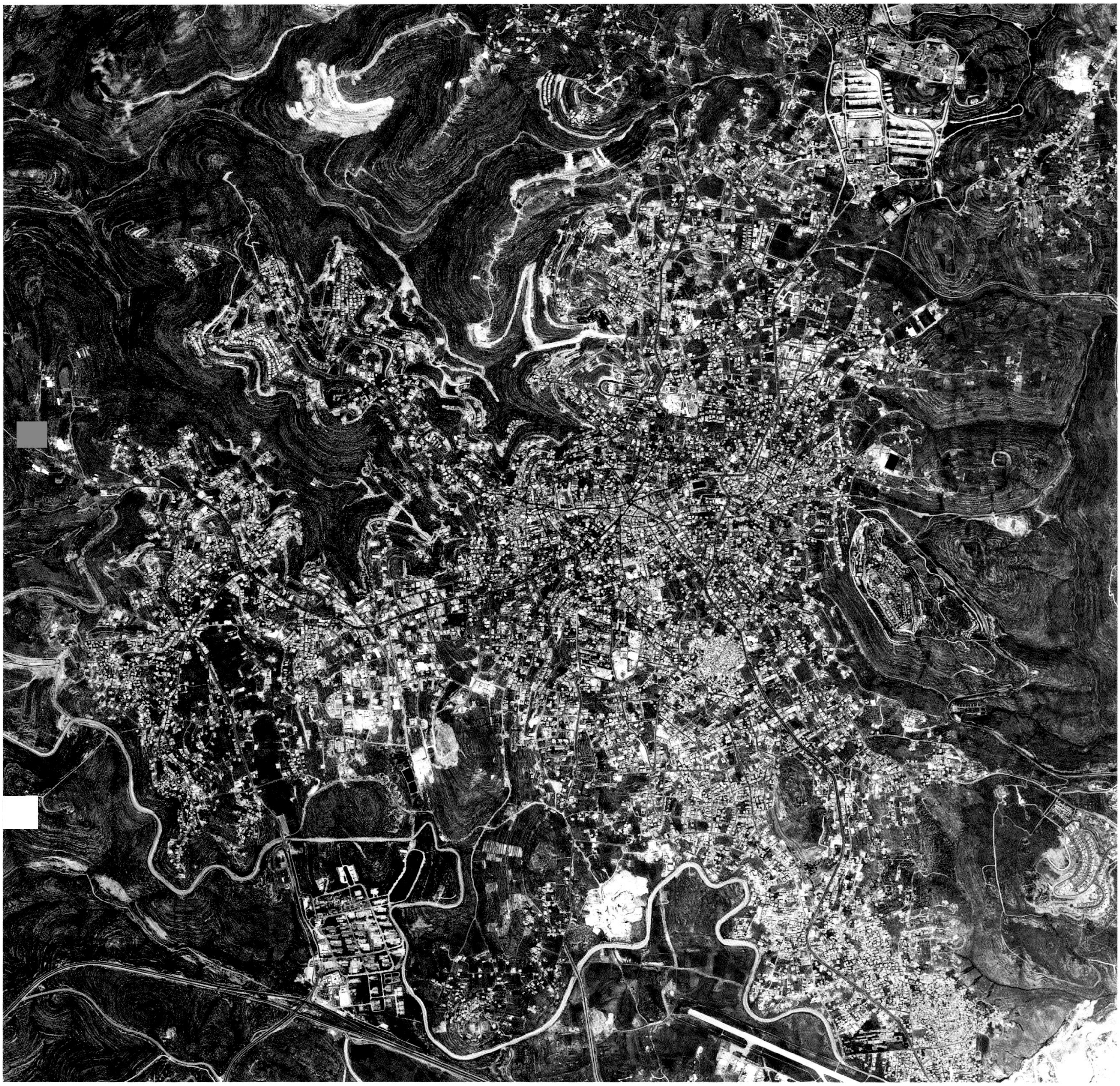

Transportation has also undergone many improvements thanks to advances in telecommunications and satellites. GPS systems make travel easier and allow us to calculate the shortest distances, which ensures the lowest energy consumption.

Today nearly everyone can have real-time access to this information with cell phones, which have become the symbol of a nomadic modernity: In late 2011, there were more than 867 cell phones per thousand people in the world.

INNOVATION

Cities are capitals of culture and technology, but also intellectual centers where multiple innovations are hatched, some of which use satellite imagery or positioning data in unexpected ways. Thanks to Google Earth, researchers were able to discover some 4,500 urban vegetable gardens—covering close to 2,800,000 square feet (260,000 sq m)—in Chicago alone and volunteered to help these gardeners better organize their plots. Others use satellite tools to control greenhouse gas emissions from energy consumption (combustion of fossil fuels) in major American cities in real time.

No doubt the combination of imagination and technology will lead to other projects using satellite data to improve quality of life in cities. Then the challenge will be to implement these plans!

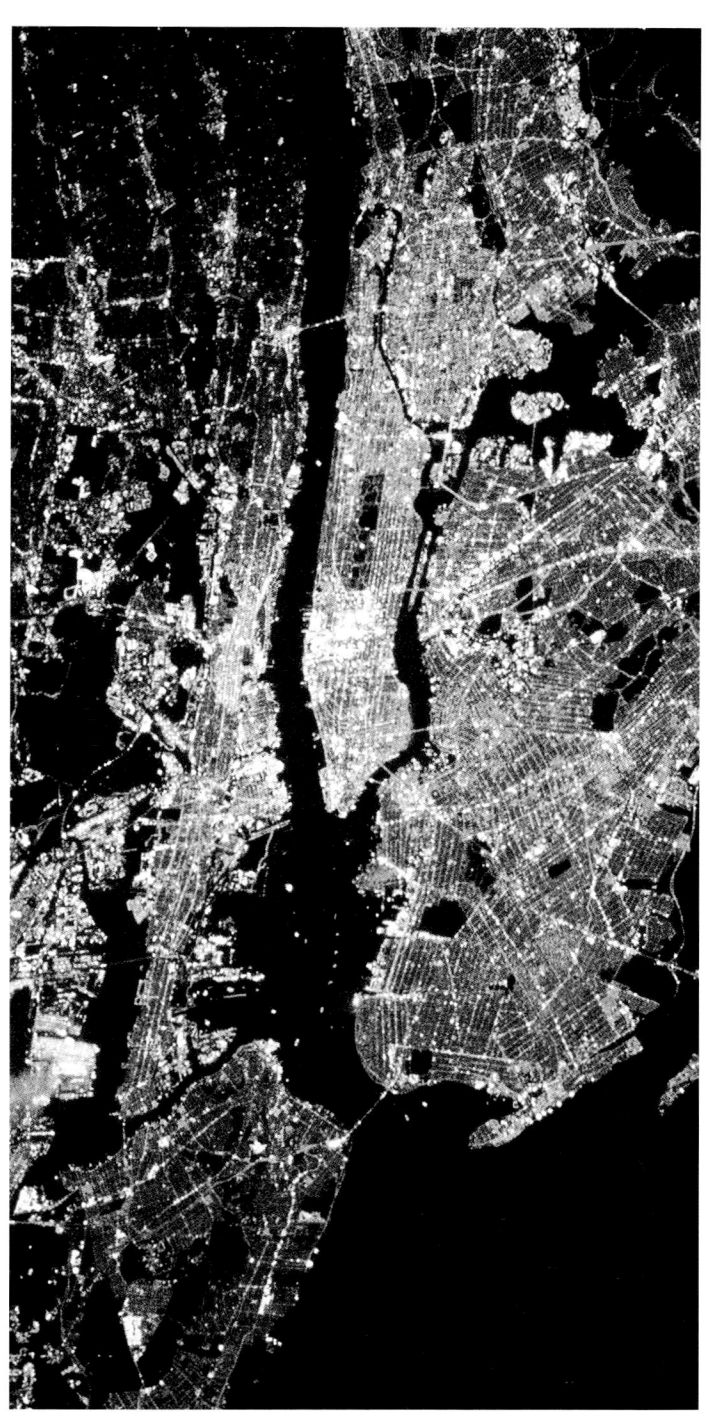

Today the human race CONSUMES THE RESOURCES OF ONE AND A HALF PLANETS IN ONE YEAR, meaning that it uses 1.5 TIMES THE RESOURCES the Earth is able to supply in one year.

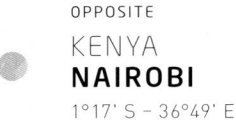

Lagos, the capital of Nigeria, is the largest city in West Africa.
It is changing quickly. And for the better.

Talking with Benoît Verdeaux

Changing the City

How has the city of Lagos evolved in recent decades?

Each year, Lagos welcomes 250,000 to 300,000 new inhabitants. This trend is partially due to Nigeria's rural exodus, but the capital also draws migrants from the rest of West Africa and Central Africa. Some neighborhoods have a population density greater than 384,000 inhabitants per square mile (150,000 inhabitants per sq km). The city is therefore increasingly vertical: Residential towers and multistory buildings are no longer limited to apartment blocks in rich neighborhoods. Traffic jams, known locally as "go-slows," are also increasing. One frequently spends hours in a traffic jam.

How do you use satellite data?

Seventy percent of Lagos's inhabitants live in slum neighborhoods. For urban planning, satellite photos allow us to study the feasibility of projects to rehabilitate slums, detect precarious housing, and estimate the number of people concerned, in order to best adapt the programs and their financing to the situation.

We also use satellite images to measure urban density and achieve a more accurate estimate of population figures. For instance, Lagos has about eleven million inhabitants, rather than the twelve to fifteen million generally stated. The same overestimations are found on a national scale: In 2012, the national population was officially estimated at 162 million inhabitants, but according to Africapolis' analyses, which cross-reference statistics and satellite images of cities, it is actually closer to one hundred million.

Are the city authorities able to control this urban growth?

The authorities are very involved, which is good news for a city undergoing such rapid growth. Security has improved and infrastructures are developing. Transportation, waste, and housing remain central concerns. Overall, one senses a political desire to move toward well-thought-out city planning and to work on rehabilitating slum neighborhoods. The agencies responsible for executing these urban policies are top notch. Some of their executives participated in London's development before coming home to Nigeria.

What are Nigerians' living conditions like?

There is a significant disparity in Nigeria between urban and rural populations and north and south—oil riches do not benefit everyone. This embodies the country's contradictions: Despite its riches—it contains half the resources of ECOWAS (Economic Community of West African States), in terms of population and economy—it is far behind neighbors such as Ghana in the fields of education and access to health care and housing.

Benoît Verdeaux, coordinator of the activities of the Agence Française de Développement (French Agency for Development) in Nigeria, heads the Africapolis study on urbanization in West Africa.

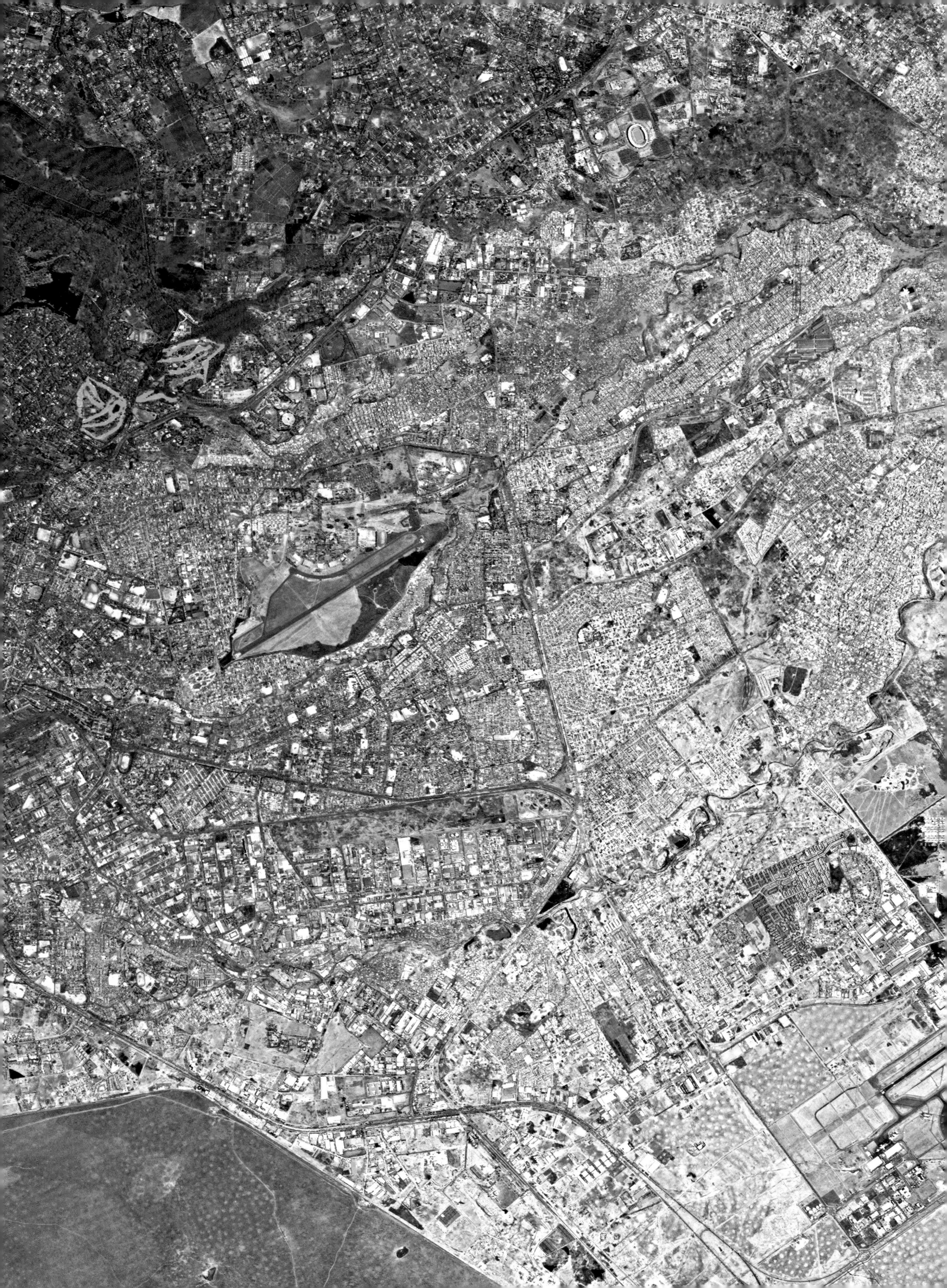

THE BIGGEST CITY IN AFRICA

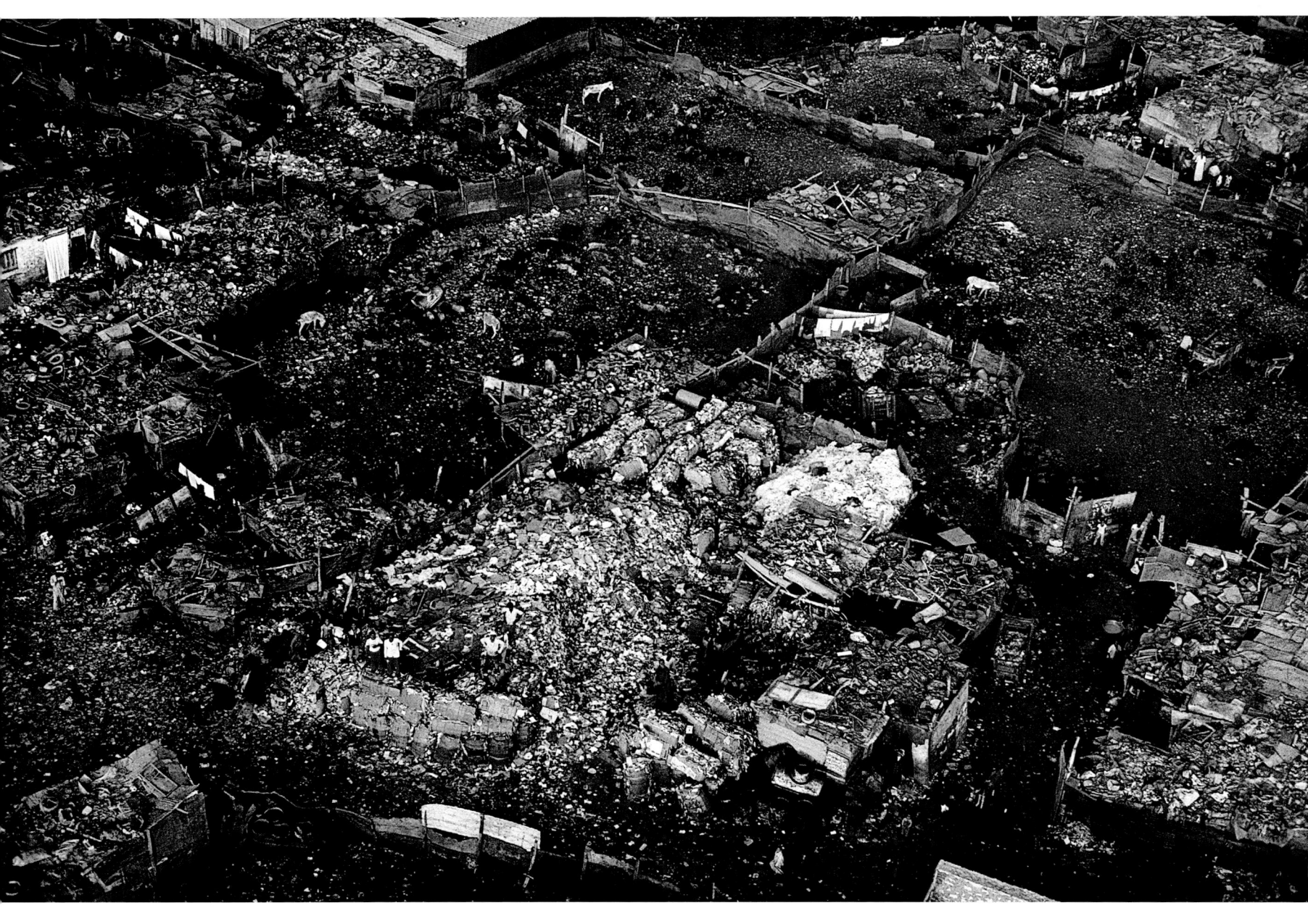

ABOVE
EGYPT
RAGMAN VILLAGE IN CAIRO
30°02' N – 31°12' E

Zabalines ("ragmen" in Arabic) are responsible for collecting, sorting, and recycling trash in Egypt's biggest city. In this country where 90 percent of the population is Muslim, the ragmen are part of the Coptic minority.

OPPOSITE
EGYPT
CAIRO
30°03' N – 31°14' E

The capital of Egypt is home to some sixteen million people, one-fifth of the country's population. The city has expanded its suburbs all the way to Giza, at the foot of the Great Pyramids—which can be seen here to the west.

With seventeen million dwellers, the greater urban area of Cairo, the biggest city in Africa, is home to one-fifth of Egypt's inhabitants. The country's agricultural land is located along the Nile on a strip of land only a few miles wide. Rapid population growth is sparking deep changes. The thirty-nine-year-old Egyptian scientist Ahmed Hassan, a specialist in geology and earth sciences, used satellite images, taken between 1984 and 2006 and provided free of charge by the Planet Action program, to show that the area of Egypt occupied by the metropolis has practically tripled—growing from 12 percent to 29 percent—while cultivated

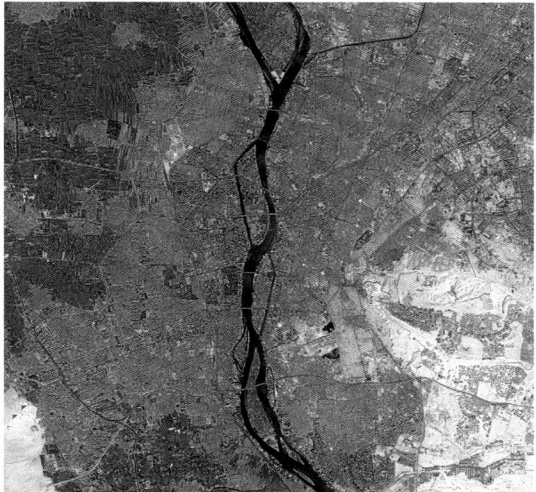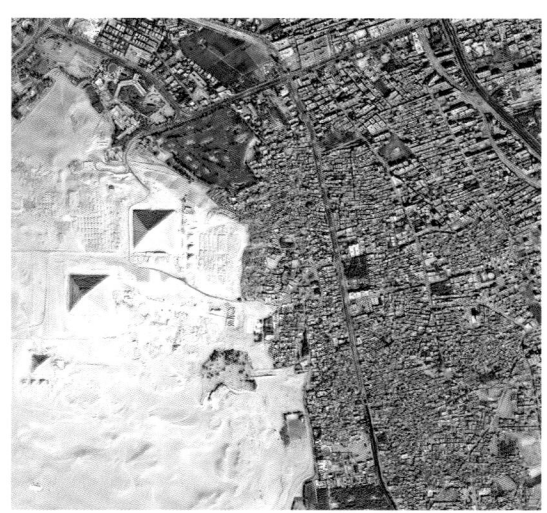

land has shrunk by half, dropping from 24 percent to 11 percent.

Hassan analyzed the color of each pixel rendered by the satellite—one pixel corresponds to several hundred square feet—to identify what the land is being used for. This enabled him to describe the evolution of land use: "The city is primarily growing on its peripheries, with the development of suburbs and slums, which encroach upon cultivated areas. Additionally, farm owners are building new houses on their land for their children and grandchildren, near the city, eating into fertile land. Sometimes these constructions are not directly intended for family members, but are built as investments."

With this research, Ahmed Hassan has shown that urban growth is taking place to the detriment of agricultural land, potentially intensifying water and food shortages. Additionally, population growth may upset the natural network of water circulation, while the artificialization of surfaces simultaneously leads to a local increase in temperatures.

PAGES 138-139
ITALY
VENICE
45°26' N – 12°20' E

The Venetian Lagoon is separated from the Adriatic Sea by a coastal strip consisting of a string of narrow islands. The city itself, with its 118 islets and 160 canals crossed by more than 400 bridges, must face, more frequently, the *acqua alta*, or rising water: Over the last thirty years, Venice has been under more than 3.2 feet (1 m) of water about one hundred times. The MOSE Project adopted in 2002 aims to close the channels; some eighty mobile dykes should be operational in 2014.

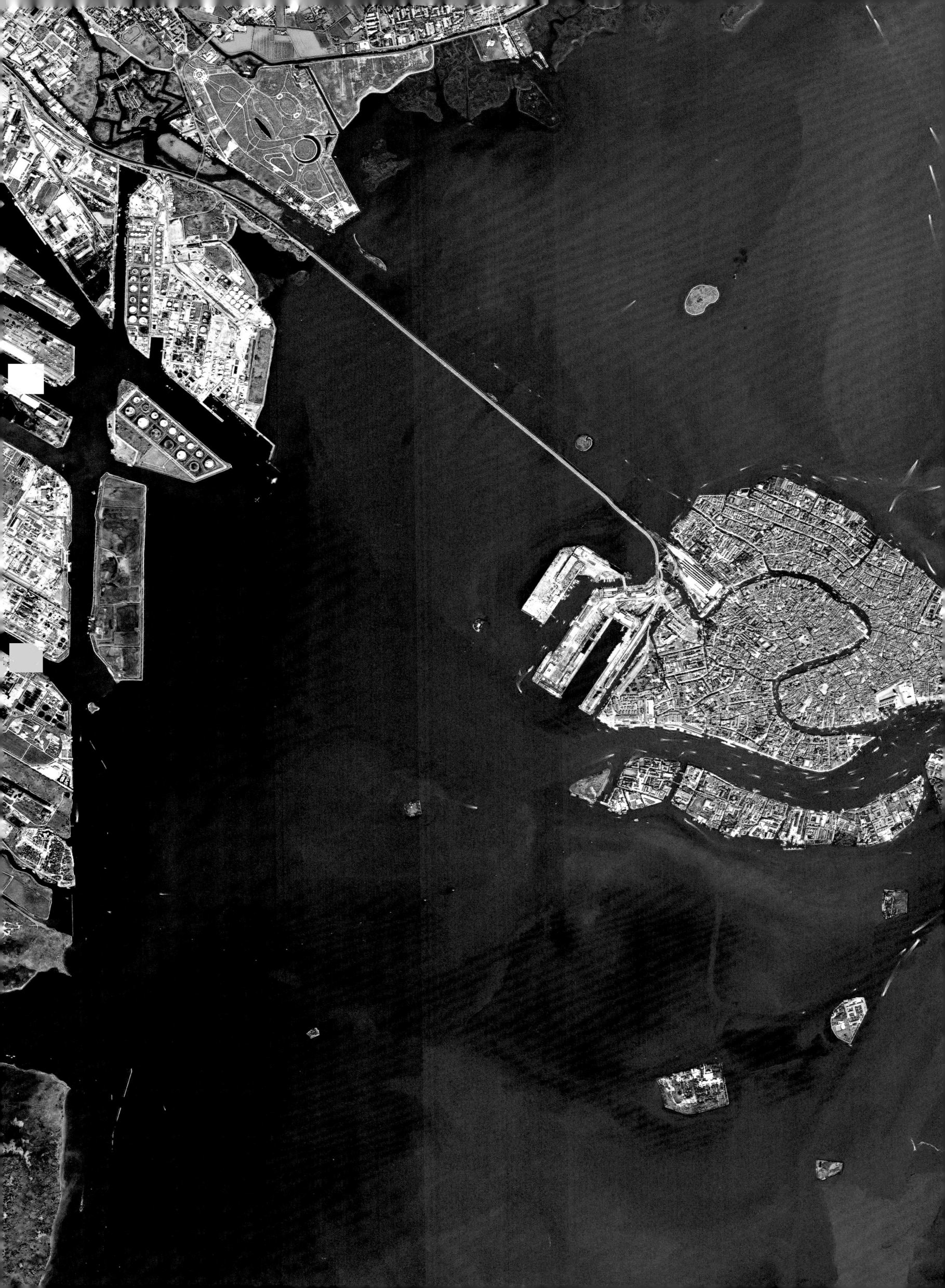

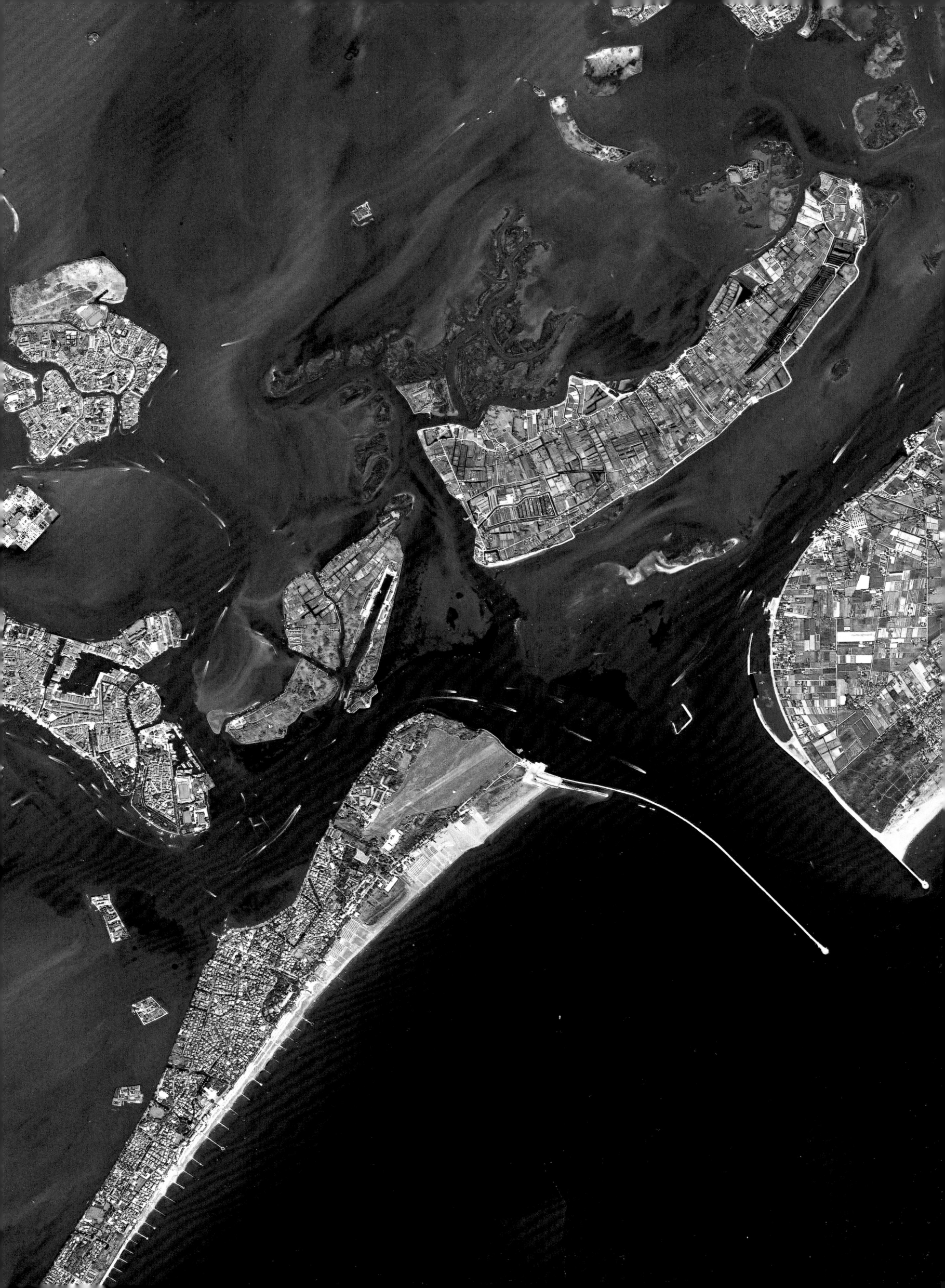

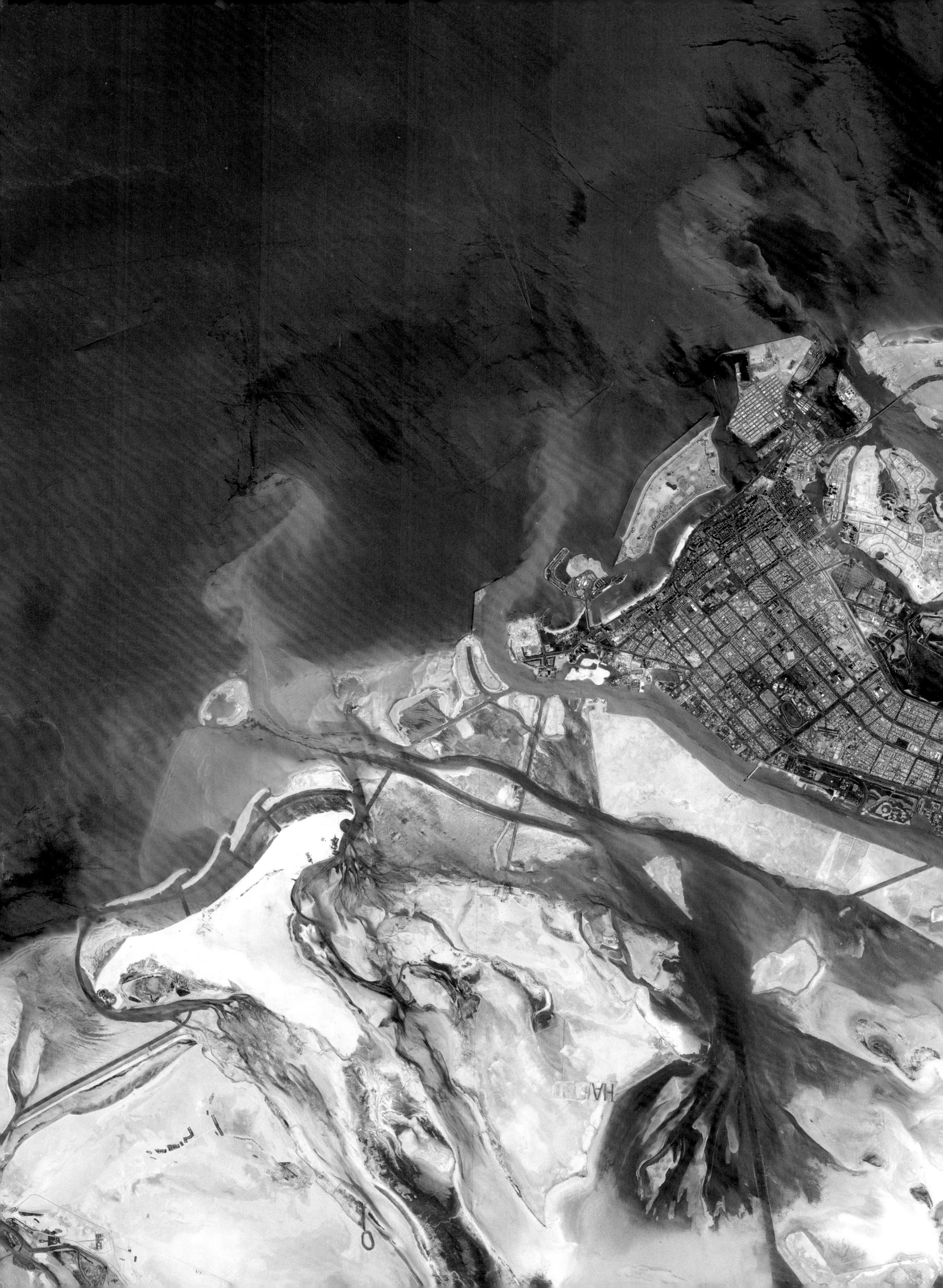

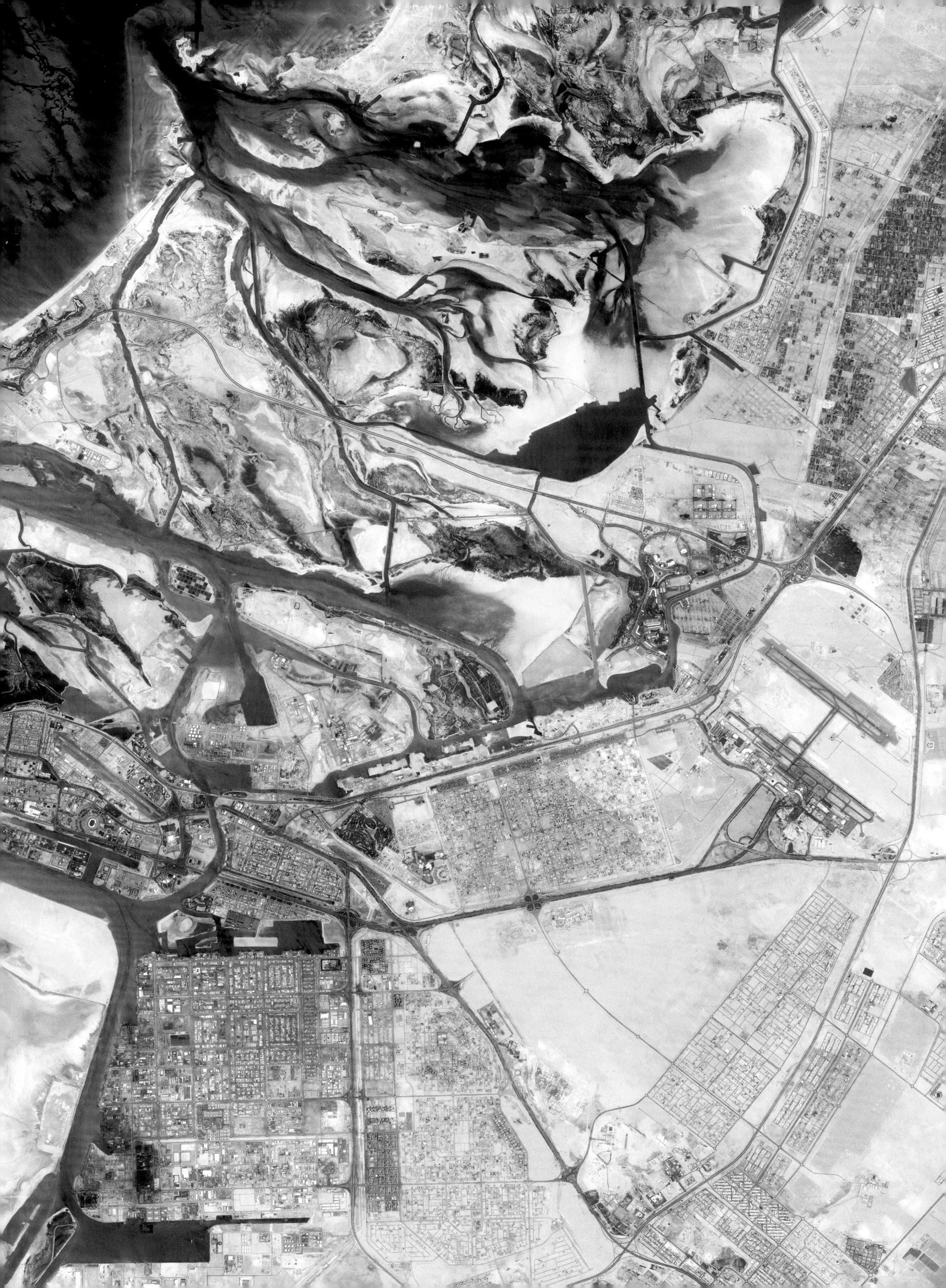

PAGES 140-141

UNITED ARAB EMIRATES
ABU DHABI
24°27' N – 54°22' E

Abu Dhabi is the capital of the United Arab Emirates, which is the seventh-largest global oil exporter. The dollars generated by oil revenue have provided the Gulf countries with a considerable annuity, bringing modern cities dense with skyscrapers, malls, and universities out of the heart of the desert. In less than a century, oil has become indispensable: It provides 32.4 percent of the primary energy in the world, principally for transportation.

BELOW

SWEDEN
STOCKHOLM
59°21' N – 18°04' E

Sweden's capital stands where Lake Mälaren drains into the Baltic Sea and consists of 14 islands connected by 57 bridges. The country's approach to foreign policy is based on pacifism and non-alignment, and the city has regularly served as a location for international negotiations and appeals, such as the 1950 Stockholm Appeal against nuclear proliferation.

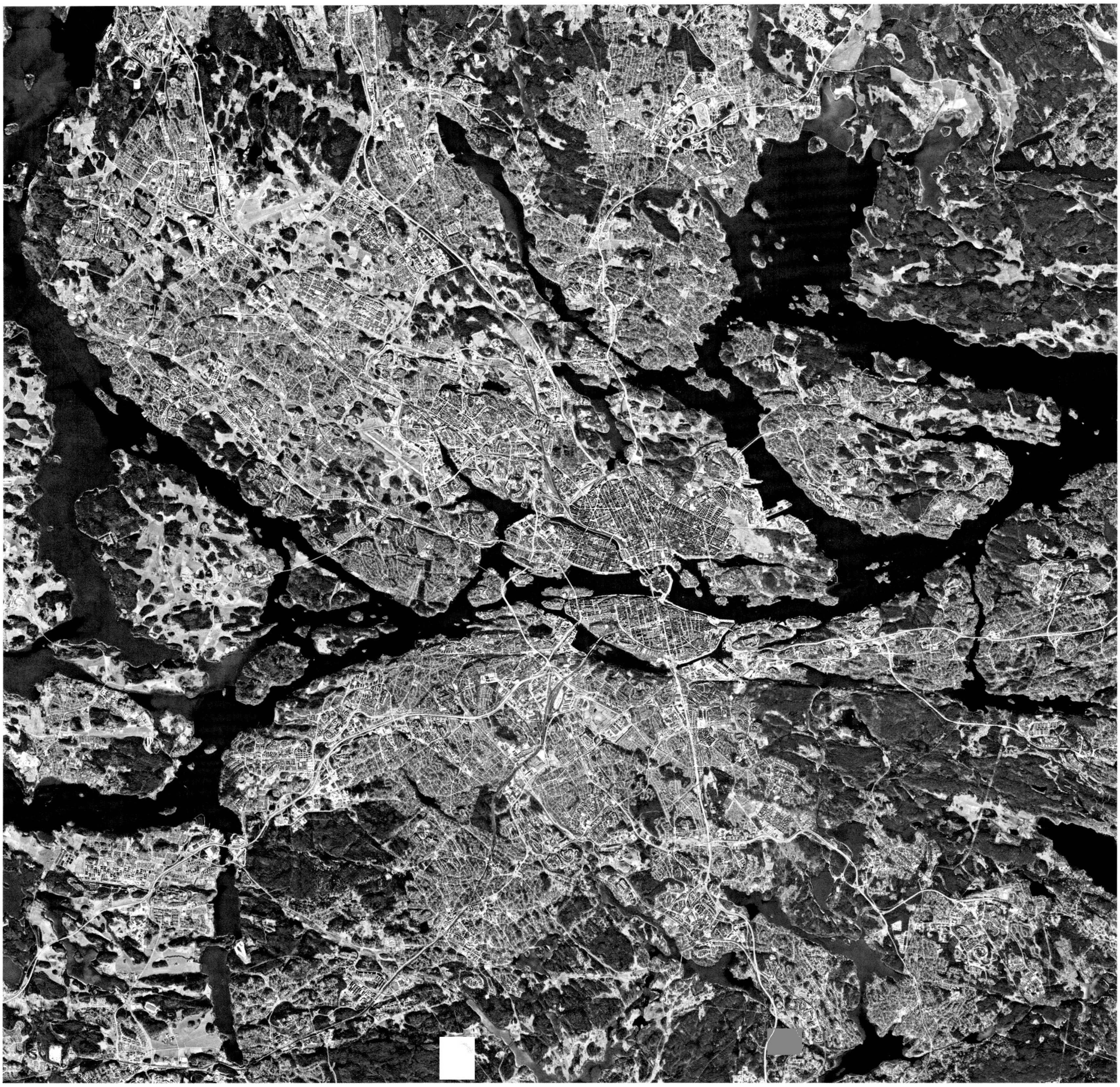

BELOW

FRANCE
LYON
45°45' N – 4°50' E

The Lyon urban area is an important commercial crossroads in the heart of Europe. While traffic often blocks the Rhône Valley—the people of Lyon spend an average of forty-six hours a year in traffic jams—the city has proven innovative with urban mobility: Lyon was the first city to use bike shares, through the Vélo'v system, funded by advertising. In Lyon and Villeurbanne, four thousand bicycles are available to the public.

PAGES 144-145

CHILE
VALPARAÍSO
33°02' S – 71°36' W

The second-largest city in Chile, Valparaíso is the country's leading port. Valparaíso consists of many little streets rising from the sea and winding their way to the tops of dozens of hills accessed by cable cars. The lower city is home to wealthy residences and businesses, while the upper city is where the poorest populations live. Valparaíso is on UNESCO's World Heritage List.

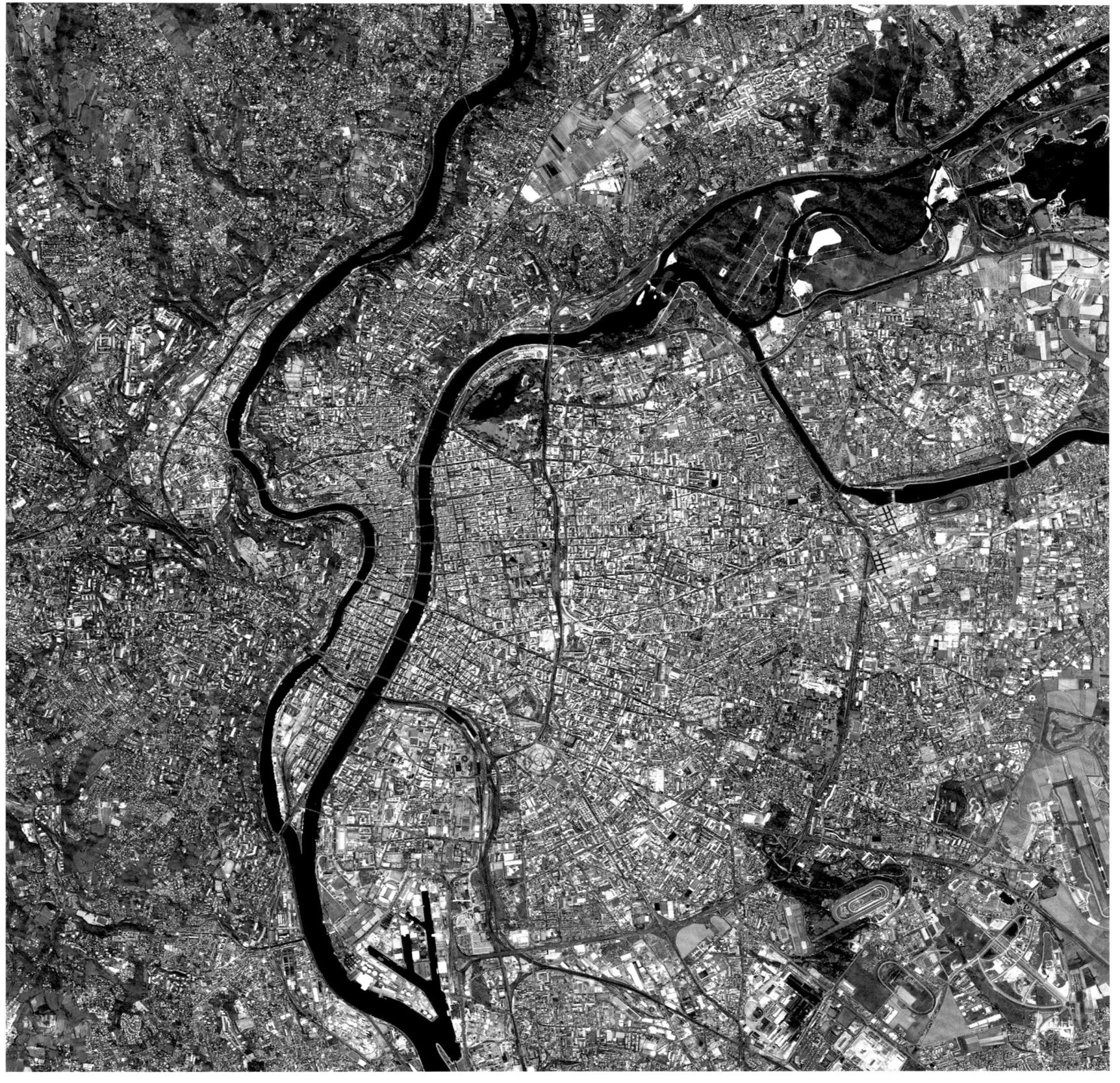

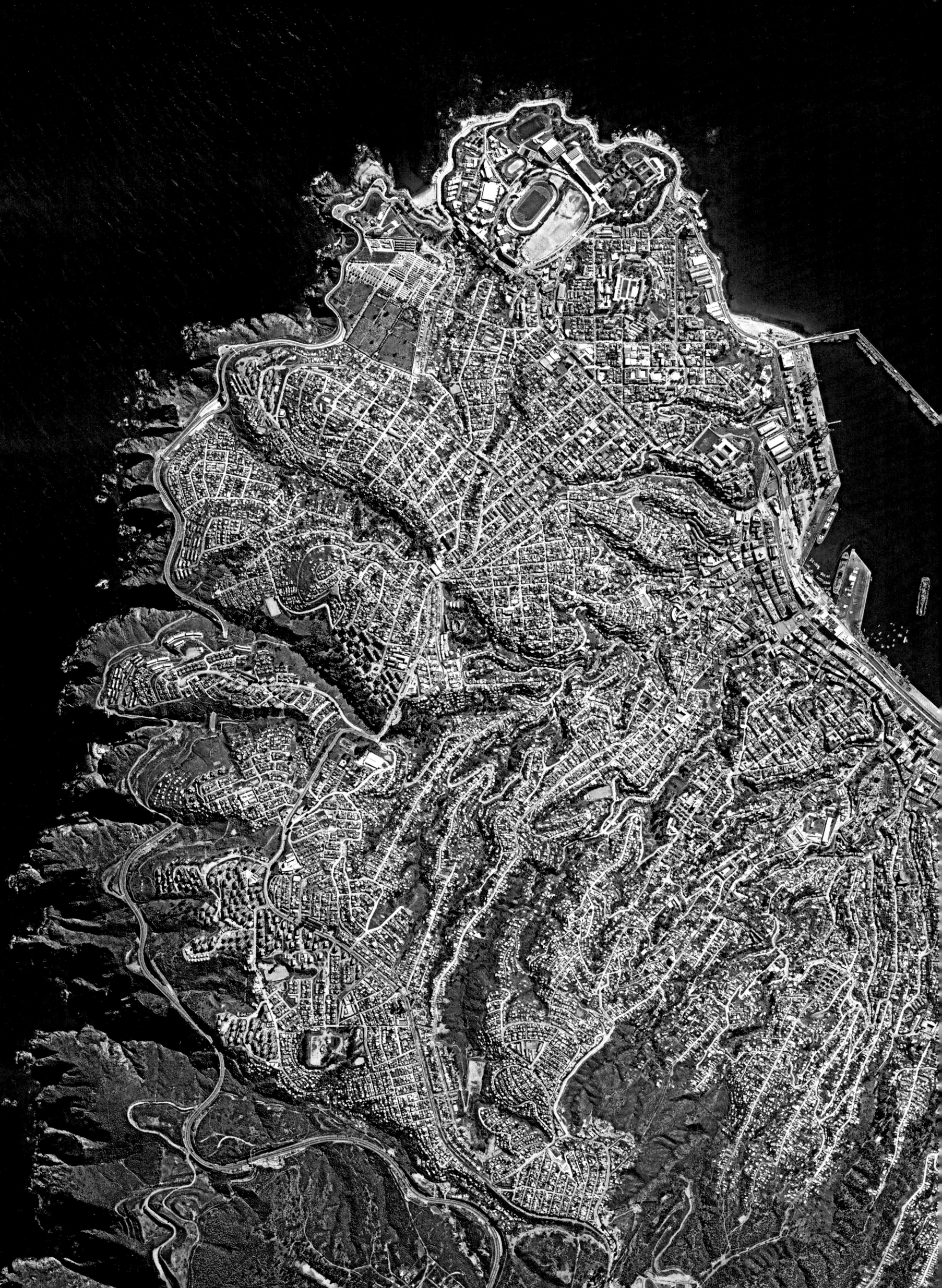

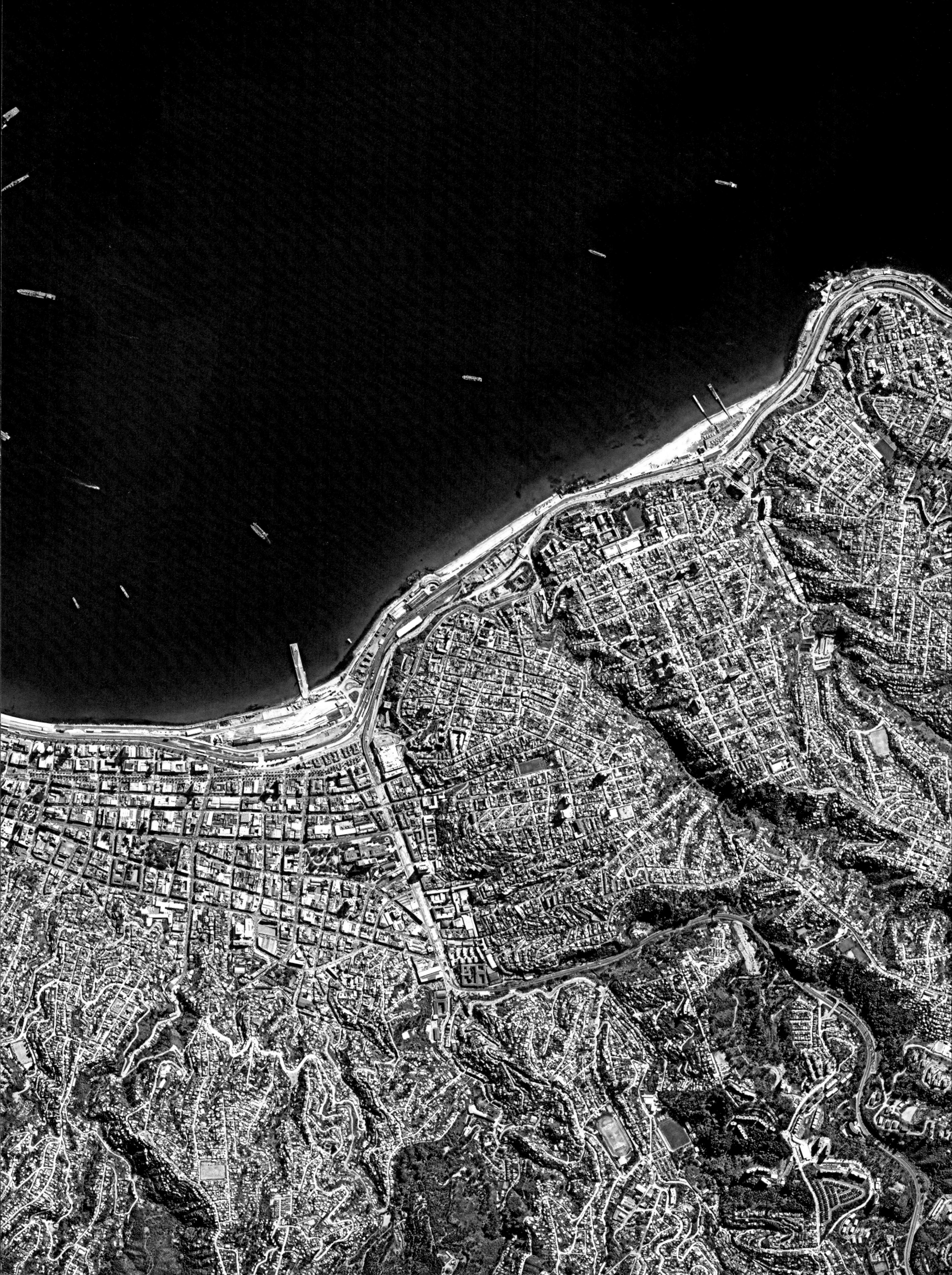

From space, you can observe tiny plantlets or gigantic ecosystems. You can also track migratory birds along their entire journey, monitor fishing boats' activities, and establish long-term plans to protect endangered species.

—

LIFE ON EARTH

—

When the oceans warm up and favorable conditions are aligned, billions and billions of miniscule plant organisms appear and proliferate. The scientific term used to describe this phenomenon is "phytoplankton bloom" or "algal bloom." Individually, one of these organisms is no bigger than 0.008 inch (0.2 mm), yet together they make up phytoplankton, which accounts for 98 percent of the oceans' biomass. Phytoplankton is at the base of the food chain on which all marine animals depend.

Surprisingly, the most powerful tool to study these microscopic organisms' dynamic is satellite observation. High concentrations of phytoplankton, with up to several million cells per milliliter, give water a green, blue, brown, or red color, which is visible from space. By analyzing the water's color, we can deduce the abundance of colored pigments in the phytoplankton, and thus the quantities of microorganisms. Satellites provide a way of detecting and monitoring the appearance of blooms due to pollution or predicting the appearance of "dead zones" where excessive proliferation of microalgae provokes the total asphyxiation of an environment.

AN UNEQUAL DISTRIBUTION OF BIODIVERSITY

Satellite imagery also allows us to draw the multicolored map of the unequal distribution of life in the oceans. The areas of the world's oceans richest in biodiversity are along coastlines—with occasional zones of heavy productivity, due to the rising of nutrient-rich waters from the deep—and at the mouths of rivers. On the other hand, the heart of the

ocean, which is incredibly transparent, is like a desert. In 2004, researchers combined satellite observation and in situ measurements to determine that the clearest water in the world is off Easter Island, in the Pacific Ocean. This exceptional transparency is indicative of the resource-poor water in this area of the world.

The unequal distribution of life is not exclusive to the marine environment. For example, Costa Rica, which occupies 0.03 percent of terrestrial surface area, shelters close to 6 percent of the world's biodiversity. Areas where life is abundant are known as biodiversity hot spots. Scientists have identified thirty-four hot spots, including the Cerrado, a savanna region in South America thought to have at least ten thousand plant species, four thousand of which are endemic; Madagascar; the mountains of Central Asia; New Zealand; and several islands in the Indian Ocean. Yet hot spots are also endangered. The concept of the hot spot is therefore used to prioritize areas that must be protected. Only 38 percent of the surface area of hot spots around the world is currently classified as protected (parks and reserves).

Satellites have proven to be incredible tools to help manage these natural environments and their biodiversity. Thanks to satellite images, we can follow the evolution of the tree cover (deforestation, for example) and representative endangered ecosystems, such as coral reefs. Coral reefs are on the front lines of climate change: With the increase in temperatures, corals bleach—a phenomenon that the Envisat satellite and its MERIS sensor can evaluate at depths of up to 33 feet (10 m). Once a coral bleaches, it often dies; it appears that 20 percent of tropical coral reefs

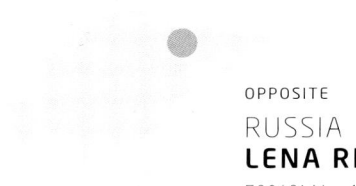

OPPOSITE

RUSSIA
LENA RIVER
72°42' N – 127° E

Running 2,734 miles (4,400 km), the Lena River is one of the longest rivers in the world. The Lena Delta Nature Reserve, an important refuge for many Siberian wild species, is one of the largest protected wild environments in Russia. Here, vegetation is shown in green, sand zones in pink, and water in blue-mauve. The delta's tundra, which is frozen during the region's seven months of winter, turns into a lush wet zone in the summer.

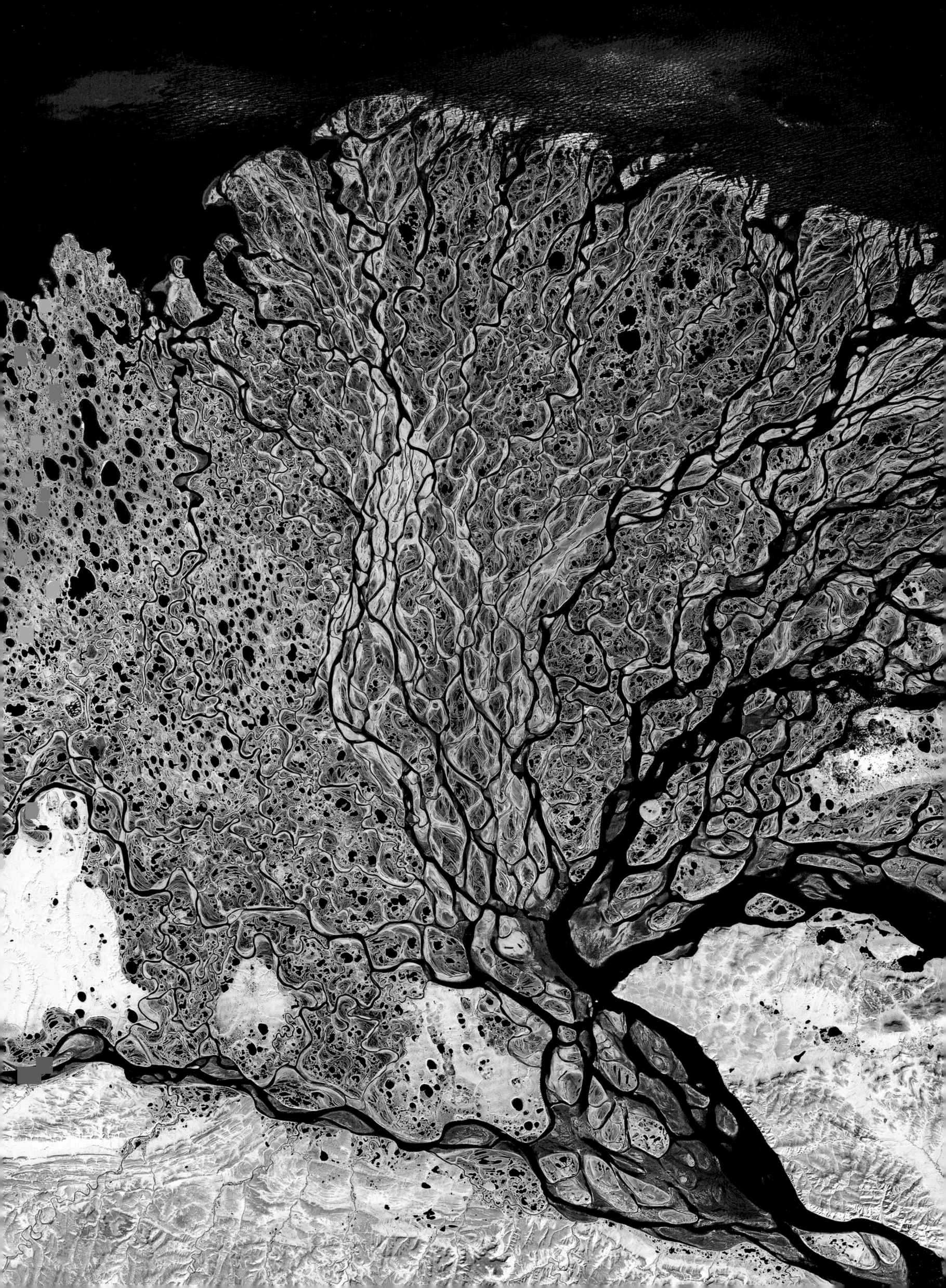

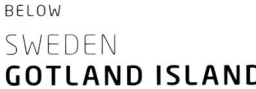

BELOW

SWEDEN
GOTLAND ISLAND

57°30' N – 19°30' E

Reminiscent of Van Gogh's famous painting *Starry Night*, this vast pattern of greenish whirls off Gotland Island in the Baltic Sea consists of billions of microscopic organisms known as phytoplankton, which proliferate when deep currents carry nutriments to the surface.

OPPOSITE

SOUTH AFRICA
NAMAQUALAND

31° S – 20°30' E

The artificial colors in this satellite image reflect the abundant biodiversity of the wild fauna and flora of this area of South Africa close to Namibia. Generally arid, the area is transformed when the many plant species it harbors bloom. Much of this area's 169,884 square miles (440,000 sq km) is now on UNESCO's World Heritage List.

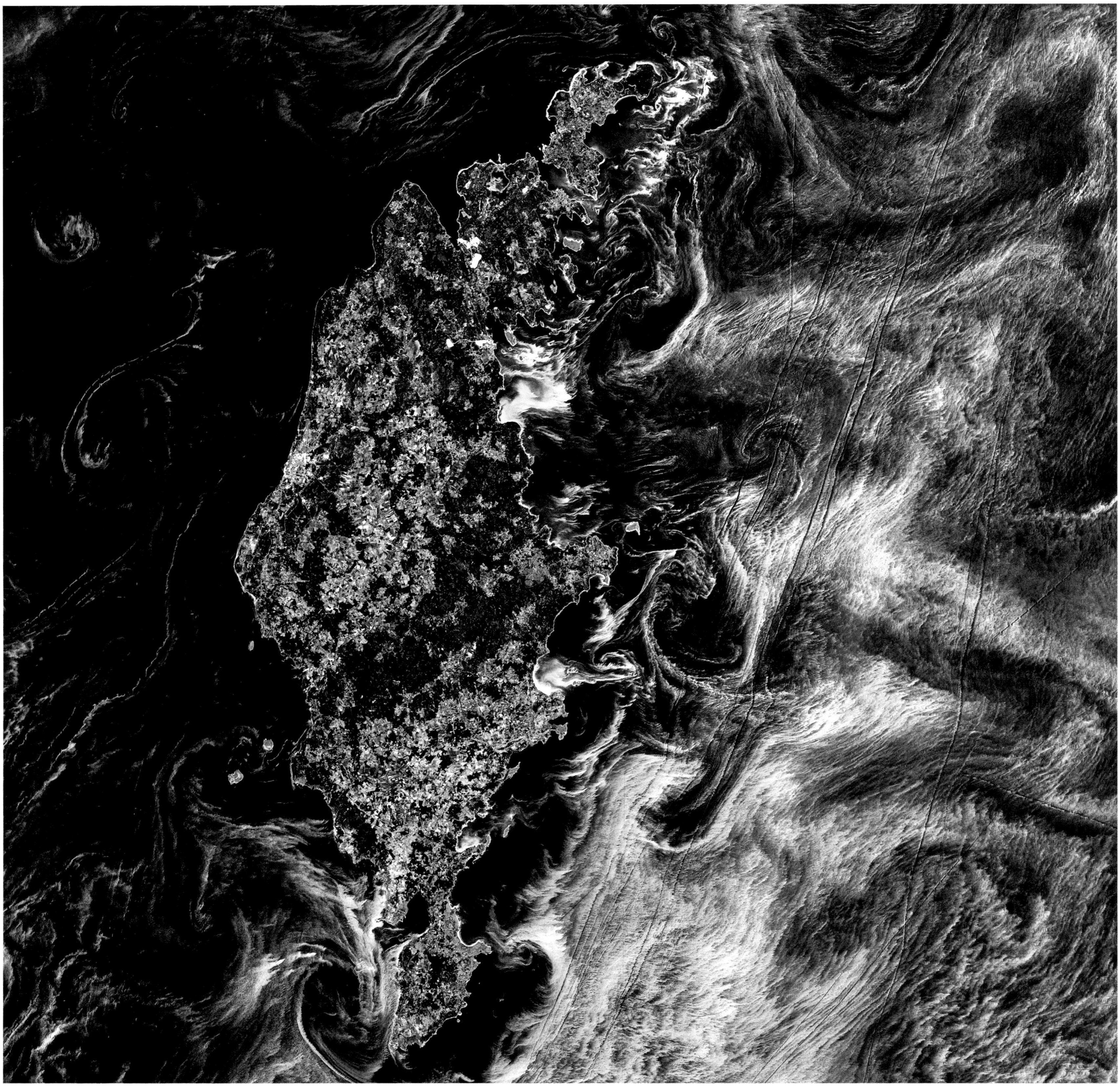

COSTA RICA, which OCCUPIES 0.03 PERCENT OF
TERRESTRIAL SURFACE AREA, shelters close to
6 PERCENT OF THE WORLD'S BIODIVERSITY.

have already been destroyed. With high resolution, regular passes, and coverage of more than 621 miles (1,000 km) per shot, satellites are ideally adapted to studying this kind of process. Indeed, bleaching happens quickly, and to observe it, one requires frequent images over stretches of ocean as vast as the Great Barrier Reef, measuring 135,000 square miles (350,000 sq km).

ARGOS BEACONS

Despite all of this, studying biodiversity from space is a difficult task, particularly in the case of terrestrial ecosystems. For example, while it is easy to measure deforestation, it is more complicated to measure its impact on species.

The famous Argos beacons devised by CNES (Centre National d'Études Spatiales, the French space agency) in 1978 are remarkable tools to track the evolution of the animals populating our planet. Now extremely miniaturized (they can be as small as a postage stamp), the beacons have been deployed on more than seven thousand animals and transmit their data to satellites in real time. In many cases, solar panel systems allow the beacons to run autonomously, which makes it possible to follow tagged animals over long periods: A stork named Max has been monitored by Argos for eleven years. Lynx, penguins, storks, bluefin tuna, and even jellyfish are currently tagged and monitored. CNES has even developed sensors that work with the beacons to collect biological data like heart rate, depth of descent, and stomach content.

Beacons are also used to protect fish stock. Many countries have implemented a VMS (vessel monitoring system) to track fishing vessels, which allows fisheries to be monitored and helps to manage activity. This type of system also improves the safety of fishing boats at sea.

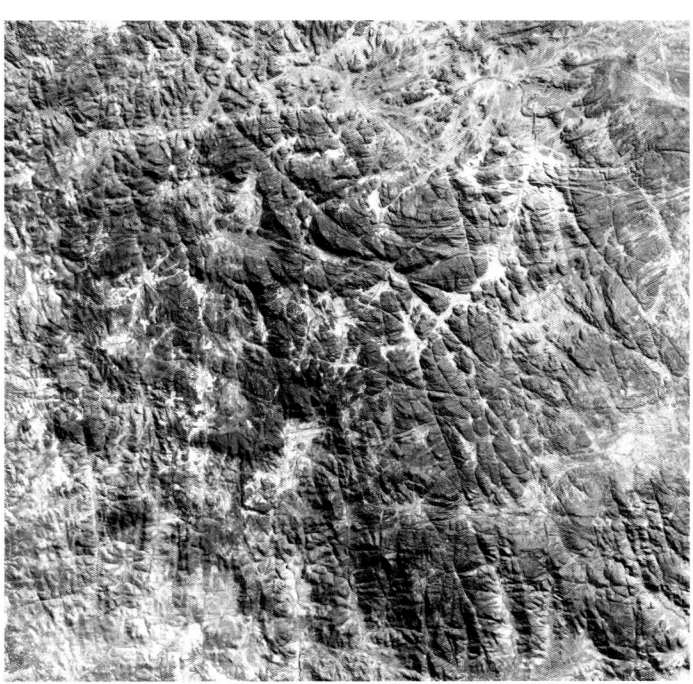

MIGRATIONS

Analysis of the distribution of life provides excellent indications of the ecosystems' health. Because of global warming, many animal and plant species slowly migrate to the poles or move to higher altitudes in search of cooler temperatures.

Certain trans-Saharan migratory species, such as the white stork and the common crane, have radically changed their behavior by increasingly wintering along the edges of the Mediterranean or along Atlantic coastlines. Insects, such as the pine processionary moth in France, also take advantage of mild winters to extend their distribution area. And

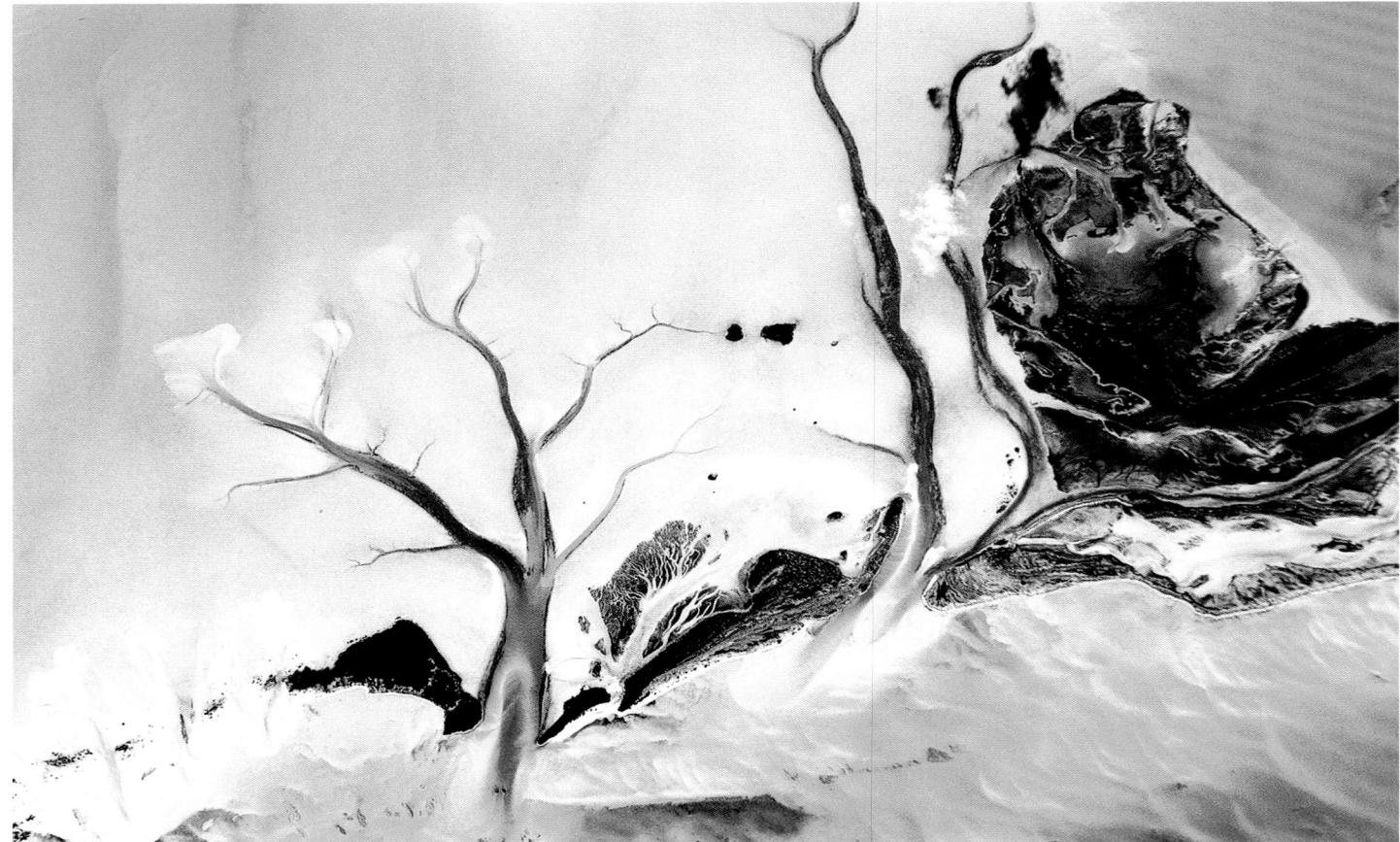

the oak tree is quietly moving north at a rate of 1,640 feet (500 m) a year.

Marine life is also affected. In the Atlantic, plankton has shifted to the north by 10 degrees of latitude, or about 620 miles (1,000 km), in forty years. Some fish have fled to cooler water, including two species of John Dorys that migrated from the south of Portugal to Ireland in the last thirty years. These movements are sometimes accompanied by the desynchronization of interactions between species, such as when a flower buds out of sync with the reproduction of the insect that pollinates it. These kinds of sudden changes in habitat range could provoke competition between several species, which are at risk of eliminating one another. Other species that are now migrating to mountain peaks, such as the woodchuck, or the extremities of a landmass, such as

the Arctic polar bear, will eventually run out of places to fall back on. Also, sometimes the obstacles standing in the way of migration are man-made: Highways, cities, and farms prevent animal and plant species from moving freely.

CORRIDORS

In order to allow animal species to migrate or simply move around—whether to reproduce, hunt, or respond to a change in the environment—researchers are working on what they call biological corridors, a kind of environmental highway reserved for life-forms that links habitats, which are often fragmented by roads and other constructions. Satellite data is crucial in determining the path of these corridors, whether on a local scale, as with the blue and green networks called

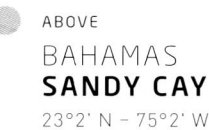

ABOVE
BAHAMAS
SANDY CAY
23°2' N – 75°2' W

The islands of the Bahamas are located on two limestone platforms, the Little Bahama Bank and the Great Bahama Bank. The islands are the emerged part of these sedimentary platforms that formed from the decomposing skeletons of microorganisms on the seafloor. This image shows the intertidal zone near Sandy Cay, along the eastern edge of the Great Bahama Bank.

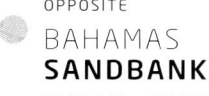

OPPOSITE
BAHAMAS
SANDBANK
23°36' N – 76°50' W

Composed of seven hundred islands, the Bahamas lies east of Florida in the Atlantic Ocean and is part of the Greater Antilles. This satellite image shows sandbanks and algae sculpted by the waves. One can even discern swimmers who have come to admire the underwater beauty around these islands.

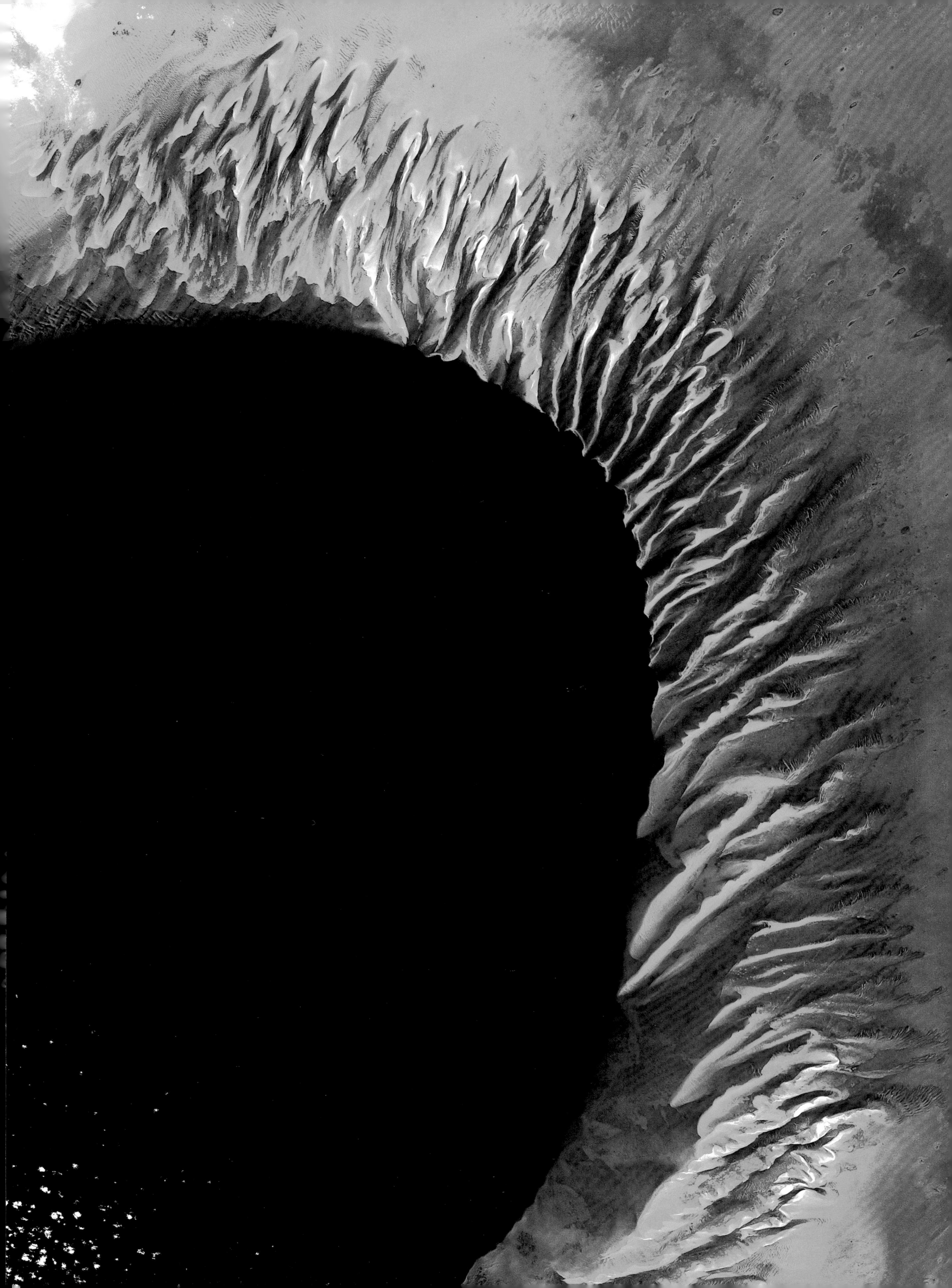

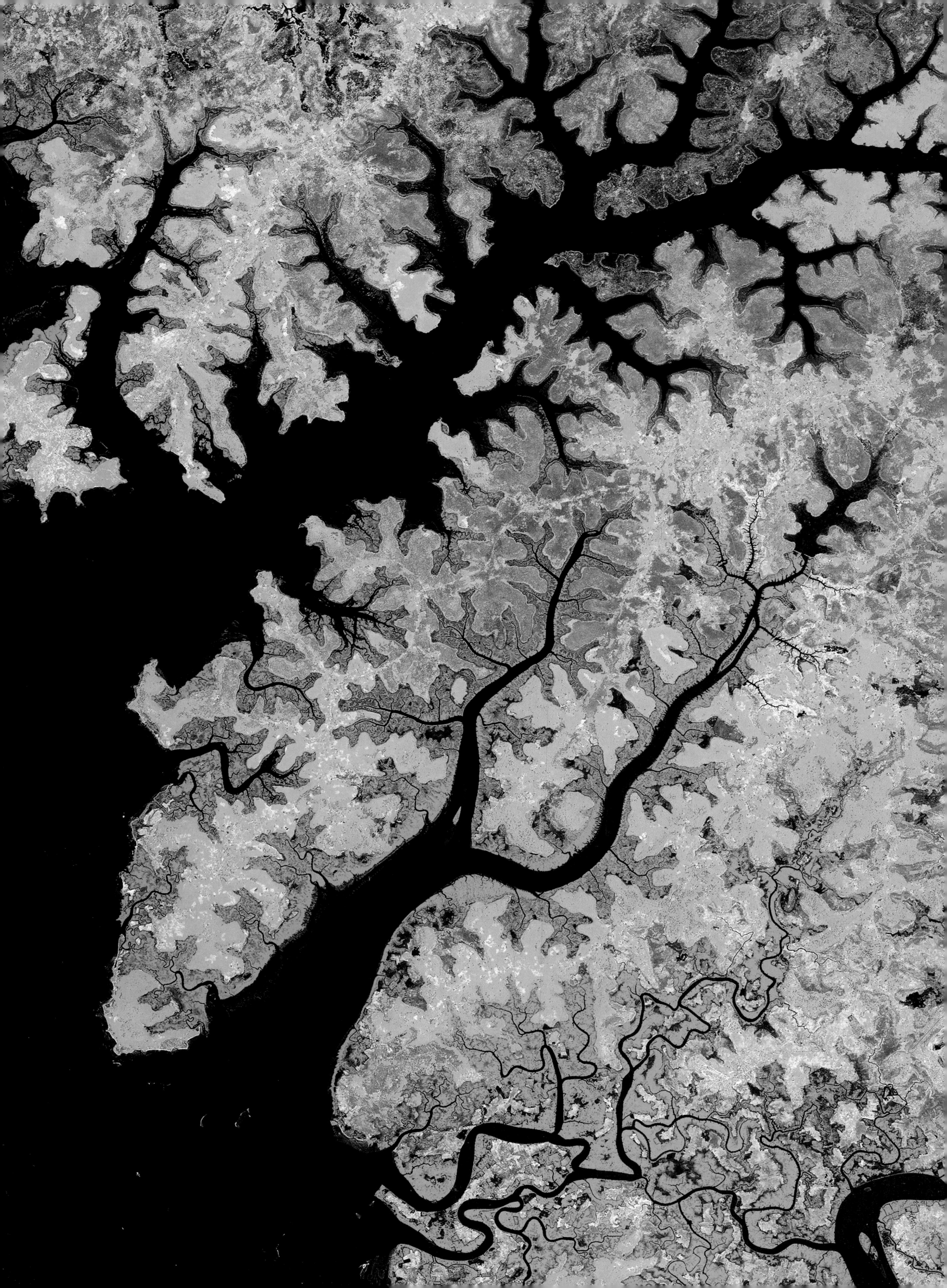

OPPOSITE
GUINEA-BISSAU
RIO POBREZA
11°39' N – 15°15' W

The Rio Pobreza is a channel covered in
mangroves, those peculiar forests found at the
meeting point between land and sea. Plants,
and particularly mangrove trees, develop a
wide range of methods for resisting the water's
salinity. Mangroves serve as nurseries to the
wide variety of aquatic animals that find food and
shelter there.

BELOW
NEW ZEALAND, NORTH ISLAND
MOUNT TARANAKI (EGMONT)
39°18' S – 174° E

The TerraSAR-X radar satellite reveals the outline of Egmont
National Park, which can be identified by the difference in color
between the slopes of the Mount Taranaki (or Mount Egmont)
volcano and the rest of the area. This circle with a radius of
5.9 miles (9.6 km) is not a natural occurrence: It separates the
park's lush forest from surrounding pasture zones. Mount
Taranaki stands 8,261 feet (2,518 m) tall and has been totally
inactive since 1755. Its peak is covered in perennial snow, while
nearby Fanthams Peak has none at all.

for in France by the Grenelle Environnement, or on the scale
of a continent, as with the Paseo del Jaguar (Path of the Jaguar) in Central and South America and the Trans-Australia
Eco-Link, which stretches about 2,175 miles (3,500 km).

A QUESTION OF MONEY

Creating these corridors is often a slow and complex process. As with any other measure to protect the environment,
it is often objected to because of its cost. Yet even the cost
of developing satellite tools, whose usefulness for observing
our planet is clearly demonstrated in these pages, pales in
comparison to the many riches offered to us daily by natural environments: The water we drink is filtered by soil, the
air we breathe is produced or purified by plankton and forests, the food we eat is borne of biodiversity, as are a large
proportion of our pharmaceuticals.

In order to justify the initiatives implemented to protect
nature, some specialists have attempted to provide an estimate of its economic value. In 1997, an initial study led by
Robert Costanza for the United Nations Environment Programme (UNEP) estimated the annual value of services carried out by nature around the world at $3.3 trillion. Since
then, numerous studies by the World Bank and UNEP have
continued to refine these initial findings.

Defending nature is not only a moral necessity; it is also
in everyone's best interest.

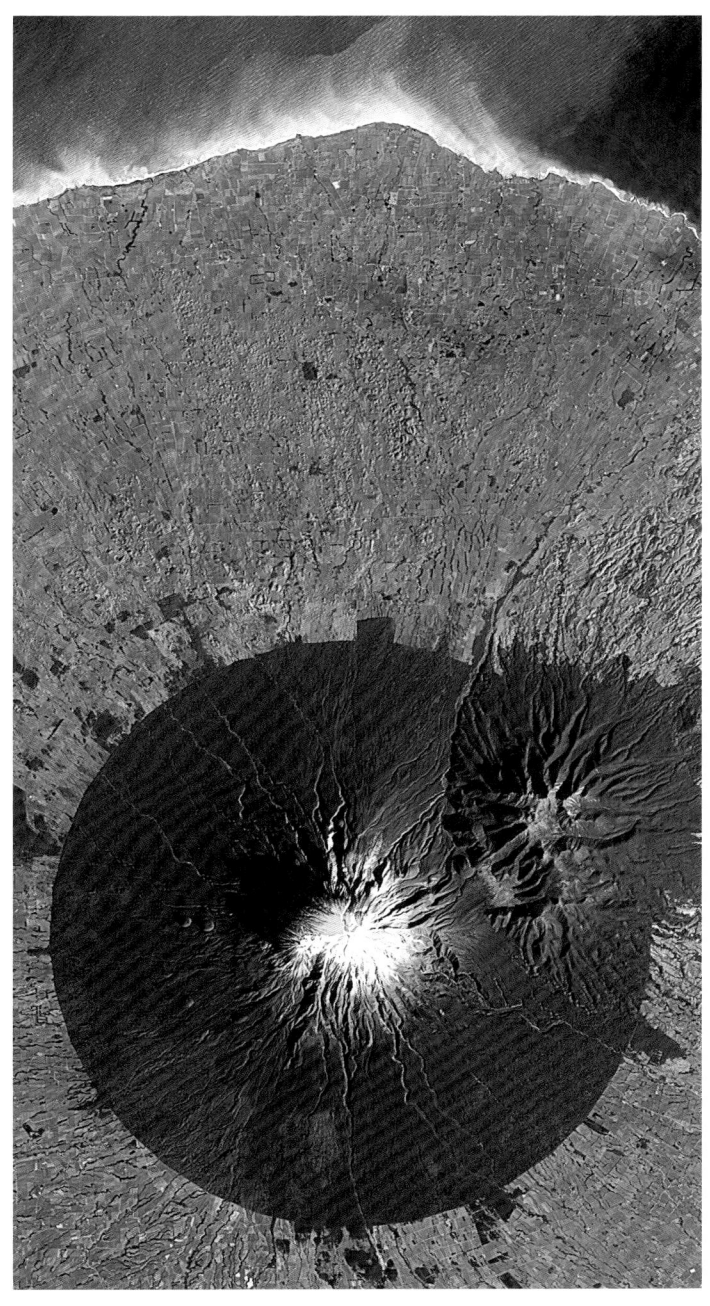

Penguins can be counted thanks to the stains left by their guano on the ice floes. In Antarctica, there are twice as many penguins as previously believed.

Talking with Peter Fretwell

Multiplying Penguins

How did you think of locating penguin colonies based on satellite images?

I made that discovery by chance. I was preparing a map of the area around the Halley Antarctic research station for the purpose of having pilots try to avoid penguin breeding grounds. But at the time, the information we had about the colonies was really vague. Then I noticed brown marks on the ice in images from the Landsat satellite where the sea of ice should have been pure white. Emperor penguins spend eight months a year raising their young in Antarctica; their guano stains the surrounding ice, making their presence noticeable. So I contacted the base to get better information about the colonies' location. The GPS coordinates my colleagues gave me coincided with the brown marks perfectly. So I decided to find the other colonies by following the same approach.

What happened next?

We found brown marks in several places where we did not know penguins lived. In an initial study published in 2009, with images accurate to 92 feet (28 m), we located most of the colonies previously identified, as well as ten new ones. By returning to the project with images accurate to 1.6 feet (0.5 m), we discovered sixteen new breeding grounds.

What did you learn about emperor penguins from these studies?

We doubled the number of known breeding sites, from twenty-six to forty-nine. These studies led to a better estimation of the emperor penguin population: about 595,000 versus 270,000 to 350,000 prior to our work.

Why observe emperor penguins from space?

Climate change is the primary threat weighing on penguins. We estimate that their population will drop by half over the next sixty years, due to the melting of the ice caps. We are perfecting a program to annually count the total emperor penguin population by satellite to verify whether our projection is borne out. This will provide an indicator, a kind of "barometer," to evaluate the implications for other species that are more difficult to study, as well as implications for the ecosystem as a whole, which is one of the most remote areas in the world.

Peter Fretwell is a cartographer with British Antarctic Survey, the United Kingdom's national research center in Antarctica, which operates five research stations, two ships, and five airplanes.

SAVING THE PANDAS

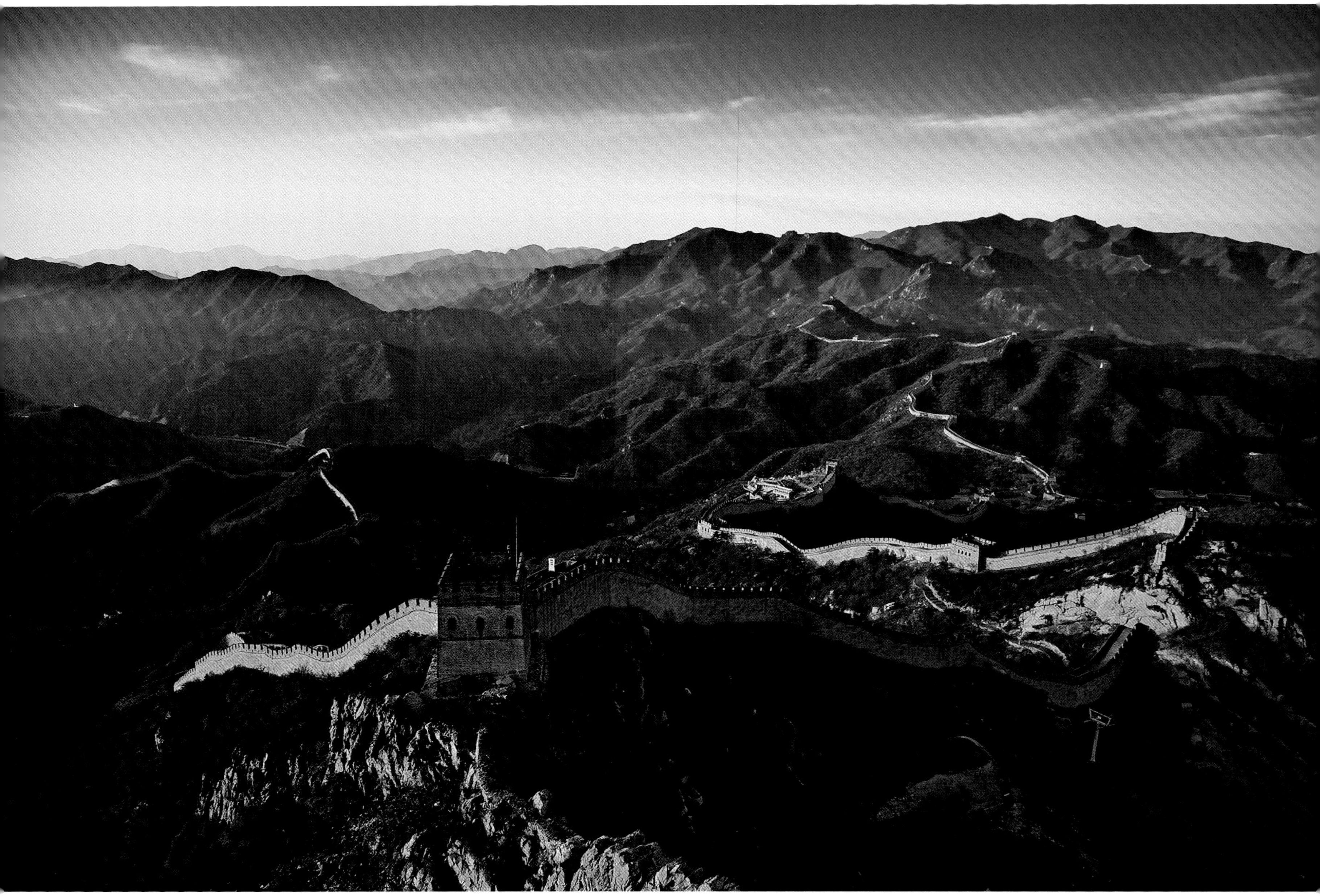

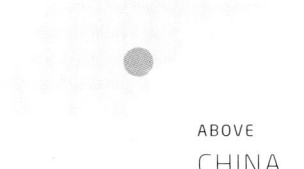

ABOVE
CHINA
**GREAT WALL,
NEAR BEIJING**
40°22' N – 116°01' E

With its sprawling surface area and wide diversity of ecosystems, China has a great deal of biodiversity. Close to 35,000 animal and plant species have been identified in China, including the panda, the country's emblematic animal. China is the third-richest country in the world in terms of biodiversity, after Brazil and Colombia. Yet industrial growth is causing multiple threats to the environment and the populations that depend on it. Sichuan, which is located in China's center, is home to the world-famous panda as well as countless other species.

The giant panda (*Ailuropoda melanoleuca*) symbolizes biodiversity's vulnerability and the efforts undertaken to preserve it—the WWF even uses the panda as its emblem. Under the combined pressure from poaching and the destruction of its habitat for agriculture, the species has seen its distribution range shrink drastically: Once widely distributed in southeast China, pandas are now believed to have dwindled to 1,600 individuals in the wild, and are primarily found in Sichuan, in the center of the country.

As this endangered species' final refuge, China is trying to protect it. But the panda's rarity, its shy behavior, and the remoteness of the regions it inhabits make it particularly difficult to observe and study. Given this context, the use of satellite images is a powerful complement to field studies. Research jointly carried out by the San Diego Zoo and the Chinese Academy of Sciences has focused on the Foping National Nature Reserve, one of the species' last bastions.

Ten giant pandas were located and tracked using GPS-equipped collars. In order to accurately understand the pandas' movements within the reserve, their itineraries were retraced in detail thanks to high-resolution images provided by Planet Action. These studies have shown that the large mammals' distribution range changes drastically according to the season: Pandas live in the valley in winter and move to the mountaintops, above 6,562 feet (2,000 m), in the warm season. Researchers also observed that, while pandas have shared distribution ranges, each individual is particularly independent and does not seek contact with its fellow creatures. Even when two individuals are close to each other, no interaction is observed.

On the basis of this study, refuge zones have been more clearly identified and conservation measures can be reinforced for those areas. China has also significantly increased the number of nature reserves dedicated to the animal: Over the last ten years, they have increased from forty to sixty-two.

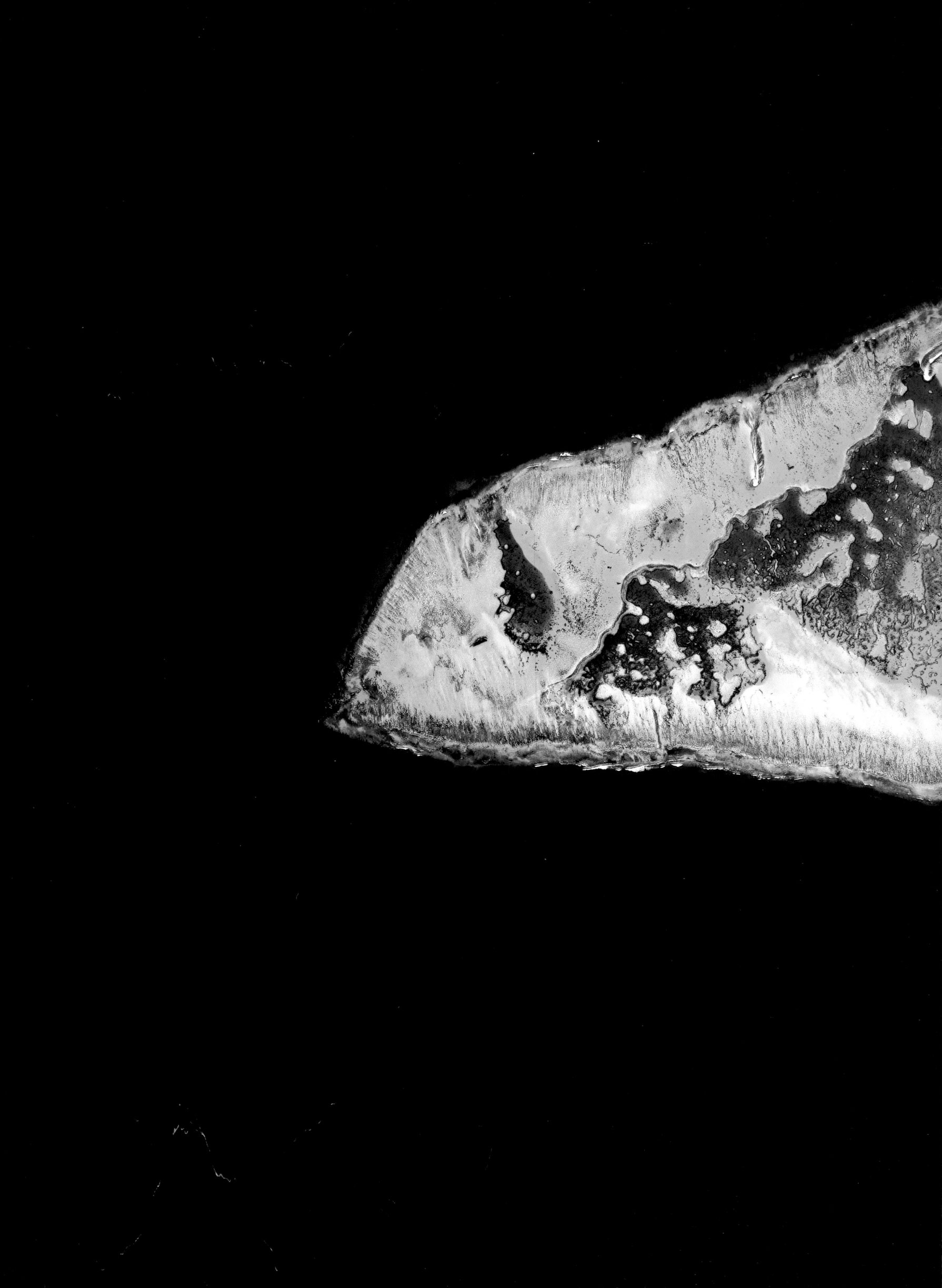

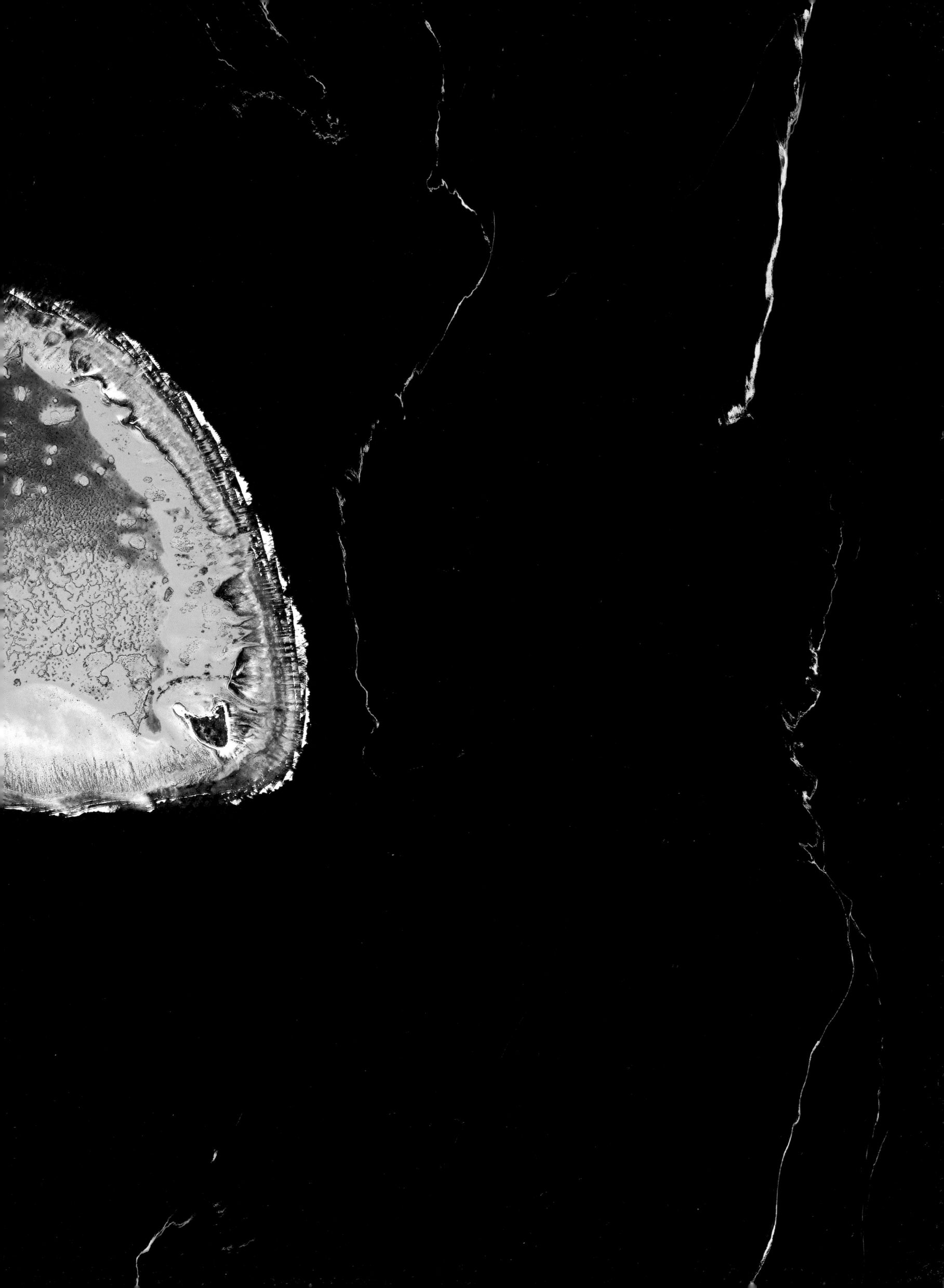

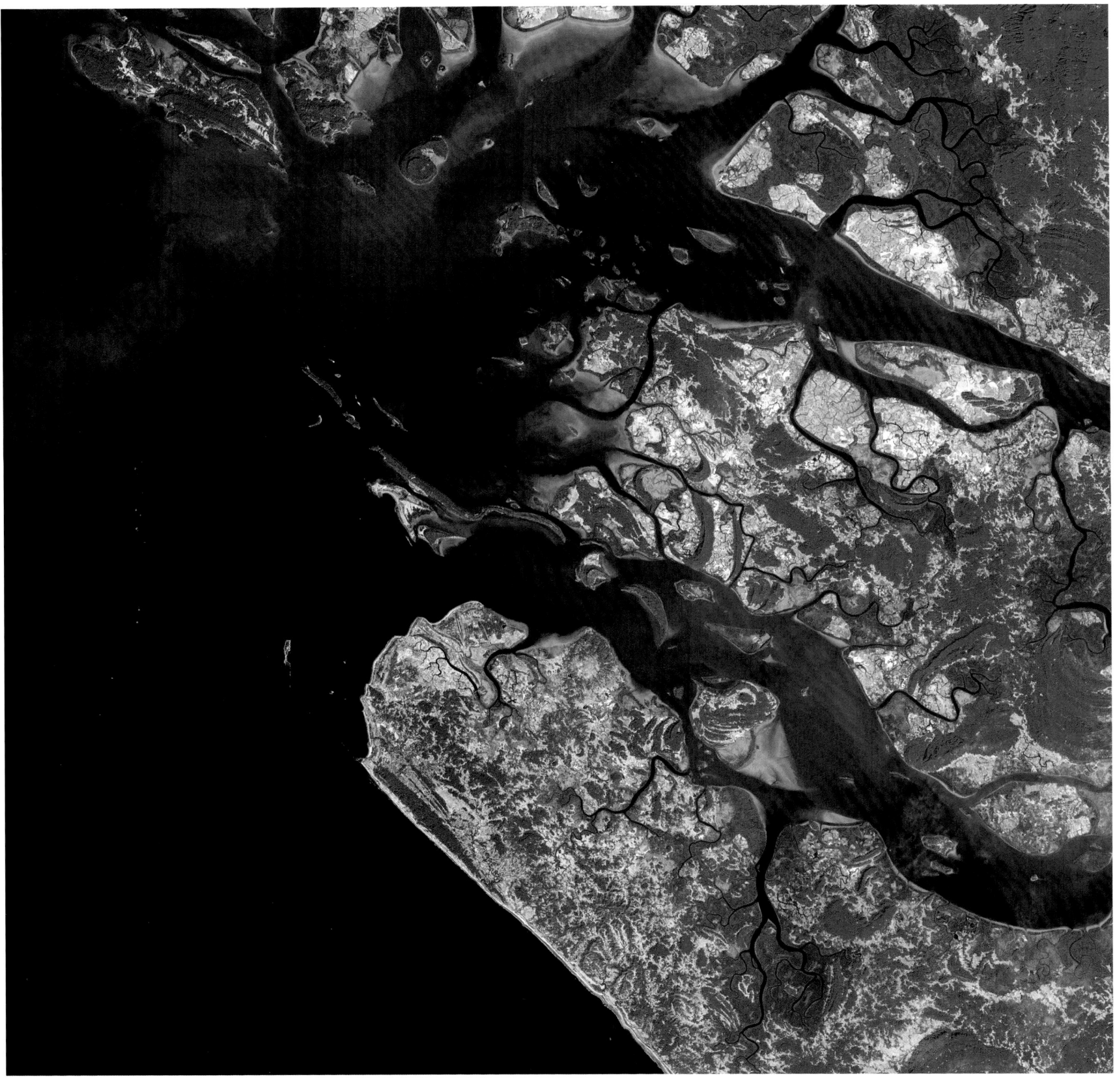

ABOVE

MYANMAR
COMBERMERE BAY
19°37' N – 93°34' E

Combermere Bay, east of the Gulf of Bengal, is a treasure
of biodiversity. The color red shows the presence of
dense vegetation, which also covers the little islands; the
mangrove is shown in brown. While the environment was
long preserved in Myanmar, the exploitation of its forests
has rapidly increased in the last thirty years due to its wealth
of rare woods, such as teak. Today the deforestation rate in
Myanmar is one of the highest in Southeast Asia.

PAGES 158-159
AUSTRALIA, QUEENSLAND
HERON REEF
23°29' S – 152°03' E

Heron Island, located to the south of the Great Barrier
Reef, is surrounded by a reef containing more than one
thousand fish species. It is a constant marvel to the
many divers and scientists who travel there.

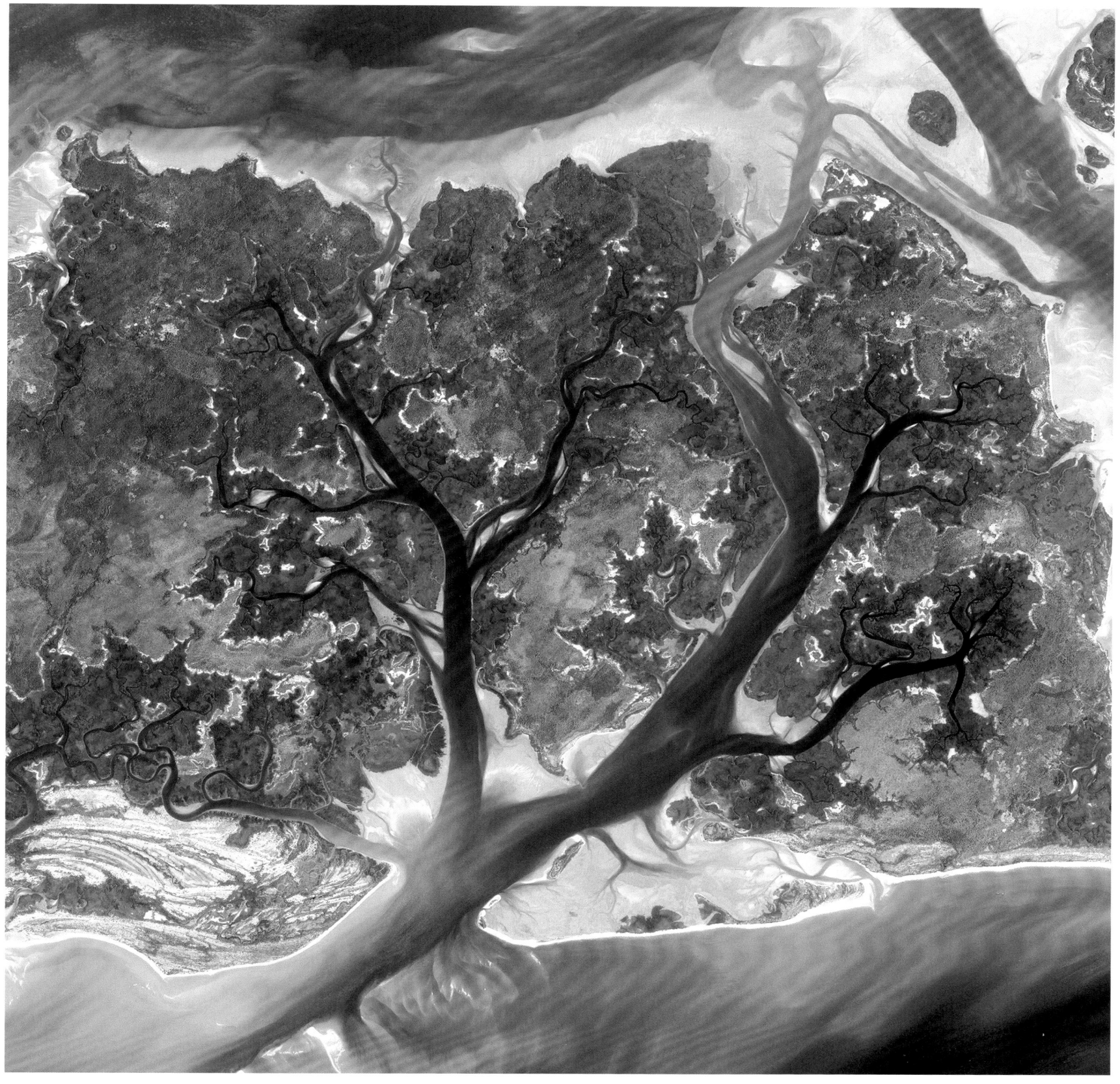

ABOVE

GUINEA-BISSAU
BIJAGÓS ISLANDS
11°09' N – 15°57' W

Only about ten of the eighty-eight islands composing the Bijagós Islands are inhabited. These islands—located at the mouth of the Geba River and listed as biosphere reserves by UNESCO—are home to exceptional marine fauna and flora. Some of the islands are covered in forests, while the mangrove often overruns the space between land and sea. With less than thirty-three thousand inhabitants spread over the archipel-ago's 1,013 square miles (2,624 sq km), the Bijagós Islands have an honest claim to the title of "deserted islands."

PAGES 162-163

BOTSWANA
OKAVANGO DELTA
19° S – 22° E

Stretching 807 miles (1,300 km), the Okavango, the third-longest river in southern Africa, originates in Angola. In Botswana, it widens into an interior delta covering approximately 5,800 square miles (15,000 sq km). Its 635.6 billion cubic feet (18 billion cu m) of water are gradually absorbed by the sands of the Kalahari Desert or evaporate in the dry air. The "river which never finds the sea" drains into a vast swampy labyrinth, populated by a stunning number of wild animals.

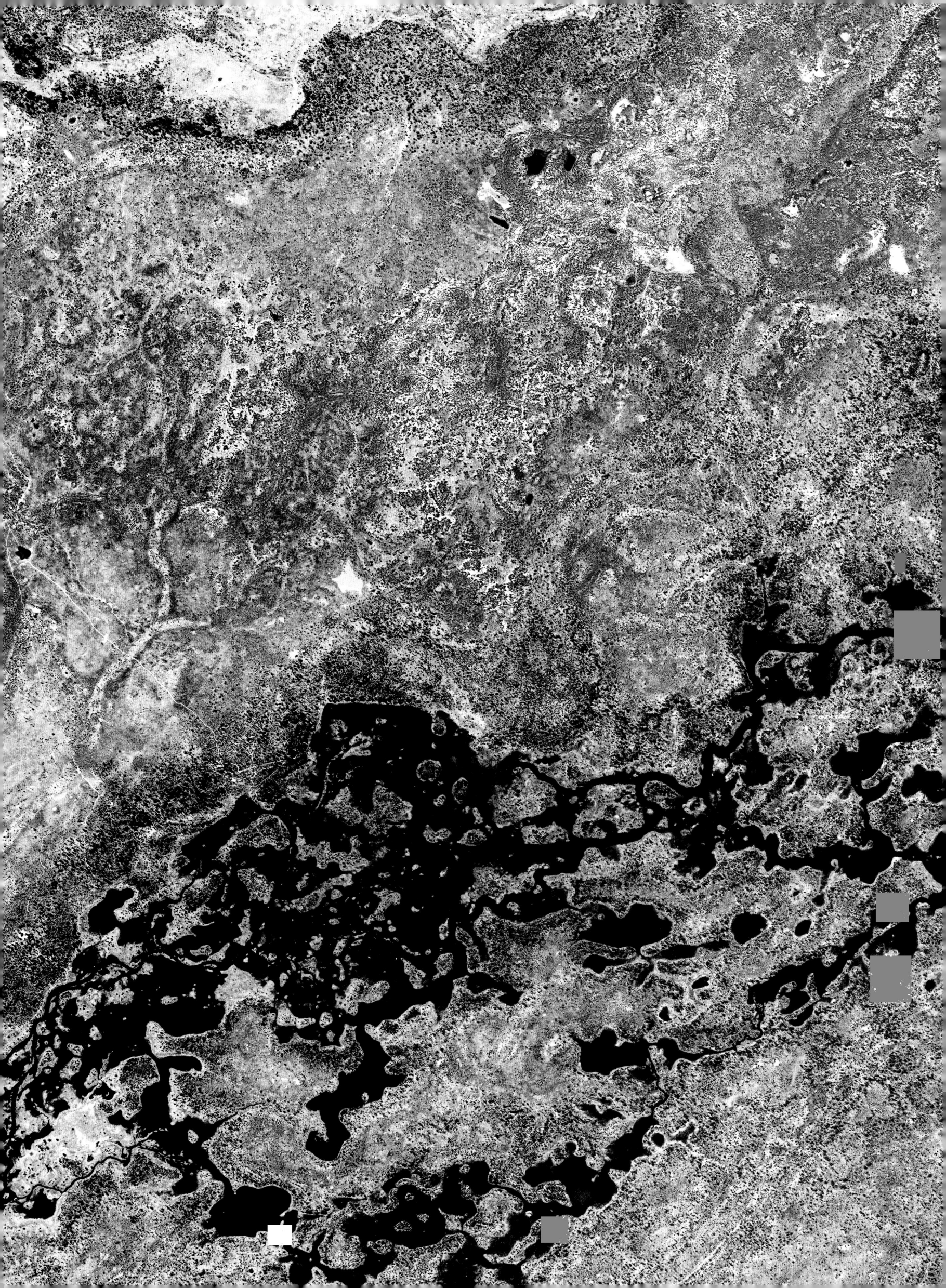

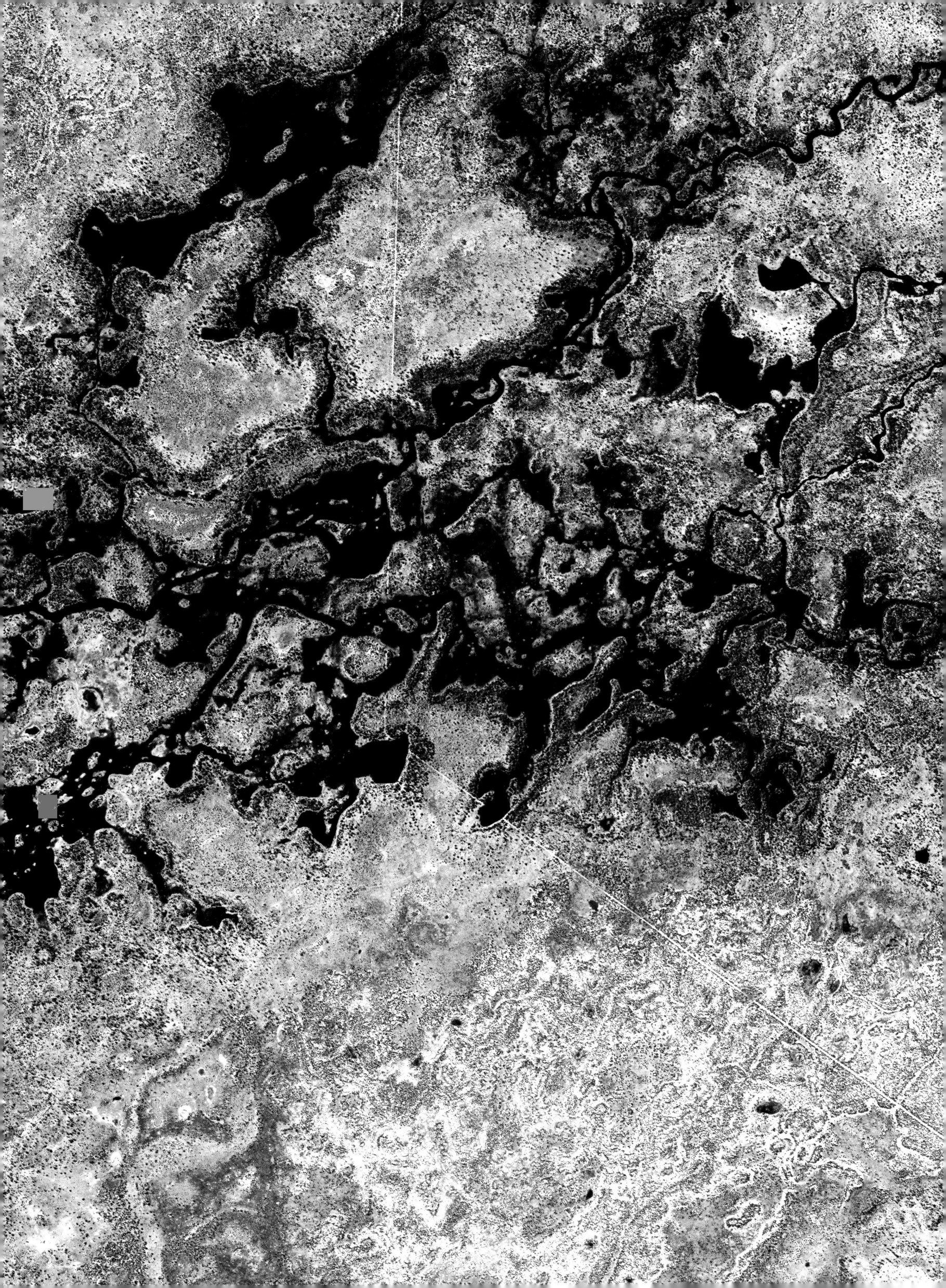

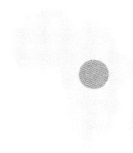

BELOW
UGANDA
LAKE KYOGA
1°30' N – 33° E

Located at an altitude of 2,998 feet (914 m) and covering an area of approximately 664 square miles (1,720 sq km), this complex shallow lake is crossed by the Victoria Nile. It is part of the network of Great Lakes though it is not one itself. From 1997 to 1998, excessive rainfall attributed to El Niño caused an exceptional rise in the water level, and a vegetal dam consisting of papyrus islands prevented the lake's waters from running off. The neighboring land was then flooded over an area of 231 square miles (600 sq km).

OPPOSITE
SWITZERLAND
GRISONS CANTON
46° N – 8°45' E

In 1914, Switzerland created a unique national park in the Grisons Canton in the southeast of the country: Today it covers 66 square miles (172 sq km) and its alpine flora and fauna flourish at altitudes of 2,600–10,500 feet (1,400–3,200 m) around the sources of the Rhine and the Inn rivers. It is a so-called strict natural reserve, where nature is entirely left to itself.

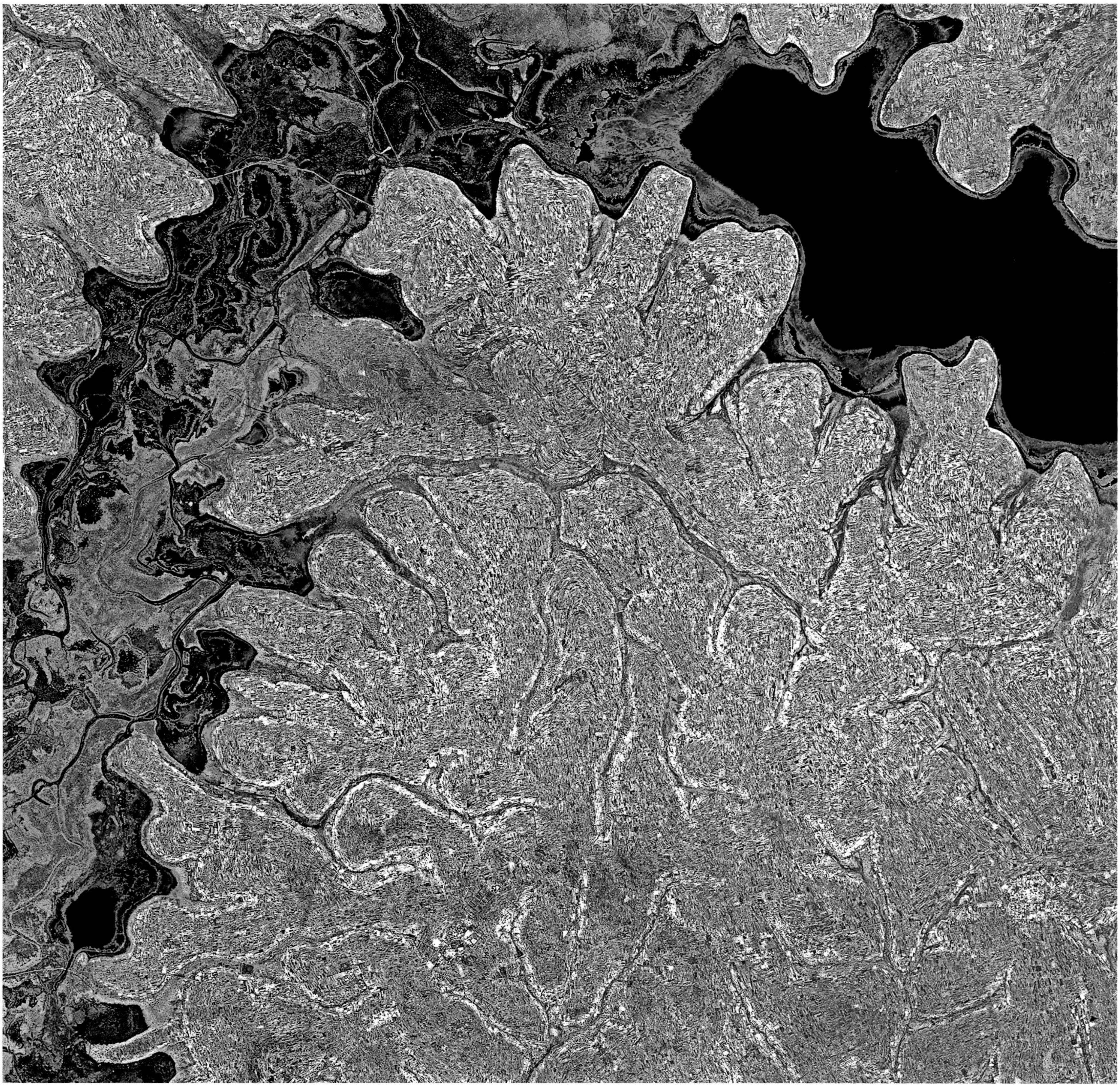

PAGES 166-167

INDIA, WEST BENGAL/BANGLADESH
SUNDARBANS
22° N – 89° E

The mangroves of the Sundarbans in the Gulf of Bengal cover about 3,860 square miles (10,000 sq km) of land. The largest of this type in the world, they also shelter a rich biodiversity: smooth-coated otters, 260 bird species, sika deer, wild boars, sand fiddler crabs . . . Some endangered species live here, too, including the estuarine crocodile, the Indian python, and the Bengal tiger.

PAGES 168-169

BRAZIL, MATO GROSSO
PANTANAL
17°43' S – 57°23' W

The Pantanal is a huge alluvial plain that, for much of the year, is transformed into a vast freshwater swamp and one of the largest wetlands of this type on Earth. This ecosystem has the richest collection of aquatic plants in the world. They are threatened by deforestation, excessive agriculture, urban development, and the increase in hydroelectric dams upriver.

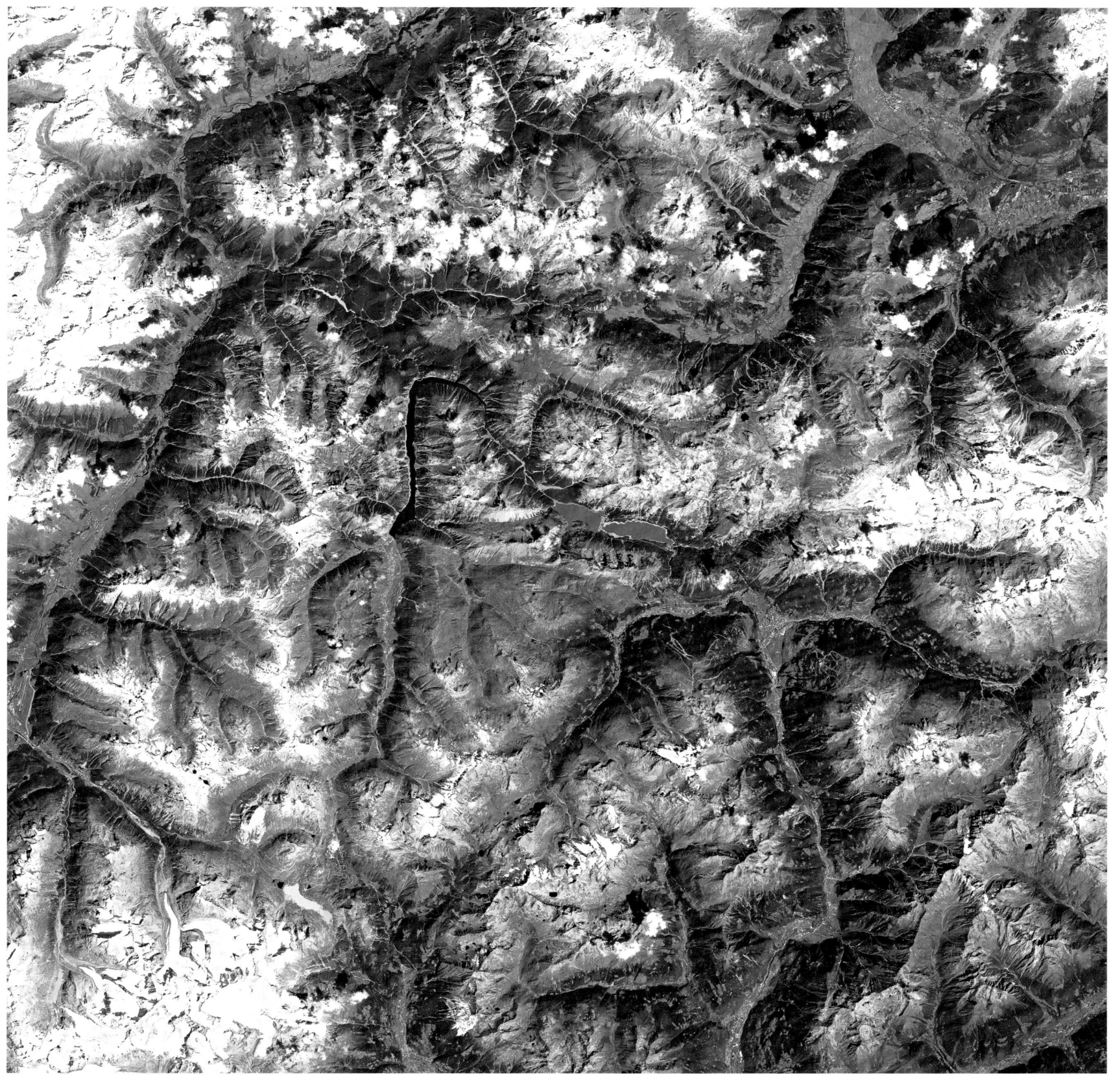

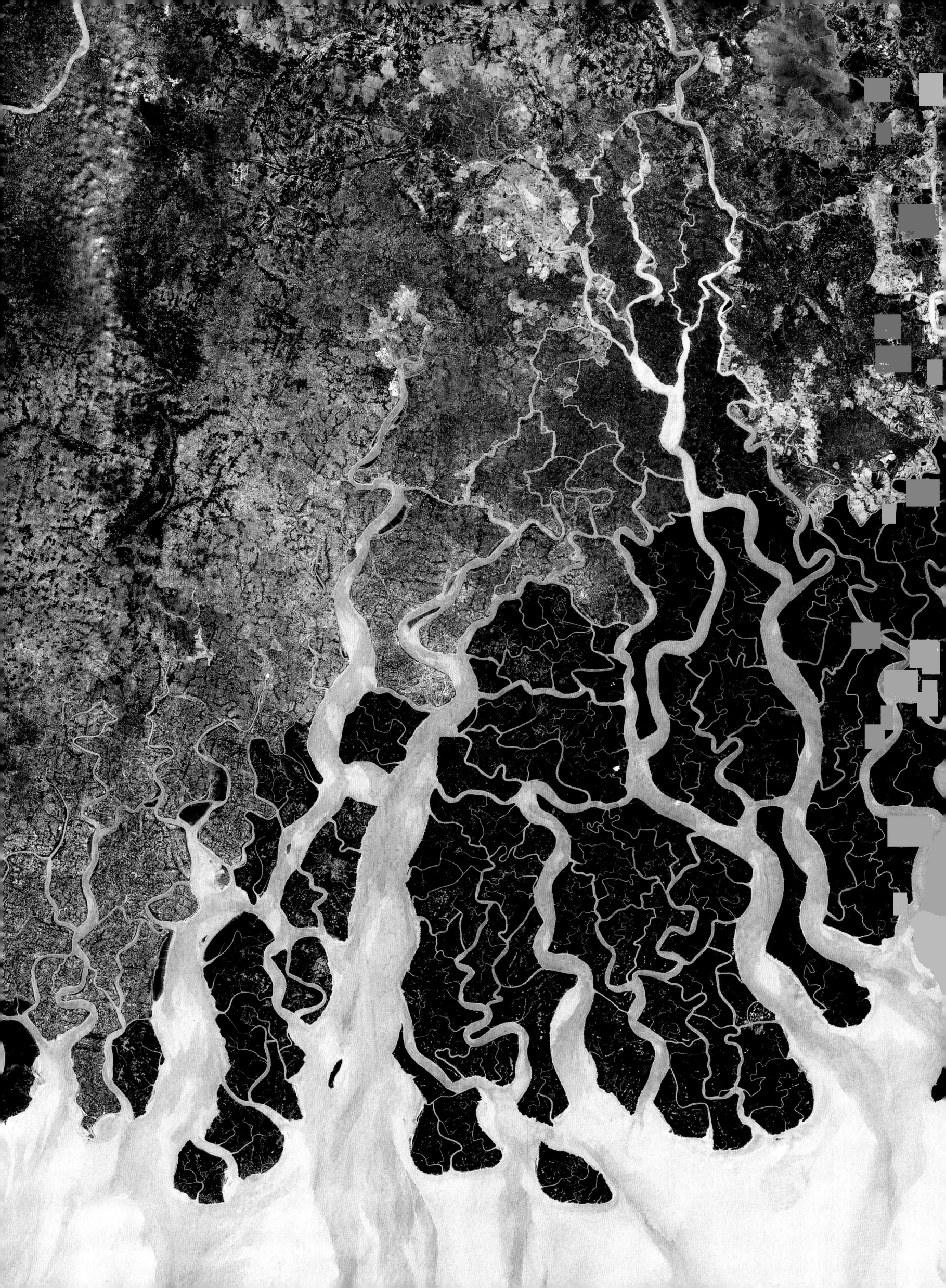

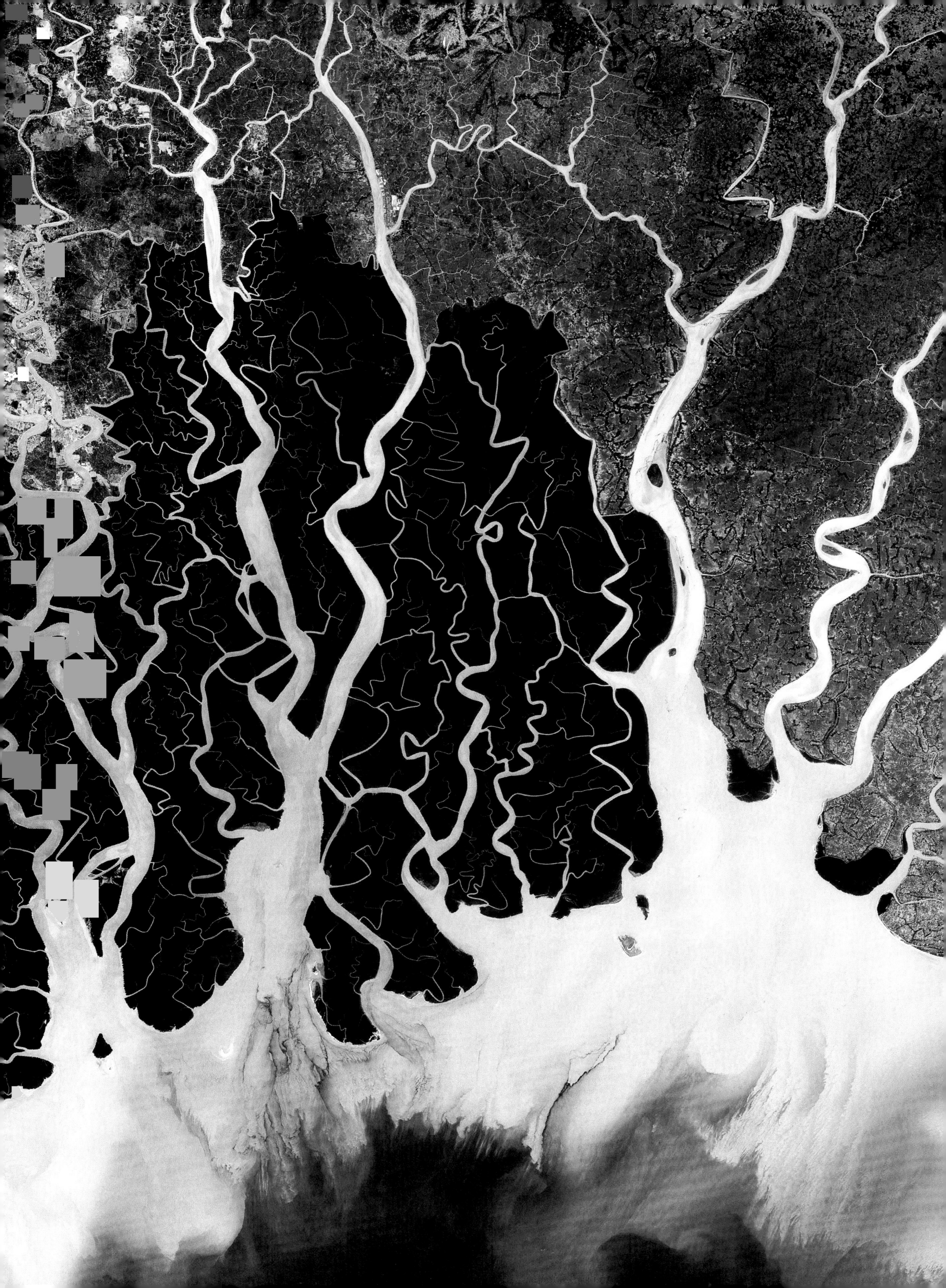

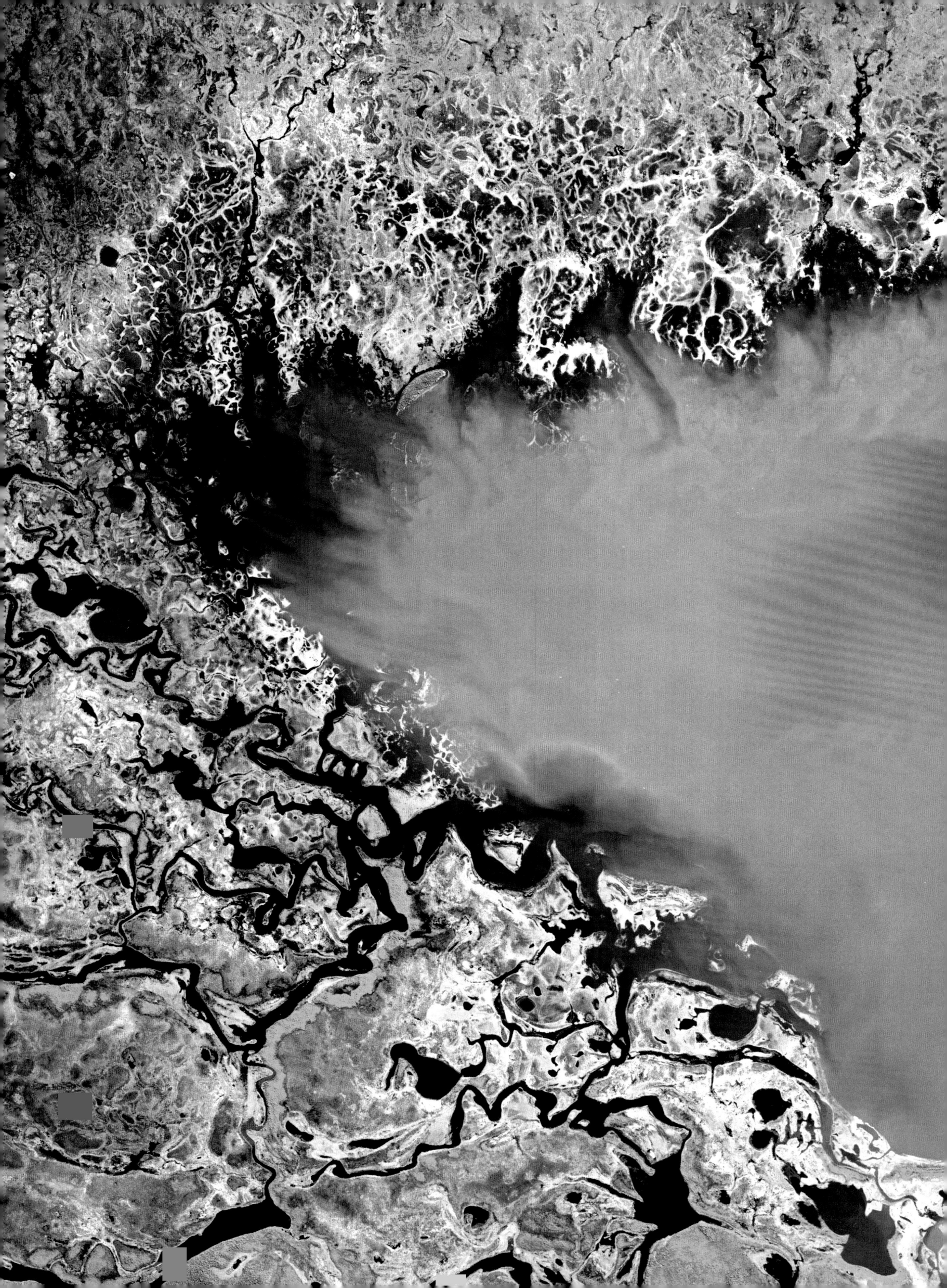

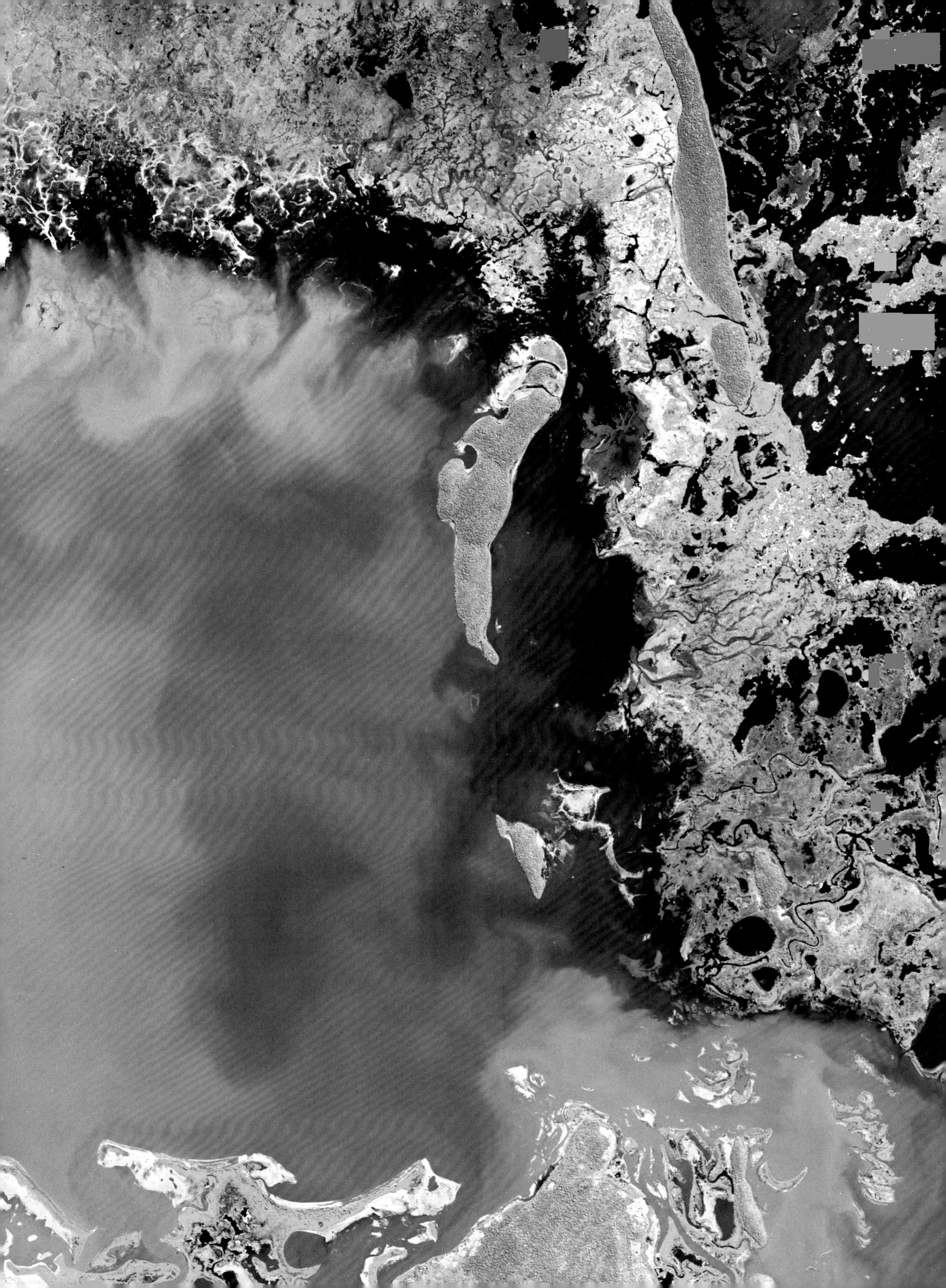

Anticipating disasters, organizing relief efforts, and evaluating the damage: Satellites can aid in many ways as we experience increasingly frequent and deadly natural threats.

——

FACING DISASTERS

——

December 26, 2004. Off Sumatra, the oceanic plate sinks under the continental plate, setting off an earthquake measuring 9 on the Richter scale. A tsunami forms and giant waves batter the coasts of Southeast Asia. The entire planet feels the repercussions of the earthquake, one of the most powerful ever recorded. One of the most deadly, too: Nearly three hundred thousand people are killed or reported missing, five million are displaced, and eight Asian countries and five African countries are hit on the same day.

It was 7:59 a.m. in Jakarta and Bangkok when the earthquake that set off the Sumatra tsunami rocked the middle of the Indian Ocean; the first waves hit the coasts of Indonesia forty minutes later. By chance, two oceanographic satellites flew over the epicenter two hours later, though they only pass over any given place every ten days; they measured the sea level with accuracy to the centimeter. On December 26, the two satellites saw the waves form offshore: two "small" bumps 20 inches (50 cm) tall, both 93 miles (150 km) long. Though this anomaly appeared minuscule, it actually represented several billion tons of water. Unfortunately, by the time this was observed, two hours after the earthquake, the waves already had the time to develop and reach the coasts of Indonesia, India, Thailand, and Sri Lanka, destroying everything in their wake.

Two hours . . . Many people could have fled and sought shelter. With this kind of disaster, reaction time needs to be quick to evacuate endangered populations. Satellites are not quick enough yet to react immediately and cannot provide instantaneous images. However, twenty-four hours after the first tremors, their images make it possible to assess the damages and organize rescue efforts.

After the tragedy, twelve earth observation satellites were reprogrammed to observe, photograph, and inform. The data collected was transmitted to the countries affected so they could evaluate and plan rescue efforts. The first maps of the disaster zones revealed hundreds of potential obstacles to rescue teams (collapsing buildings, mudslides, blocked roads) and helped to find new means of access, contributing to organizing emergency services on the ground, setting up field hospitals, and preparing refugee camps. They were also useful for locating displaced people—for example, those who sought shelter on hills in fear of aftershocks, and who could have otherwise remained isolated for days. Once the crisis has passed, this kind of data is also used to assess the human loss, and the material and agricultural costs of the disaster. In 2004, more than two hundred such images were produced and transmitted to affected countries.

HURRICANES: EVALUATING THE COST OF THE DAMAGES

Loss of human life is unfortunately not the only consequence of a natural disaster. In August 2005, Hurricane Katrina hit the southeast coast of the United States. One of the most powerful hurricanes in the nation's history, it ranged over a diameter of more than 800 miles (1,300 km), with a 25-mile (40-km) eye and accompanying winds reaching 174 miles (280 km) per hour. New Orleans was pummeled by the storm. With certain parts of the city located 19 feet (6 m) below sea level, its 13-foot (4-m) dykes were swept away. Close to 1,840 people died and 250,000 were displaced. The damages were estimated at $134 billion.

OPPOSITE

UNITED STATES, ALASKA, ALEUTIAN ISLANDS
KÁRMÁN VORTEX STREET
52° N – 175° W

The Aleutian Islands volcanic archipelago stretches more than 1,110 miles (1,800 km) south of the Bering Strait, which separates North America from Asia. The swirling vortices seen in this image, which follow the typical pattern of the "Kármán vortex street," are low-altitude marine clouds that produce drizzle. Different colors indicate temperature variations and the size of the raindrops inside the clouds.

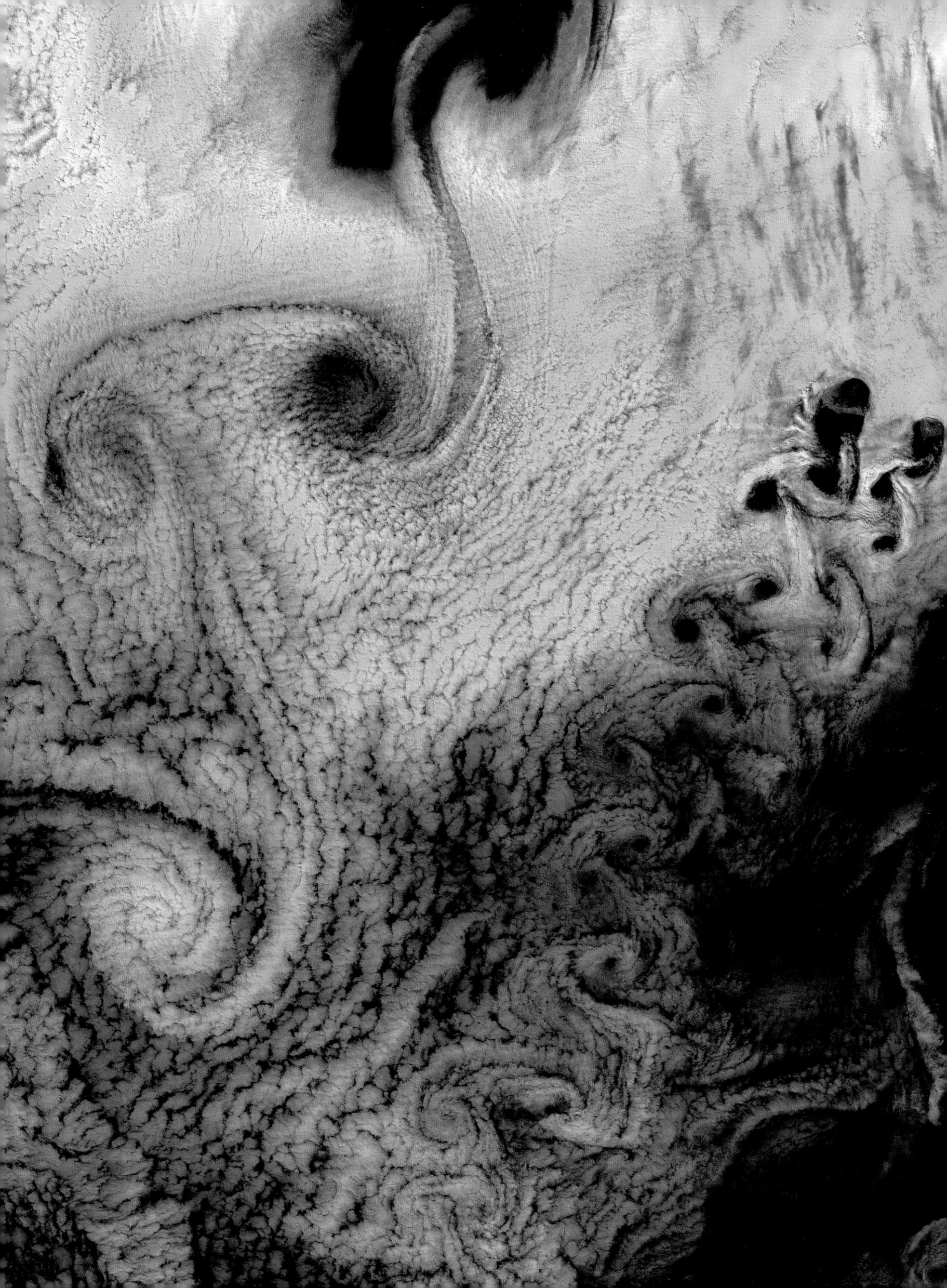

From 2000 to 2008, the **COST OF DAMAGES** associated with NATURAL DISASTERS has **INCREASED AN AVERAGE OF 11 PERCENT** per year.

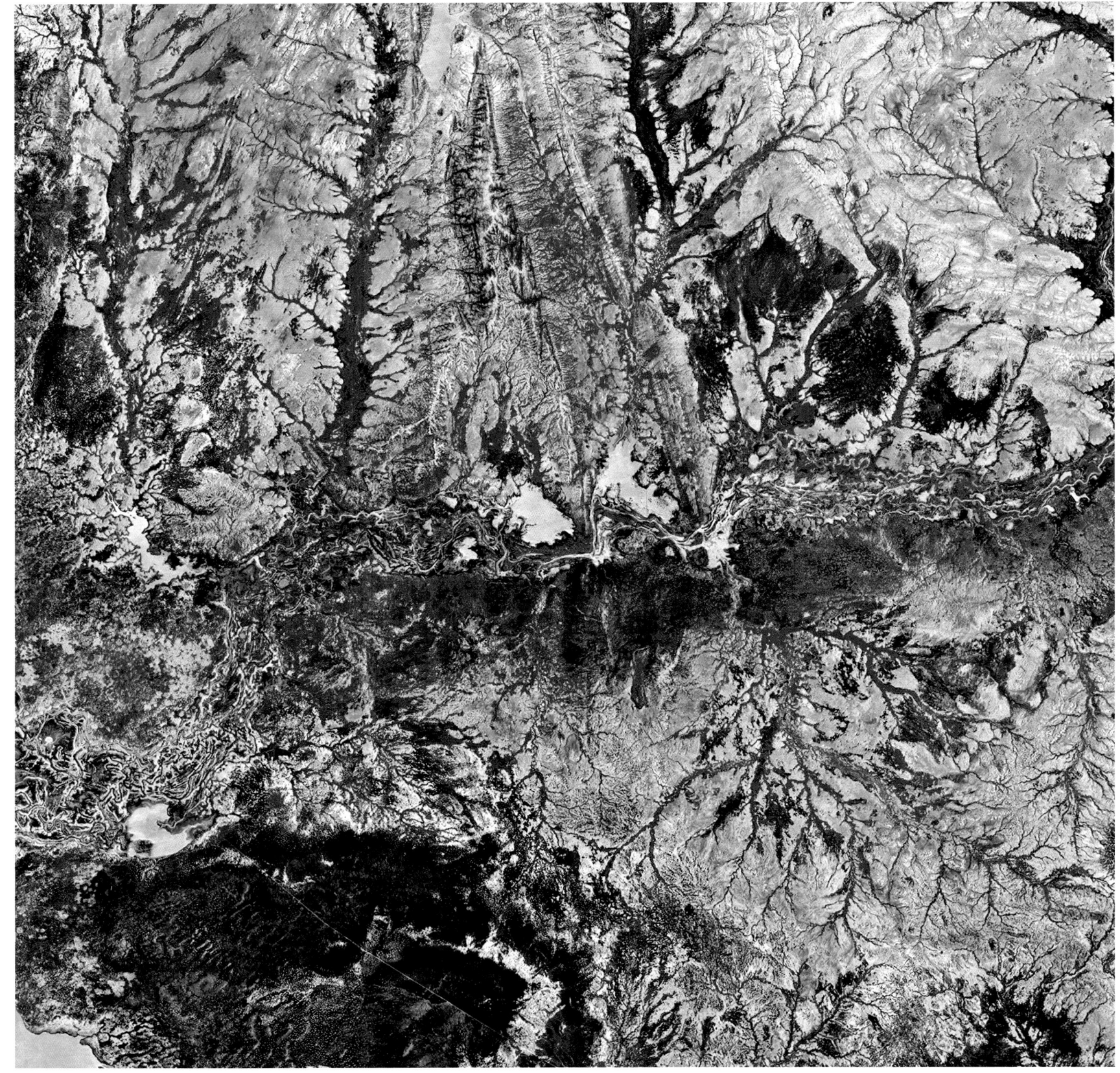

The crisis revealed significant social and racial inequality in Louisiana, then the second-poorest state in the United States: In 2004, close to 22 percent of the population lived below the poverty line. In the Lower Ninth Ward, one of New Orleans' poorest neighborhoods and one of the most badly devastated by Katrina, most inhabitants did not have flood insurance, which was not compulsory because the area was not classified as a flood risk.

Once the emergency had passed, satellite data helped to establish an initial estimate of the human and material losses by allowing a comparison of past images with those taken immediately after the disaster. This information also benefited those inhabitants who were insured by helping to establish the required inventories.

Insurance companies collect this kind of data annually and use it to evaluate the cost of natural disasters around the world. A particularly bleak year was 2005, when natural disasters—including the after-effects of the Asian tsunami and Hurricane Katrina—caused total damages of some $225 billion, of which approximately $80 billion were covered by insurance. Total damages for 2012 will probably be between $140 billion and $160 billion, with 67 percent of the total losses caused by Hurricane Sandy.

From 2000 to 2008, the cost of damages associated with natural disasters has increased an average of 11 percent per year—with a peak in 2005. And the trend seems to be continuing. The increase can principally be explained by two factors. First, populations are increasingly dense, particularly in zones that are highly exposed to natural risks. So disasters in these areas affect an increasing number of victims.

Secondly, the frequency and intensity of extreme weather phenomena (which do not include tsunamis) is tending to increase, most likely due to climate change. Some experts even see the increase in damages noted by insurance companies as one of the first signs attesting to these major climate upheavals.

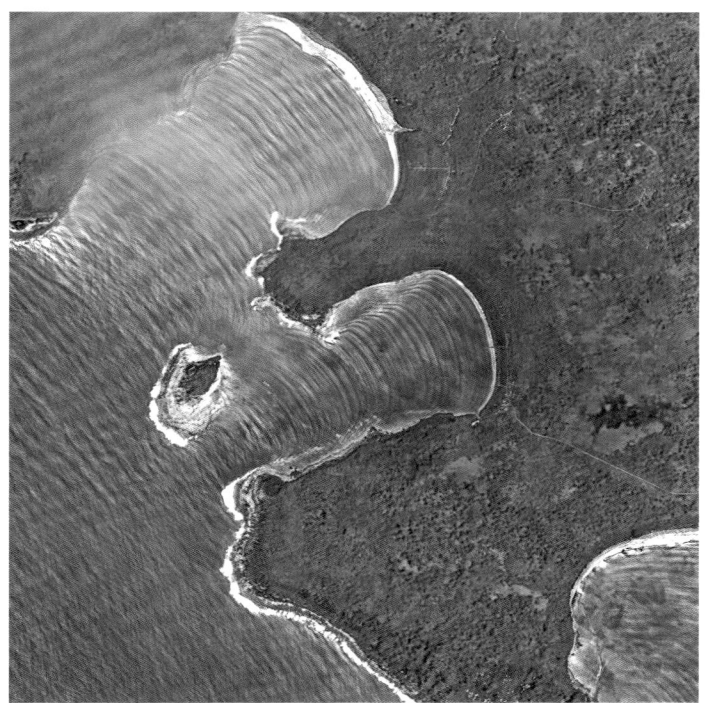

MAURITANIA
WADI GARFA
15°48' N – 12°54' W

The meanders of Wadi Garfa, an affluent of the Senegal River, flow from east to west. In this picture, some zones are covered in water (light blue), others attest to an explosion of vegetation (red and pink), while water-saturated and still-emerged land (black) bears witness to an intermediary stage. The gray-green of the steppe is mottled with bright red and the brown of the gallery forest that covers the small water-courses, displaying the entire hydrographic network.

INDONESIA, MENTAWAI ISLANDS
SOUTH PAGAI
3°24' S – 100°15' E

On October 25, 2010, an underwater earthquake measuring 7.2 on the Richter scale took place off the coast of Indonesia, setting off a tsunami that killed more than four hundred people. This image was taken by the SPOT-5 satellite only a few days after 6.5-foot (2-m) waves hit the shores of South Pagai Island in the Mentawai Archipelago off Sumatra. The earthquake occurred on the fault line previously responsible for the devastating 2004 earthquake and tsunami.

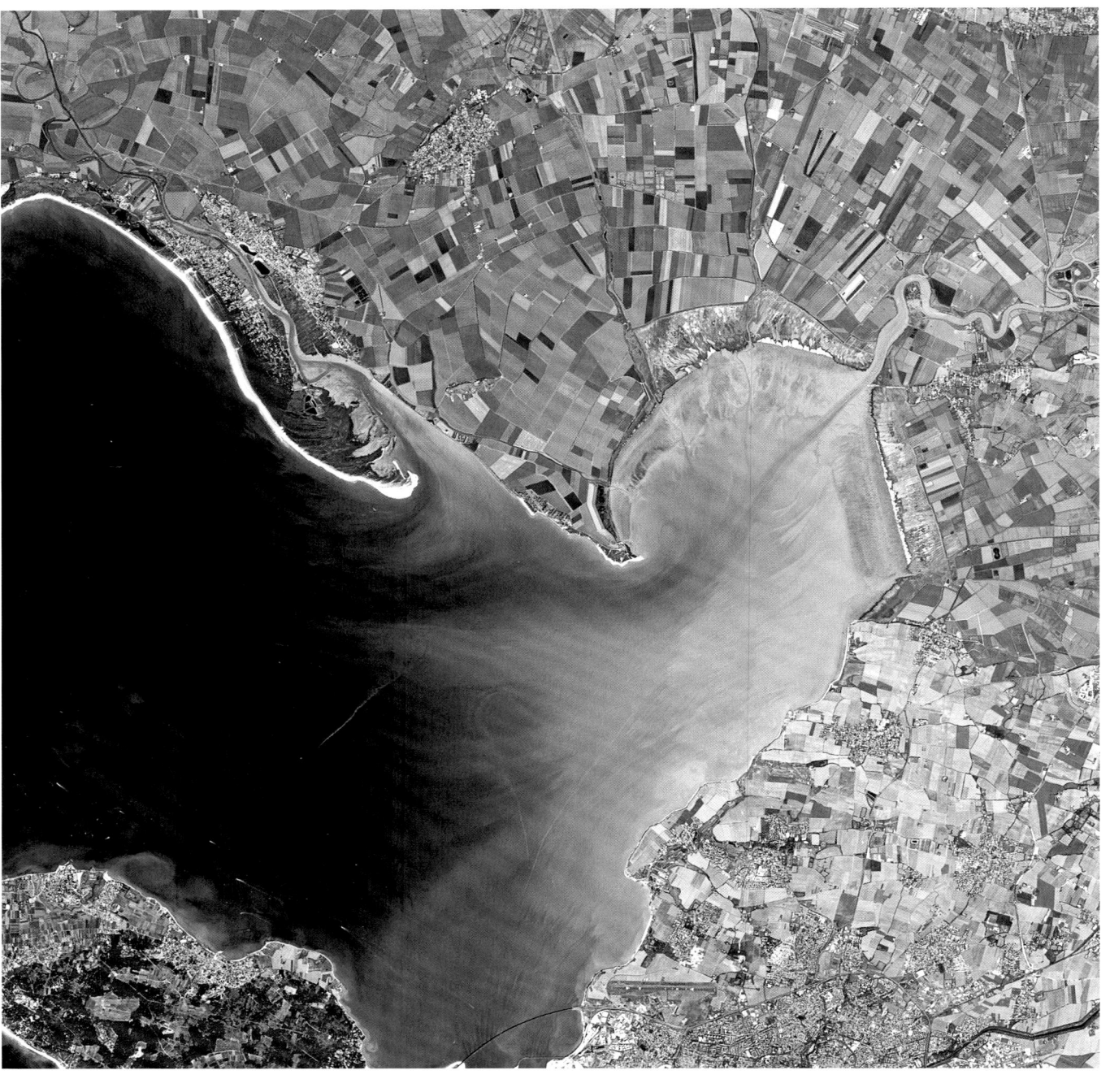

1984 2009

FRANCE
LA FAUTE-SUR-MER
46°19' N – 1°19' W

Cyclone Xynthia hit French coasts on
the night of February 27–28, 2010, killing
fifty-three people. La Faute-sur-Mer, in
Vendée, occupies a peninsula between the
Atlantic Ocean and the Lay River estuary. The
small town had been protected from flooding
by a dyke since the 1930s, and in the 2000s the
municipality authorized the construction of
two housing projects in a naturally floodable
depression 3.2 feet (1 m) below sea level.
When the storm hit the coasts of Vendée, the
tide was high with a strong coefficient. The
dyke broke and the area was devastated;
twenty-nine inhabitants died. This disaster's
consequences reiterate the problems posed
by built-up coastlines.

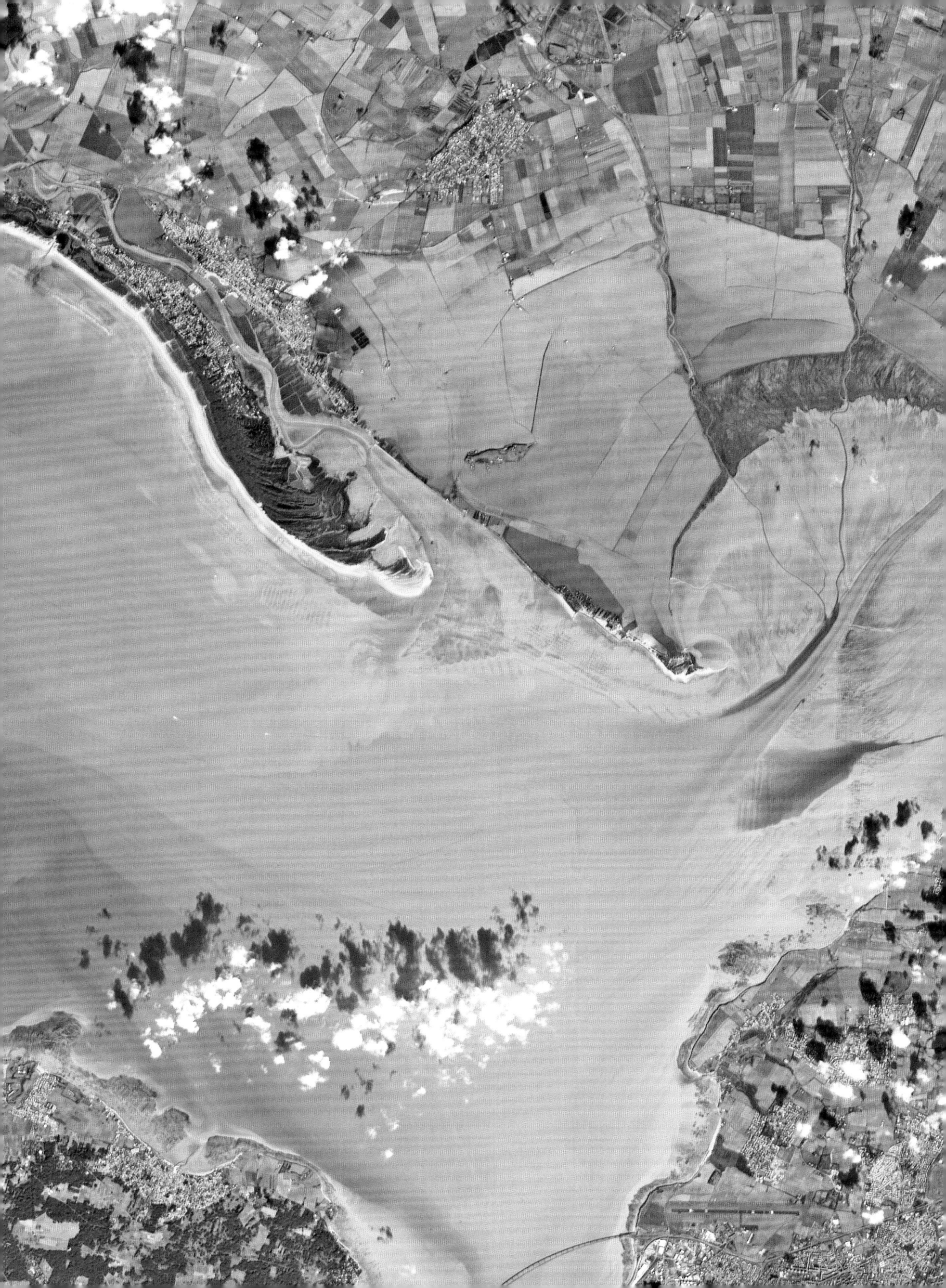

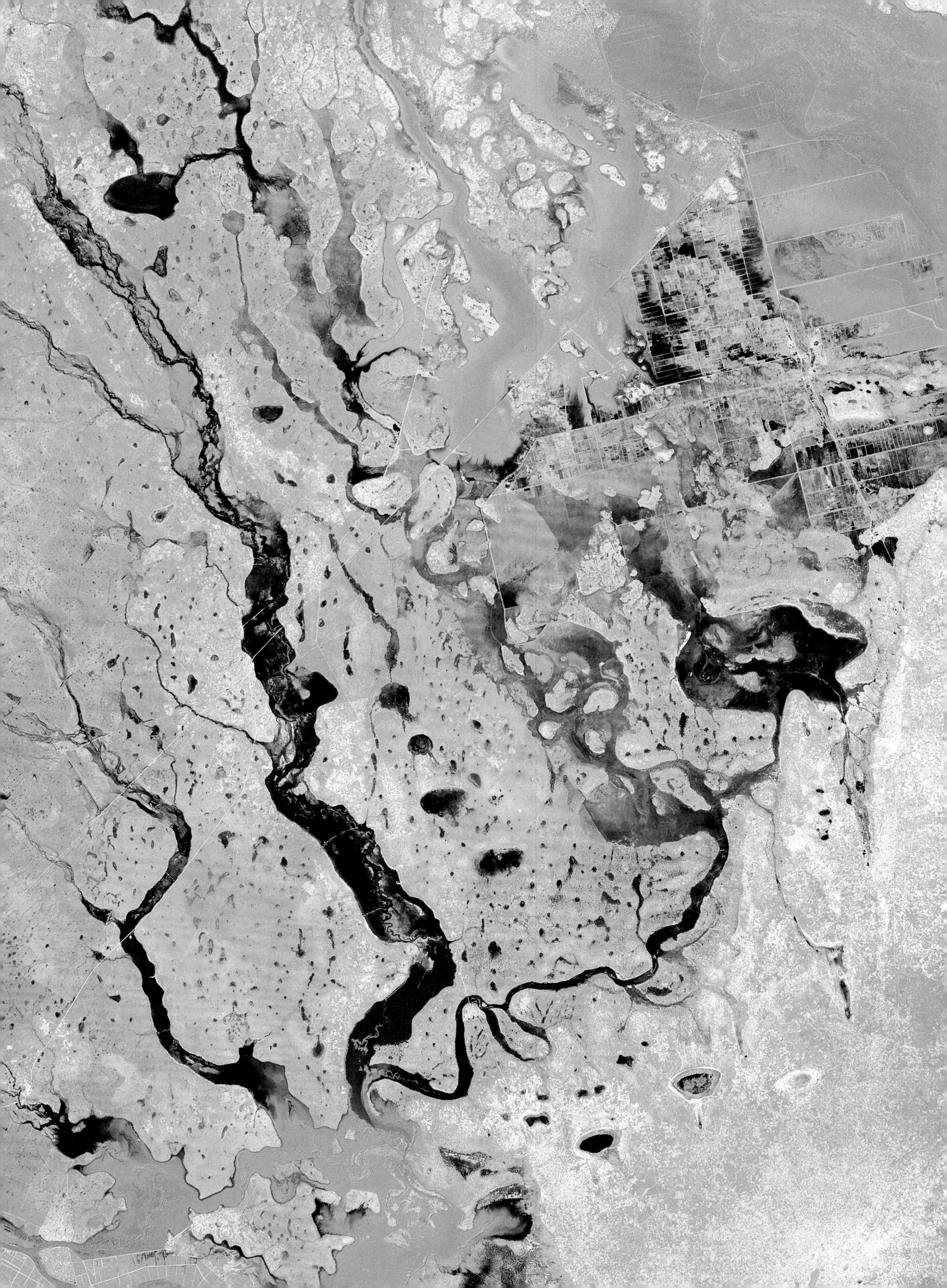

FLOODS: PREVENTING DAMAGE

In some cases, satellite data can help avoid losses and diminish the damages and costs of a natural disaster.

In January 2011, torrential downpours battered Queensland, Australia. The area was the victim of a particularly intense episode of La Niña—a weather phenomenon that is the opposite of El Niño and is characterized by unusually numerous masses of hot, humid air—that was pushed to Australia and Asia by the trade winds of the western

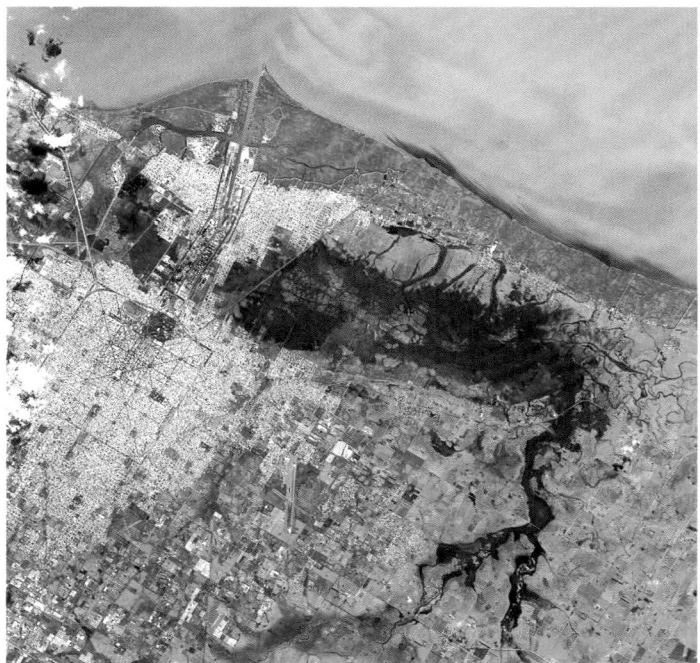

Pacific. Thirty people were killed, more than two hundred thousand were displaced, and twenty-two cities remained completely isolated for several weeks, stranded in the middle of an ocean of mud and brown water. Some specialists even stated that the effects on the country's economy could be more devastating than those of Hurricane Katrina in the United States in 2005.

While satellites belonging to the international community can be repositioned over a disaster zone within a few days to observe, photograph, and inform, they are also used before the disaster even takes place. They make it possible to map flood zones by providing information about the area's physiography and hydrology, which is data local authorities can use to understand and anticipate the evolution of the water's rise and the damages caused. This enables them to protect the population by banning construction in flood zones. In the long term, satellite images provide the means to follow topographic changes due to heavy flooding and to update risk-zone maps, thereby anticipating future floods in order to reduce the losses and costs related to this kind of natural disaster.

FOREST FIRES: OBSERVING TO BETTER PROTECT

Summer 2010: Russia is devastated by more than twenty thousand fires. About fifty lives are lost. An unprecedented heat wave is blamed: The initial fires start in a peat bog then spread uncontrollably, destroying more than 1.7 million acres (700,000 hectares) of forest and eventually getting threateningly near the Sarov military nuclear power station.

OPPOSITE
MOZAMBIQUE
CHOKWE DISTRICT
34°31' S – 32°59' E

In January 2013, a region-wide torrential downfall and the opening of a dam downstream of the Limpopo River caused devastating floods in the south of Mozambique. Some 80 people died, 180,000 were displaced, and agriculture, the only source of revenue for the region's inhabitants, was devastated. In this image, vegetation is shown in green, low-sediment water is navy blue, and high-sediment water stands out in pink.

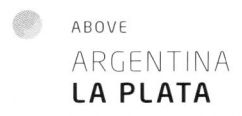

ABOVE
ARGENTINA
LA PLATA
34°55' S – 57°57' W

Located 37 miles (60 km) from Buenos Aires, the Argentine city of La Plata was severely flooded in April 2013. The zones affected, visible in purple in this image, remained under more than 6.5 feet (2 m) of water for several days; forty-six people died. A few days earlier, the city of Buenos Aires had been flooded as well.

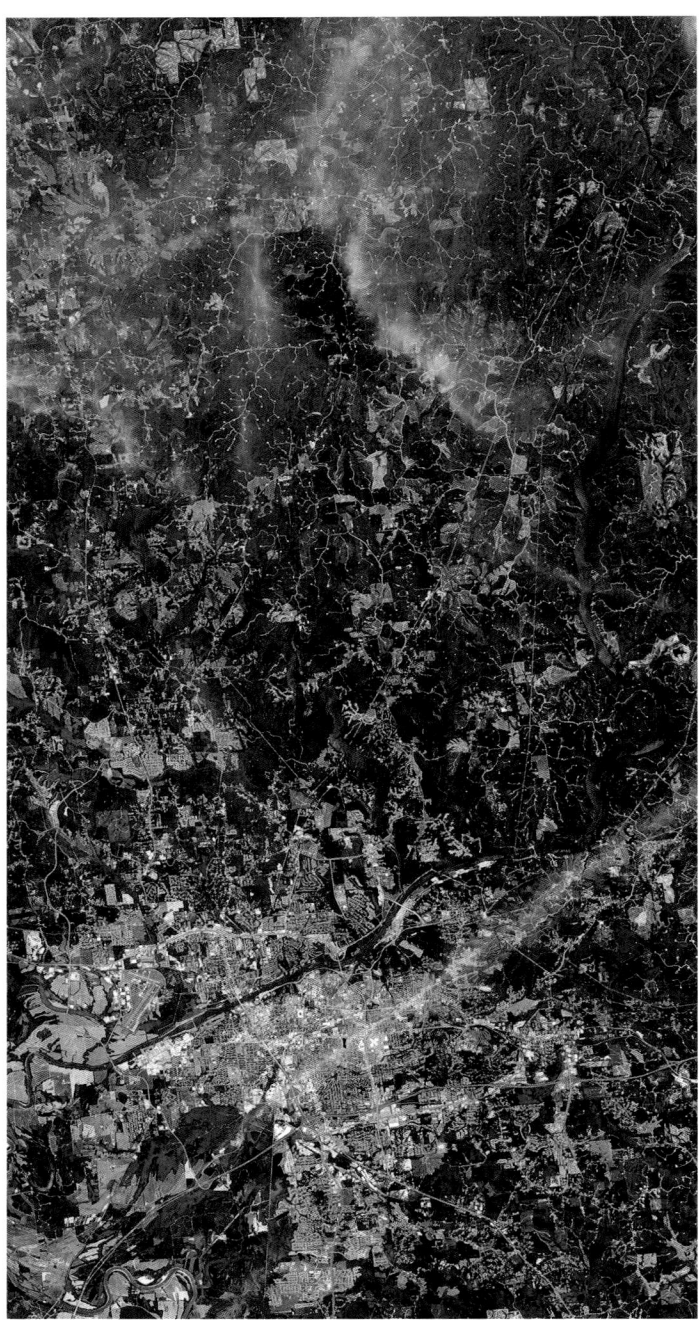

Heat sources can be rapidly detected and monitored from outer space. On the basis of satellite data, firefighters locate the fire, know its extent, and try to control it before it reaches human populations. In the case of the exceptional fires that struck Russia in 2010, this data was also used to anticipate the direction in which the fires would spread and to organize evacuations.

Russia is not the only area affected by fires, of course. Each year, forest fires destroy 864 million acres (350 million hectares) around the world, half of which are in Africa. In the summer, the Mediterranean basin is also highly vulnerable to fires. In the last few years, satellite data has provided new tools to better evaluate the post-disaster period in the Mediterranean. Once a fire is extinguished, the damages have to be assessed. Since 2006, a participatory data input system has inventoried any fires affecting more than 12 acres (5 hectares) in the region. An emergency data acquisition procedure is launched in order to determine the fire's exact contour, with the information transmitted bearing on the area's biophysical parameters: water, soil, and vegetation.

Local authorities can thus identify the state of biodiversity and determine the varieties to be replanted, particularly those that are less likely to spark a future fire. Over the years, this data has also helped local authorities position firefighting crews in the most vulnerable zones, those likely to burn the fastest.

Even more so than other disasters, fires embody the contradictions of the term "natural disaster," which seems increasingly less appropriate in the modern world. Fire is one of the natural ways in which the forest renews itself, but it is considered a disaster when it causes damage to human societies. And while it affects the natural environment, it is often caused by man (in 80 percent of cases, according to certain estimates).

Be that as it may, the increase in the number and magnitude of these disasters only makes our use of satellites more essential—to monitor and anticipate but also, whenever possible, to prevent.

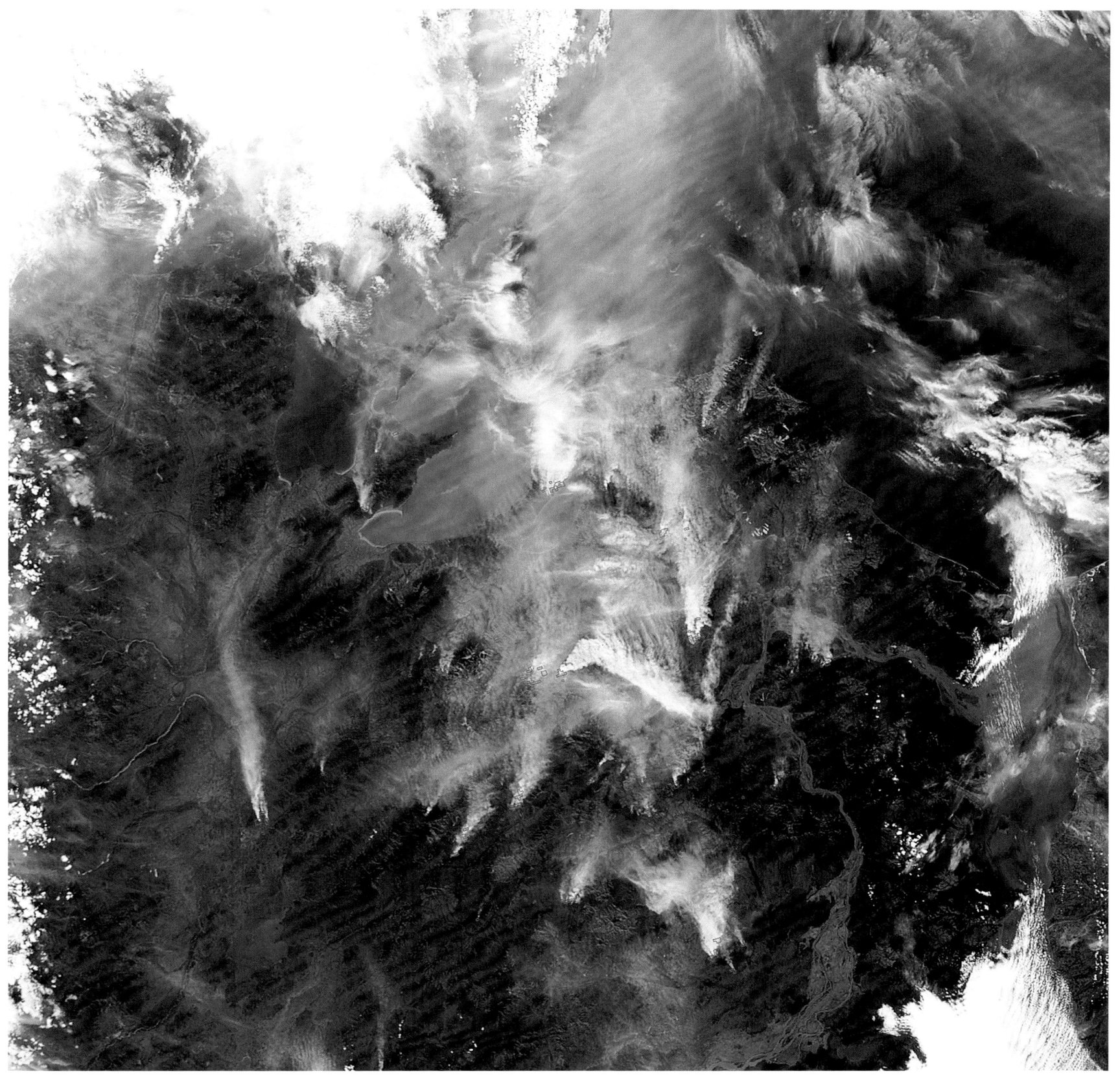

UNITED STATES, ALABAMA
BERRY
33°36' N – 87°30' W

The Cordova tornado, one of dozens of lethal tornadoes that hit the United States in April 2011, bisected Alabama, destroying everything in its path. Its trajectory is clearly visible in the satellite images: a corridor 7 miles (12 km) wide and close to 125 miles (200 km) long. According to the American weather services, winds moving at more than 170 miles (275 km) per hour stripped the area of vegetation.

RUSSIA
FOREST FIRES IN SIBERIA
53°45' N – 138°30' E

Forest fires such as the one seen here in the area of Khabarovsk, in the Russian Far East, are increasingly common during the summer in the north of Siberia, which is undergoing rapid climate warming. In 2010, close to 2.4 million acres (1 million hectares) of forest went up in smoke due to extraordinary fires.

The earthquake that shook Haiti in 2010 lasted only thirty-five seconds. But it was one of the deadliest in history. It killed more than 250,000 people—more than the fatalities caused by both atomic bombs dropped on Hiroshima and Nagasaki in 1945. Ten months after the earthquake, the UN peacekeepers who had deployed to help the stricken population were relieved by a Nepalese battalion. Unbeknown to them, these soldiers were carrying the most dangerous form of cholera: an infectious strain that would affect 660,000 people and claim 7,500 lives from 2010 to 2012.

Talking with Ivan Buendia Gayton

Fighting Disease with Satellites

How did you start fighting the cholera pandemic in Haiti?

Immediately after the earthquake, Google sent a team to take pictures of the area from a plane. Together, we perfected computer tools to allow us to map the areas inhabited by people with cholera and those coming to treatment centers. We weren't surprised to see that this data pointed to a correlation between the zones most affected by the epidemic and the presence of rivers or unsanitary water sources. Then we were able to target our efforts and reduce our intervention time.

How did you access the most remote areas?

The area of Baradères is a remote zone in the west of Haiti. We got a call for help from a nun who no longer knew what to do about the epidemic. So I sent a team there. But without a GPS, the only way to guide them was by satellite telephone. We would do a telephone update every two hours. Each time we talked, the team members would tell me: "We found it, the emergency is here." But thanks to the map, I knew that wasn't the case. They had to leave medical supplies and keep moving to get to the center of the emergency. After several phone calls and many hours on foot, they got to the end of the trail, at the edge of the water, and they continued by boat. When they found the mouth of the river, I called the nun, who sent them a boat small enough to fit through the meanders. When they finally reached the area upstream of the river, they called me back. The port was full of dead bodies piled one on top of another, like firewood!

How many people were saved thanks to your intervention?

It's hard to estimate the exact number. In Baradères, more than six hundred people with cholera came out of the forest in three days. Without treatment, the death rate is 40 percent. It drops to 1 percent with treatment. So our team's arrival most probably saved a few hundred of the people in this specific place, and overall tens of thousands of people on the entire island.

Ivan Buendia Gayton, a mission leader for Médecins Sans Frontières (Doctors Without Borders) in 2010, was rushed to Haiti to organize relief efforts and prevent a second humanitarian disaster: the cholera pandemic that was affecting the entire island, including in the Dominican Republic. How could its true extent be detected? How could those still alive be saved?

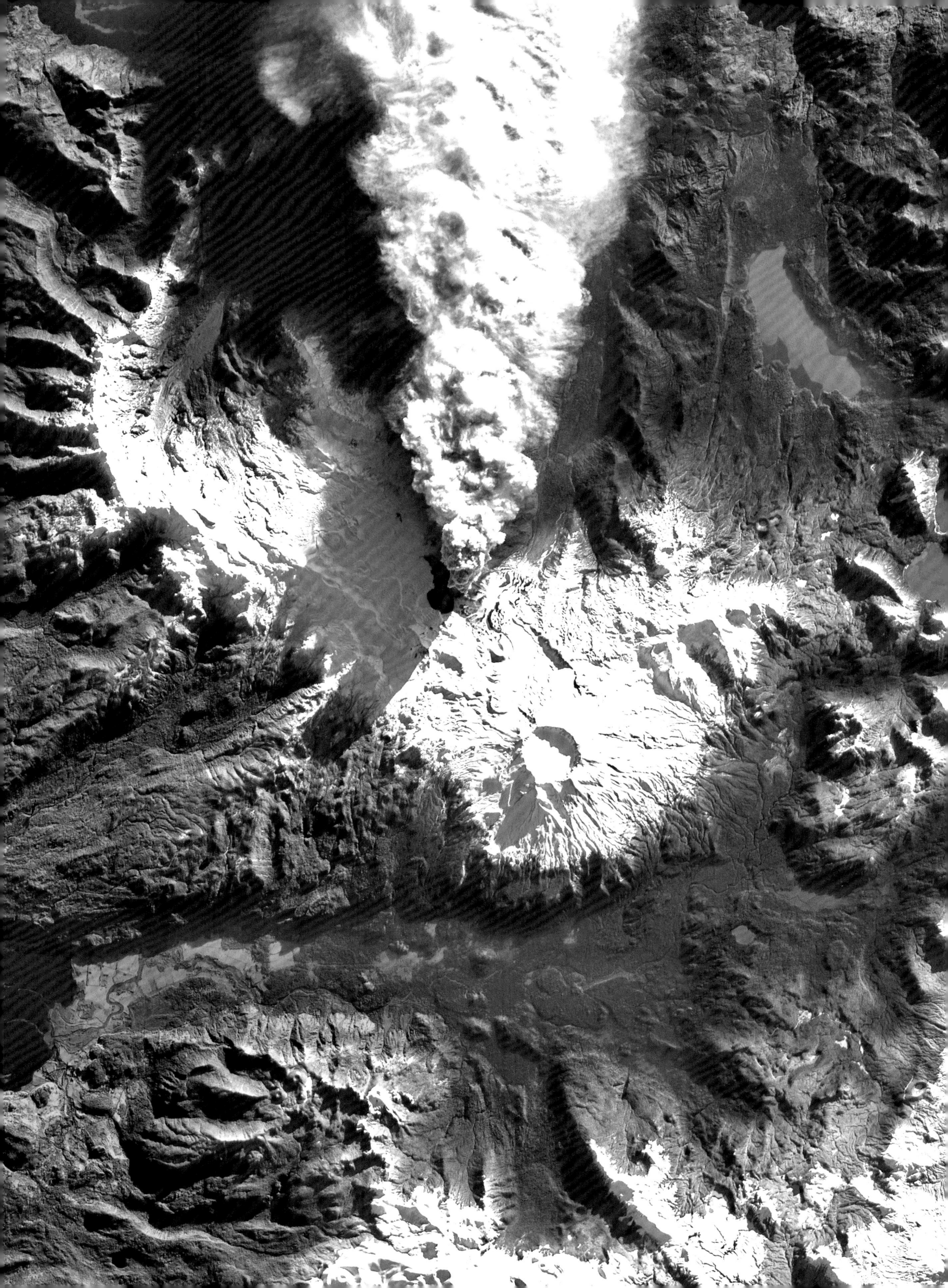

THE INTERNATIONAL CHARTER ON SPACE AND MAJOR DISASTERS

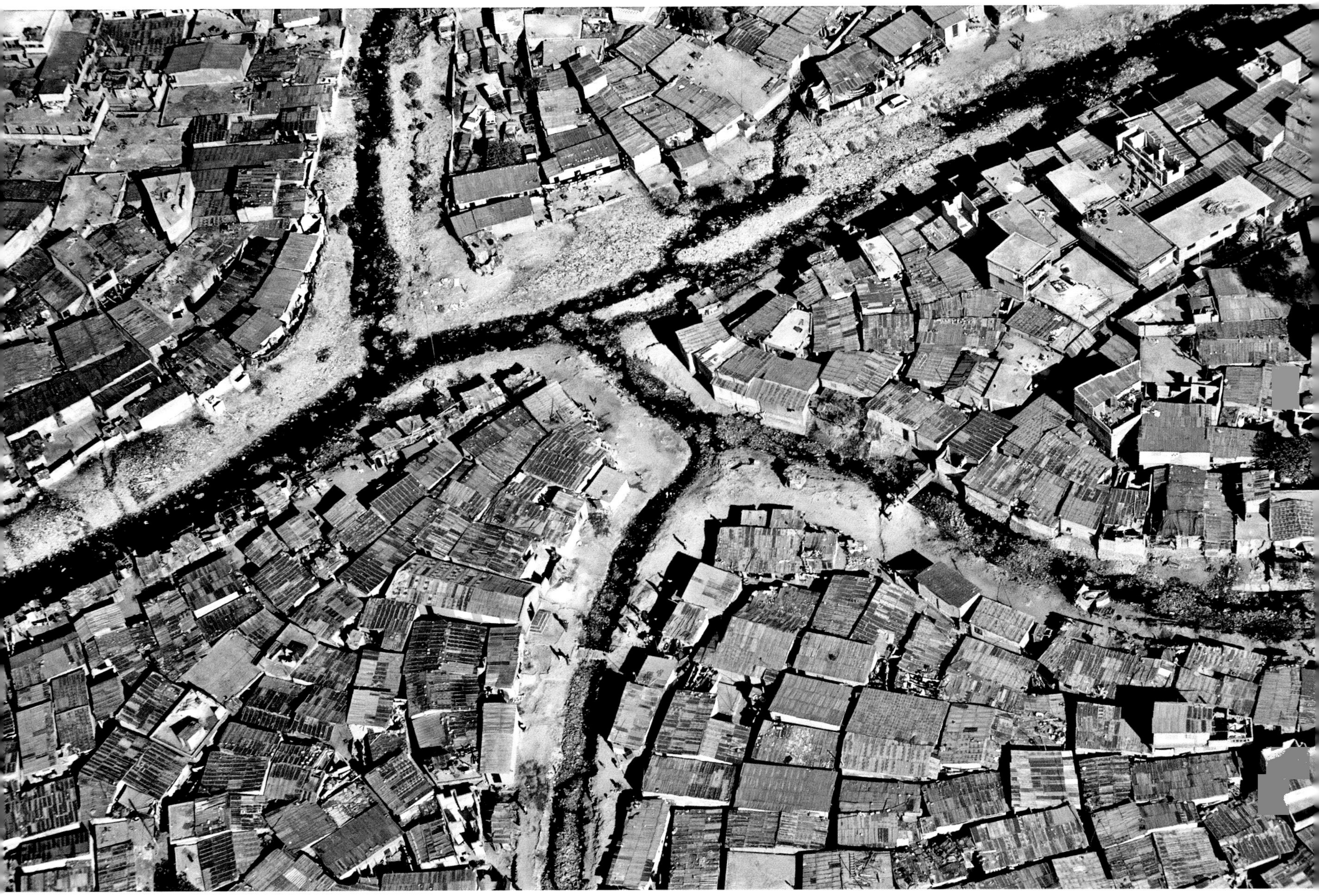

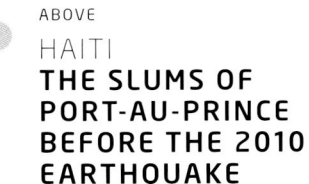

ABOVE

HAITI
THE SLUMS OF PORT-AU-PRINCE BEFORE THE 2010 EARTHQUAKE
18°32' N – 72°22' W

Even before the earthquake, Haiti was the poorest country in Latin America: a small, densely populated nation in which 70 percent of the population lived below the poverty line. The situation has further deteriorated since.

n March 2011, an earthquake measuring 9 on the Richter scale provoked a gigantic tsunami that struck Japan's northeast coast. The damage was devastating, with more than 15,860 people dead and 3,000 missing. Smashing everything in its path, the tidal wave also struck the Fukushima Daiichi nuclear power plant.

One hour after the earthquake's first tremors, a solidarity movement took shape in space. At the request of the Japanese space agency, the International Charter on Space and Major Disasters was activated. Established in 2000, the charter serves to rapidly provide free satellite images to countries struck by natural disaster, allowing them to assess the extent of the damages and to determine where to go to help the survivors.

By sharing their resources, the fourteen participating countries can mobilize their twenty-five satellites to help countries in emergency situations. These satellites photograph the affected country and send images to it within forty-eight hours and at no cost. A single call to a secret number from one of the signatory countries is all that is required for all the satellites to be repositioned.

Since it was launched, this charter has already been activated almost four hundred times due to cyclones, earthquakes, fires, floods, mudslides, tsunamis, oil spills, and volcanic eruptions.

By using this charter and the satellite images provided, authorities can rapidly determine the extent of the disaster area, the nature of the damage, the number of people displaced, and places where people are taking refuge. They can also organize emergency services, establish access roads, and start planning for the recovery.

PAGES 184-185
JAPAN, HONSHU
FUKUSHIMA
37°25' N – 141°01' E

Taken more than a year after the 2011 tsunami and the accident at the Fukushima Daiichi nuclear power plant, this image shows the work in progress at a site that has become hostile to man. The roof of reactor number 1, which was blown out, is covered with a white structure. The roof of reactor number 2's surrounding wall is undamaged. Red cranes clear the containment buildings for reactors 3 and 4. The roofing's metal girders are visible: Those in number 3 were torn apart and those in number 4 were burned.

PAGES 186-187
LIBYA AND EGYPT
PLUMES OF DUST ON MEDITERRANEAN COAST
32° N – 20° E

Only 1 percent of the land in Libya is farmable, with hundreds of square miles covered in sand. Sandstorms from the Sahara regularly cross the country, and sometimes even the Mediterranean.

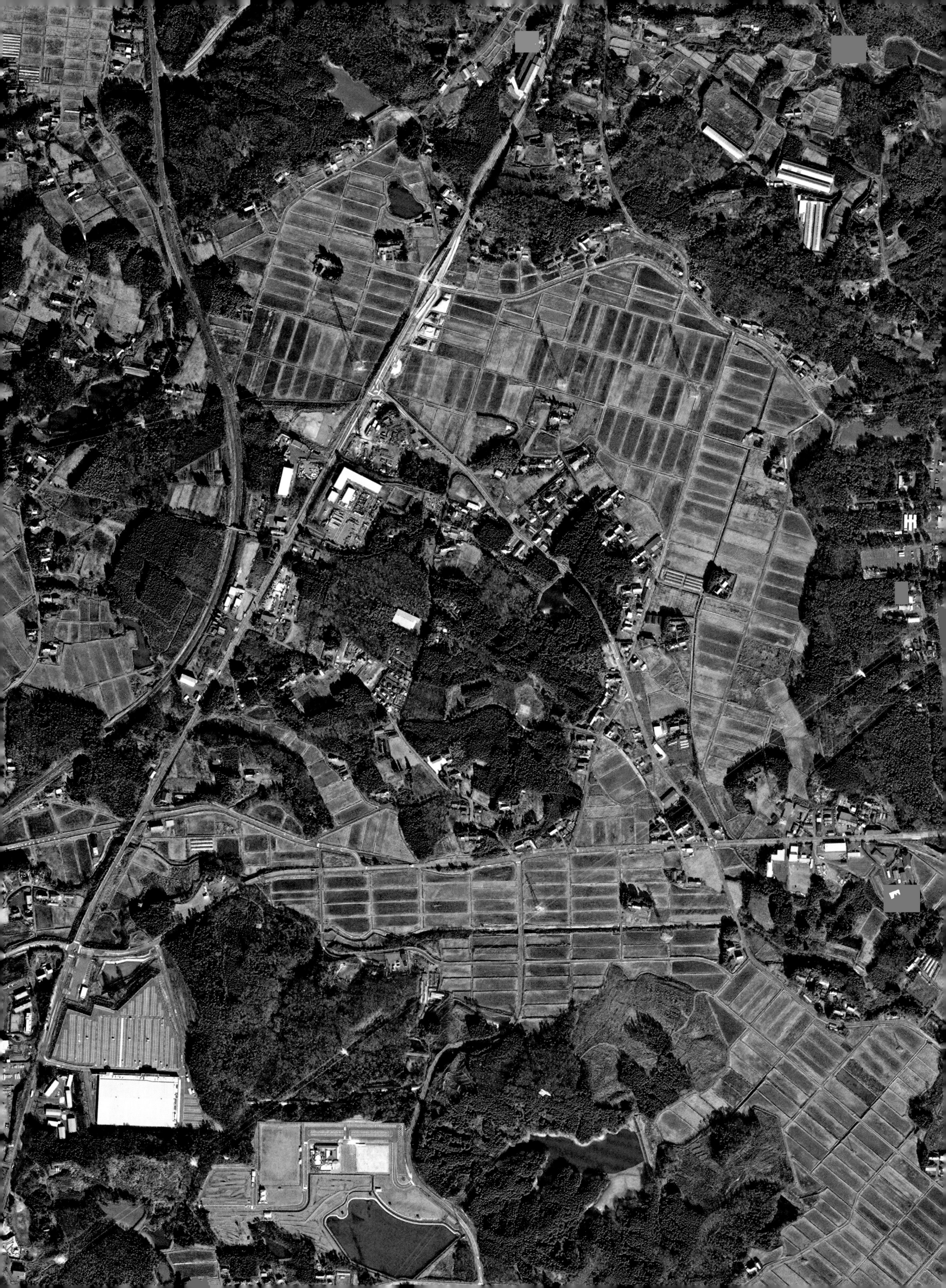

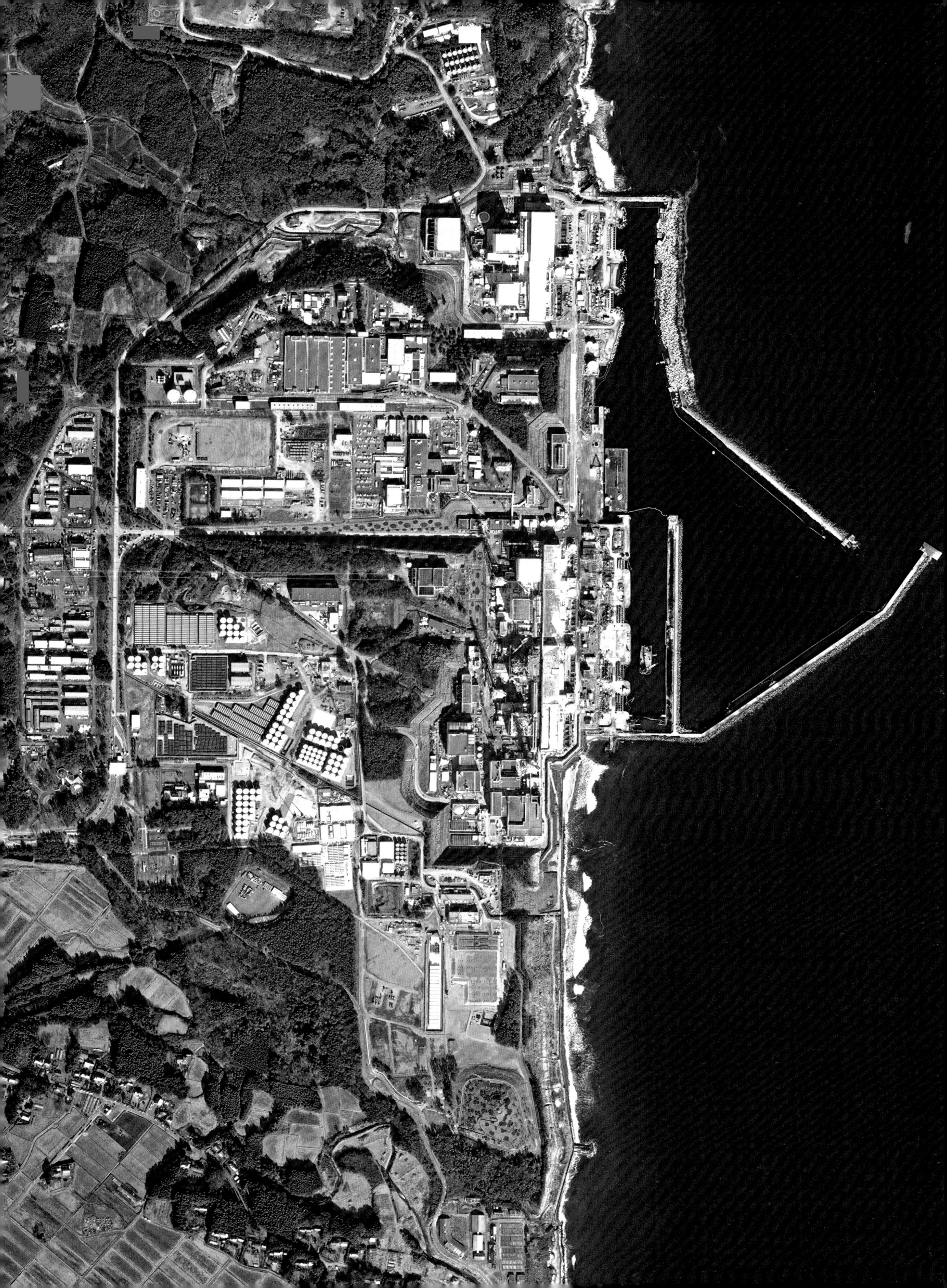

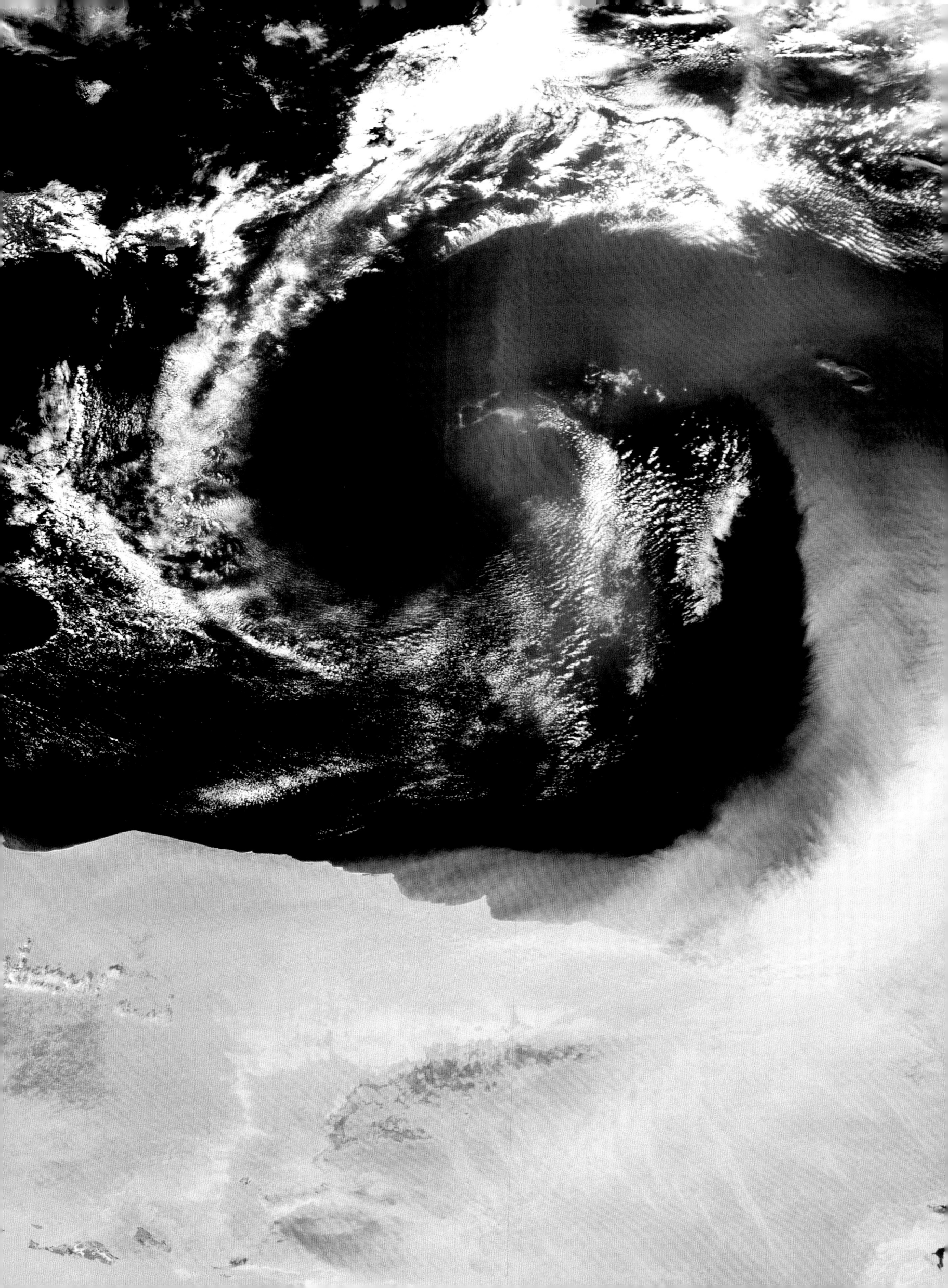

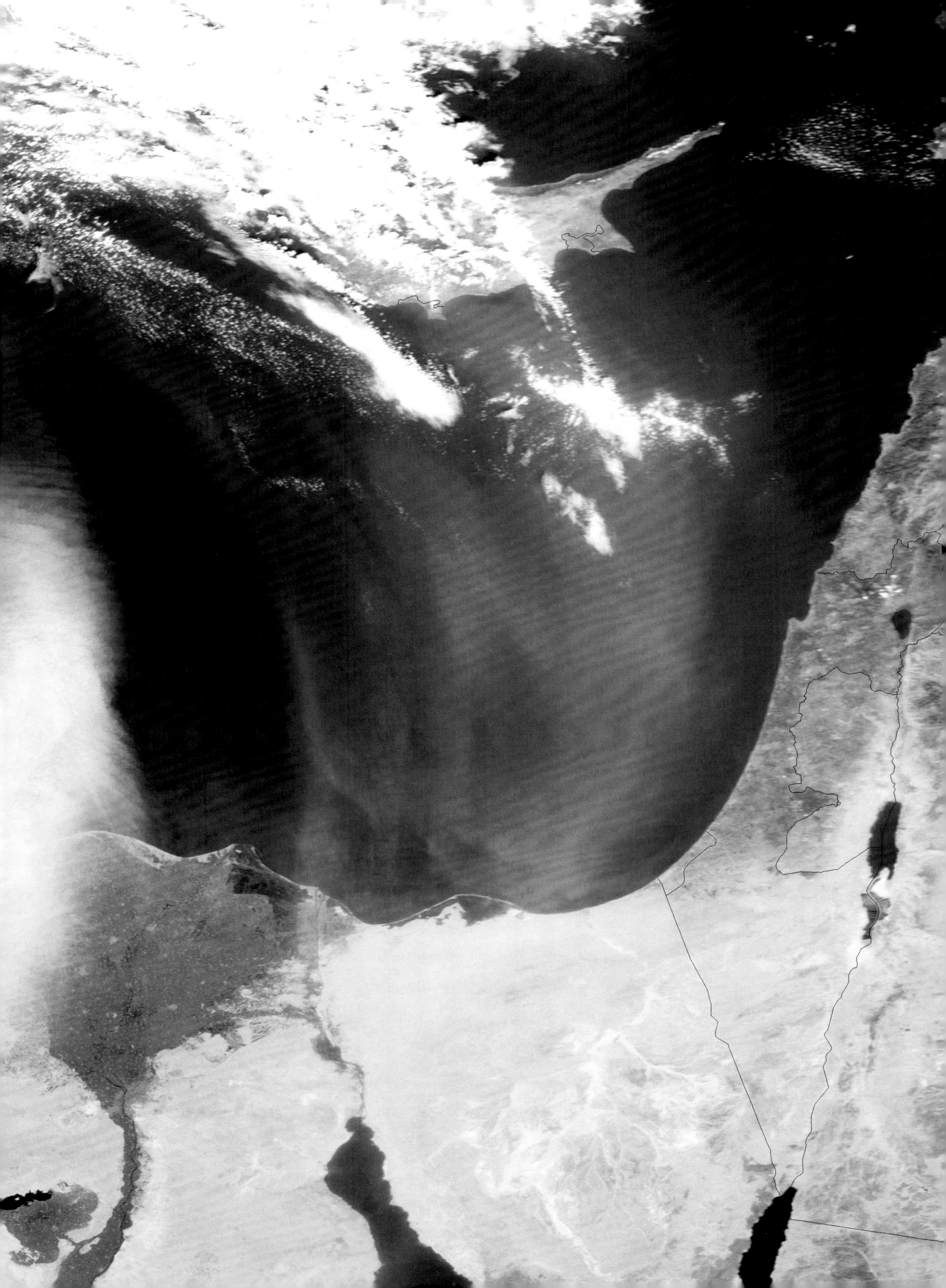

Only 0.01 percent of the water on Earth is both available and drinkable, and most of
that is used by farmers to feed the planet. Demand is increasing every day. Clearly,
it is essential for us to regulate our use of this precious resource.

____

BLUE GOLD

____

The discovery of traces of water on Mars gave credence to the idea that conditions favorable to life once existed on the planet. That's how important the precious molecule of water is. This discovery, which was largely enabled by space imagery, also demonstrated the effectiveness of remote sensing tools.

Water is far more abundant on Earth than on Mars. However, in terms of volume, freshwater accounts for only 3 percent of the water on the planet. Indeed, about three-quarters of freshwater is trapped in polar ice and most of the remaining quarter consists of subterranean water. Only a miniscule proportion of freshwater is available at the surface; the volume of freshwater directly usable is less than 0.01 percent of the total volume of water on Earth.

THE WATER CYCLE

Yes, water is a renewable resource. The total amount of water on Earth has generally remained the same for millions of years, and water is naturally purified and renewed over the course of what we refer to as the water cycle: It evaporates into the atmosphere, condenses there, falls back to Earth as precipitation (rain, snow, hail), and runs off and infiltrates soil. And then the cycle repeats again, over and over.

This cycle can sometimes last for more than a thousand years. But the rate of withdrawals has accelerated and water is starting to be scarce in many places. To address this situation, it has become necessary to better manage water, particularly given the unequal distribution of resources on the planet. Combined with other data, images provided by satellites can prove to be precious allies.

CONTROLLING DEMAND

Agriculture is currently the human activity that consumes the most water, accounting for about 70 percent of the freshwater used globally. In the latter half of the twentieth century, irrigation considerably improved agricultural yields; contributing to guaranteeing the food safety of millions of people, it was pivotal in reducing poverty in several areas of the world. But when practiced in an uncontrolled manner, it can cause genuine ecological disasters.

The most notorious case is that of the Aral Sea. Until the 1960s, this saltwater lake had two major rivers flowing into it, the Amu Darya and the Syr Darya. Then both rivers were diverted to irrigate cotton crops. This massive loss of water caused such a significant drop in the rivers' flow that the Aral Sea's surface area shrank by 75 percent and its volume by 90 percent. The fish gradually disappeared, along with the fishermen. The neighboring agricultural land became arid. The entire region was devastated.

Elsewhere in the world, the surface area of Africa's Lake Chad has shrunk by 10 percent in thirty years, while the Colorado River in the United States and the Huang He (the Yellow River) in China have experienced major episodes of dryness during which they no longer reached the sea.

These phenomena can be observed from space. Imagery satellites provide information about the surface of a body of water, while so-called altimetry satellites measure the level of the largest lakes and watercourses, making it possible to calculate their volume or flow.

OPPOSITE

KAZAKHSTAN/
UZBEKISTAN
ARAL SEA
45° N – 60° E

Located in Central Asia, with Kazakhstan to its north and Uzbekistan to its south, the Aral Sea is slowly dying. Once considered the fourth-largest landlocked sea in the world, it has been shrinking at an alarming rate over the last fifty years. The Aral Sea has gradually dried up due to the lack of water supply caused by the diversion of the two main rivers that once flowed into it, the Syr Darya to the north and the Amu Darya to the southwest. Recent initiatives in Kazakhstan, supported by the World Bank, have led to some improvement, at least in the north.

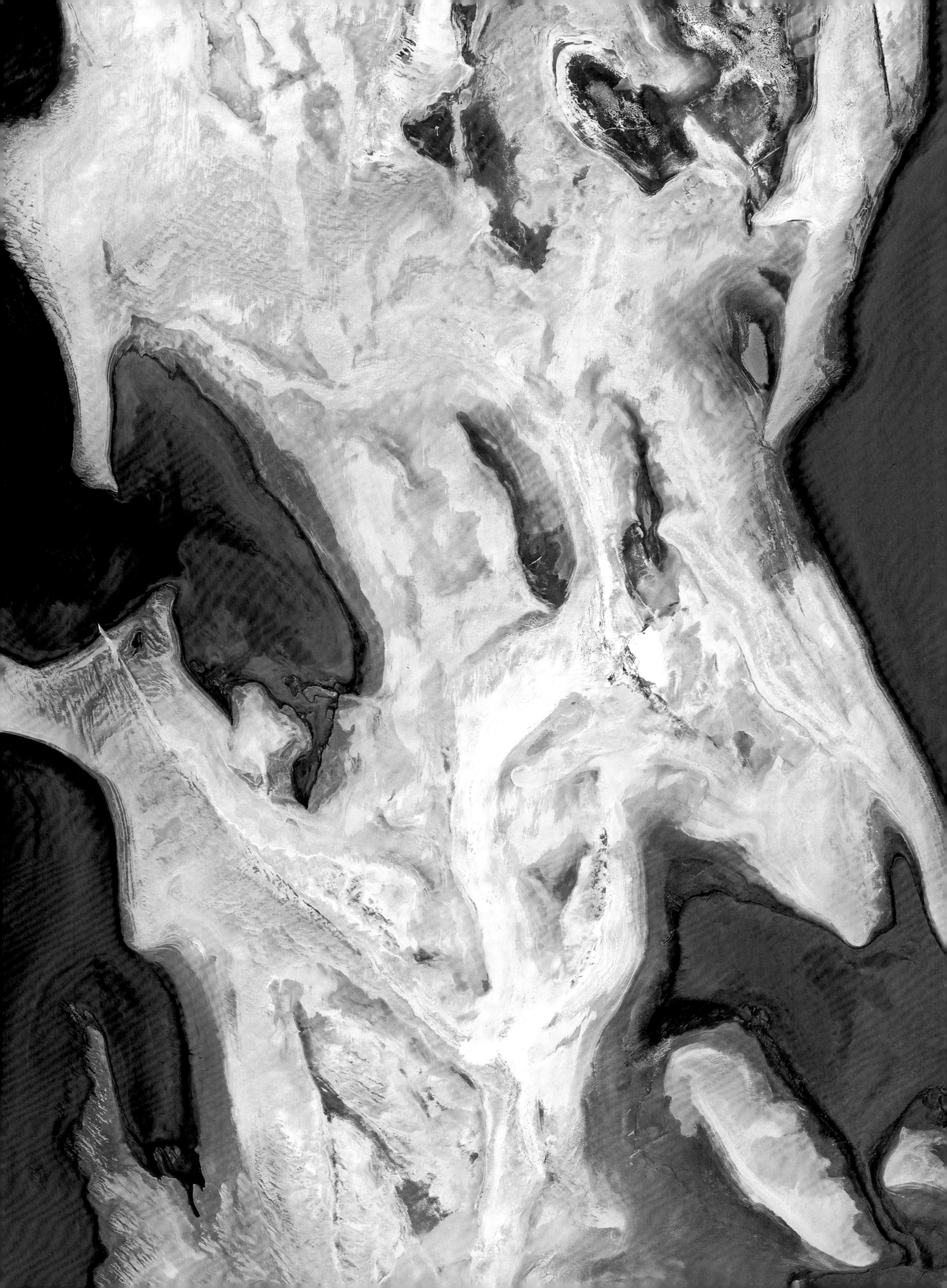

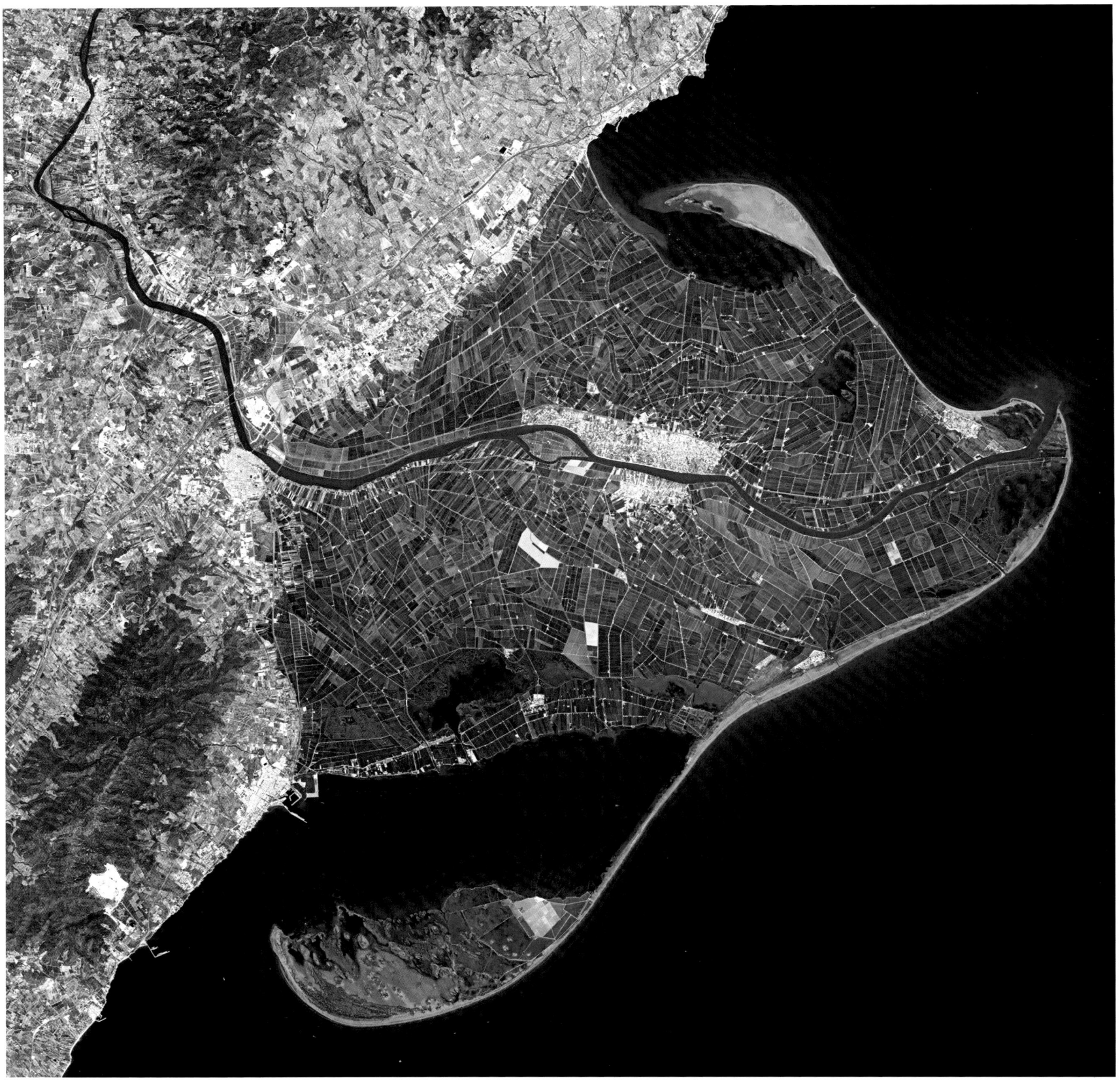

SPAIN
EBRO DELTA
40°43' N – 0°44' E

The Ebro is Spain's most powerful river. Stretching for more than 550 miles (900 km), it flows into the Mediterranean through a vast delta that juts far into the sea. Abundant alluvium and an ingenious system of irrigation channels favor rice plantations, notably of the famous bomba rice. The Ebro Delta Nature Park, founded in 1983 and covering 19,279 acres (7,802 hectares), is the largest wet zone in Catalonia.

JORDAN
DEAD SEA SALTERN
31°10' N – 35°26' E

This is the lowest point on the planet's surface: The Dead Sea, a vast salt lake between Israel and Jordan, is currently 1,387 feet (423 m) below sea level. Here we see its southern side, which is equipped with dykes delimiting evaporation ponds for extracting salt from this particularly high-salinity water.

Only **30 TO 50 PERCENT OF THE WATER USED FOR IRRIGATION** is truly useful to crops, with **THE REST BEING LOST** due to defective pumping equipment or water networks.

HELPING FARMERS

Some satellites evaluate soil moisture to help farmers better use water resources. While soil contains only 0.01 percent of the water on our planet, this water is obviously essential to plant life. Only the first few inches of soil are currently accessible to imaging technology, while roots can go as deep as three to six feet (or one to two meters). We need to establish a model to estimate what goes on at greater depths, which is only possible based on accurate analysis of the soil in the field. Farmers can thus identify which areas need

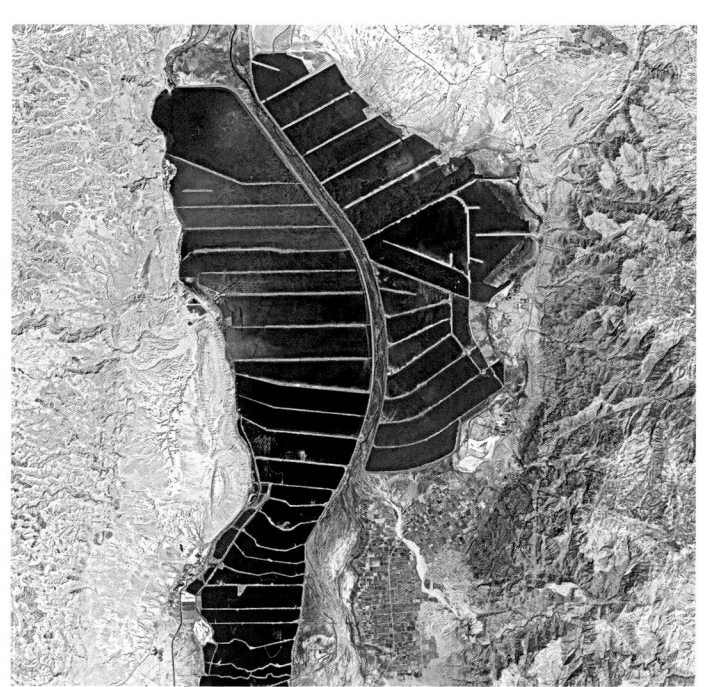

more water and more precisely determine the best time to water crops. Or rather—they will soon be able to do so, for these techniques are still being developed. This is a particularly important task. According to the Food and Agriculture Organization of the United Nations (FAO), only 30 to 50 percent of the water used for irrigation is truly useful to crops, with the rest being lost due to defective pumping equipment or water networks.

DAMS

Canals, flood-routing reservoirs, dams: A great deal of infrastructure has been built specifically for irrigation. Half of the major dams in the world were built exclusively or principally for this purpose.

Some dams are among the largest structures built by man. The Tarbela Dam in Pakistan, for instance, was built with 3.7 billion cubic feet (90.9 million cubic meters) of earth and stone, or forty times the volume of the Great Pyramid of Giza in Egypt. The Itaipu Dam in Brazil required the equivalent of 380 Eiffel Towers in steel and iron. Each of these dams generates as much electricity as several nuclear power stations combined. As for the Three Gorges Dam in China, it is expected to single-handedly produce twenty thousand megawatts, the equivalent of twenty nuclear reactors.

Long considered signs of development and economic progress, dams were built on watercourses in great numbers throughout the twentieth century. There are approximately eight hundred thousand of them around the world, of which fifty-two thousand are considered "large dams," which

According to a report of the World Commission on Dams,
FORTY TO EIGHTY MILLION PEOPLE were displaced
over the last century following the **CONSTRUCTION OF DAMS.**

means they are higher than 49 feet (15 m) with a reservoir capacity of more than 105 million cubic feet (3 million cm).

However, this kind of infrastructure causes significant social, environmental, and geopolitical problems. The social effects are staggering: According to a report of the World Commission on Dams, forty to eighty million people were displaced over the last century following the construction of dams. From an environmental perspective, dams transform watercourses and disrupt their proper functioning, which affects ecosystems—for example, by preventing fish (particularly salmon) from traveling or blocking the transport of alluvium. It is estimated that 60 percent of the rivers in the world have been altered this way. The energy produced by dams is also not entirely without effect on the climate, for the plants that decompose in reservoirs release methane,

particularly in tropical areas. From a geopolitical perspective, the construction of a dam by a state upstream of a river modifies the supply of water to the states downstream and can lead to tensions, as can be seen in Turkey's disagreement with Iraq and Syria after dams were built on the Tigris and Euphrates on Turkish soil. Most experts now argue for more modest structures, with less impact.

CONTROLLING THE RISKS

Nonetheless, dams' water retention systems give them the advantage of diminishing the risk of flooding and attenuating the consequences of droughts. Yet with the effects of climate change, extreme weather phenomena have become increasingly frequent in recent years.

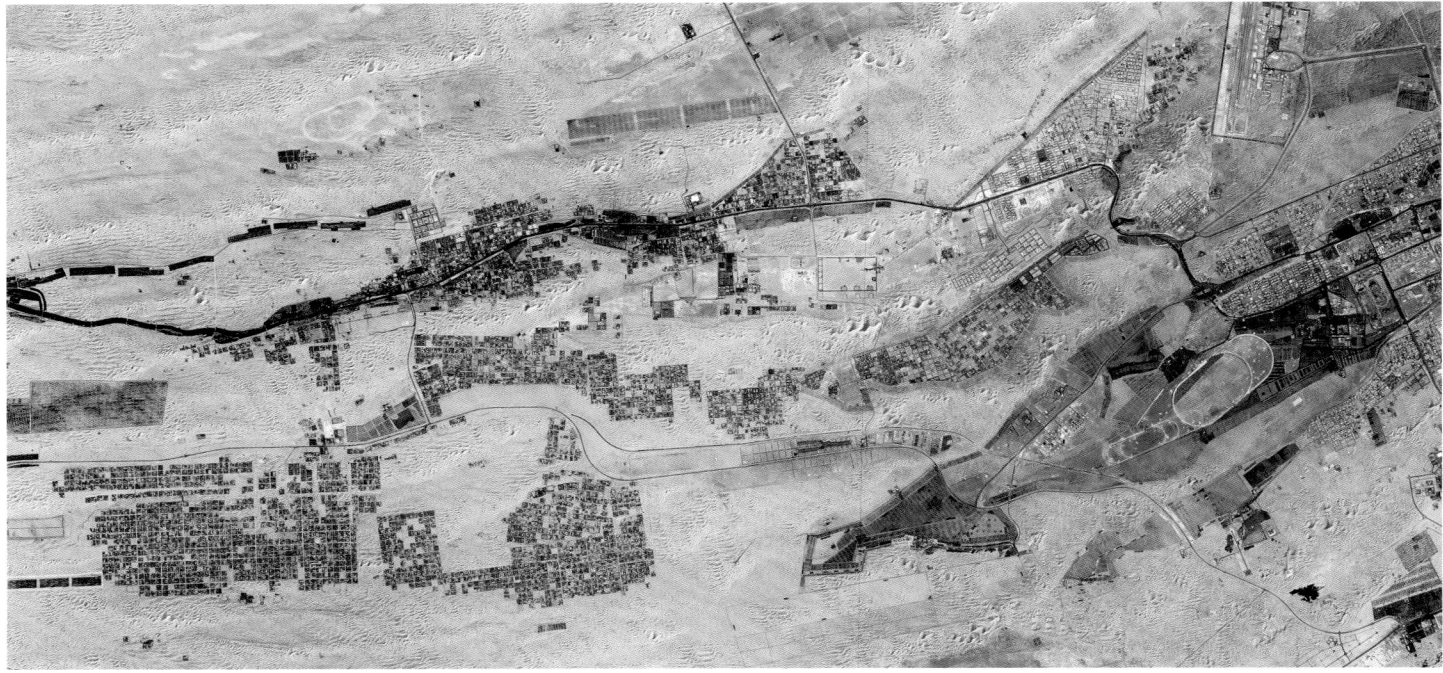

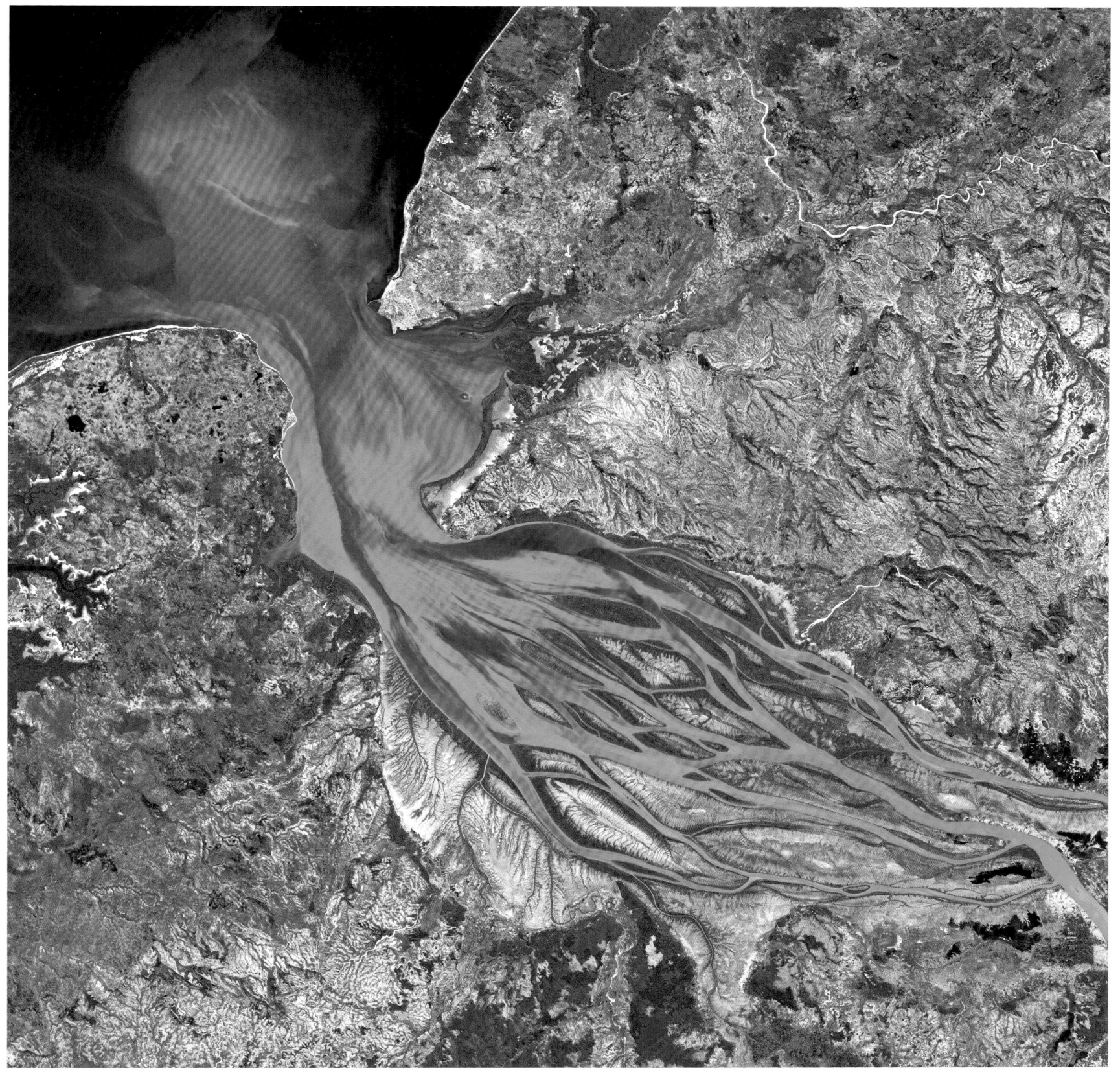

UNITED ARAB EMIRATES, ABU DHABI
AL AIN OASIS

24°12' N – 55°44' E

Underground springs and irrigation tunnels, known as *falaj*, which have been operating since antiquity, supply water to the garden city of Al Ain, the fourth-largest city in the United Arab Emirates. Satellite observation attests to the fact that palm-date plantations are developing year after year in the neighboring oases. These plantations threaten the *ghaf*, a desert tree that is used to feed cattle, as a combustible, and for its medicinal qualities.

MADAGASCAR
BOMBETOKA BAY

15°48' S – 46°16' E

Bombetoka Bay, the Betsiboka River's sludgy estuary, spans some 6 miles (10 km) on Madagascar's west coast. Alluvial islands stretch out in the direction of the current, preceded and extended by sandbanks; they are partially covered by mangrove trees, seen in red in the image. In the surrounding area, the heavily cleared forest gives way to brush and cassava and rice plantations.

BELOW

INDIA, ASSAM
BRAHMAPUTRA RIVER
26° N – 94° E

The Brahmaputra River flows from a source at an altitude of about 16,400 feet (5,000 m) in the Himalaya Kailash range of Tibet. The river valley is one of the richest sites of biodiversity in the world.

Heavy floods fertilize the soil, but can also be devastating. During the monsoon, from June to October, the river is in its high-water period, while the low-water period lasts from January to March.

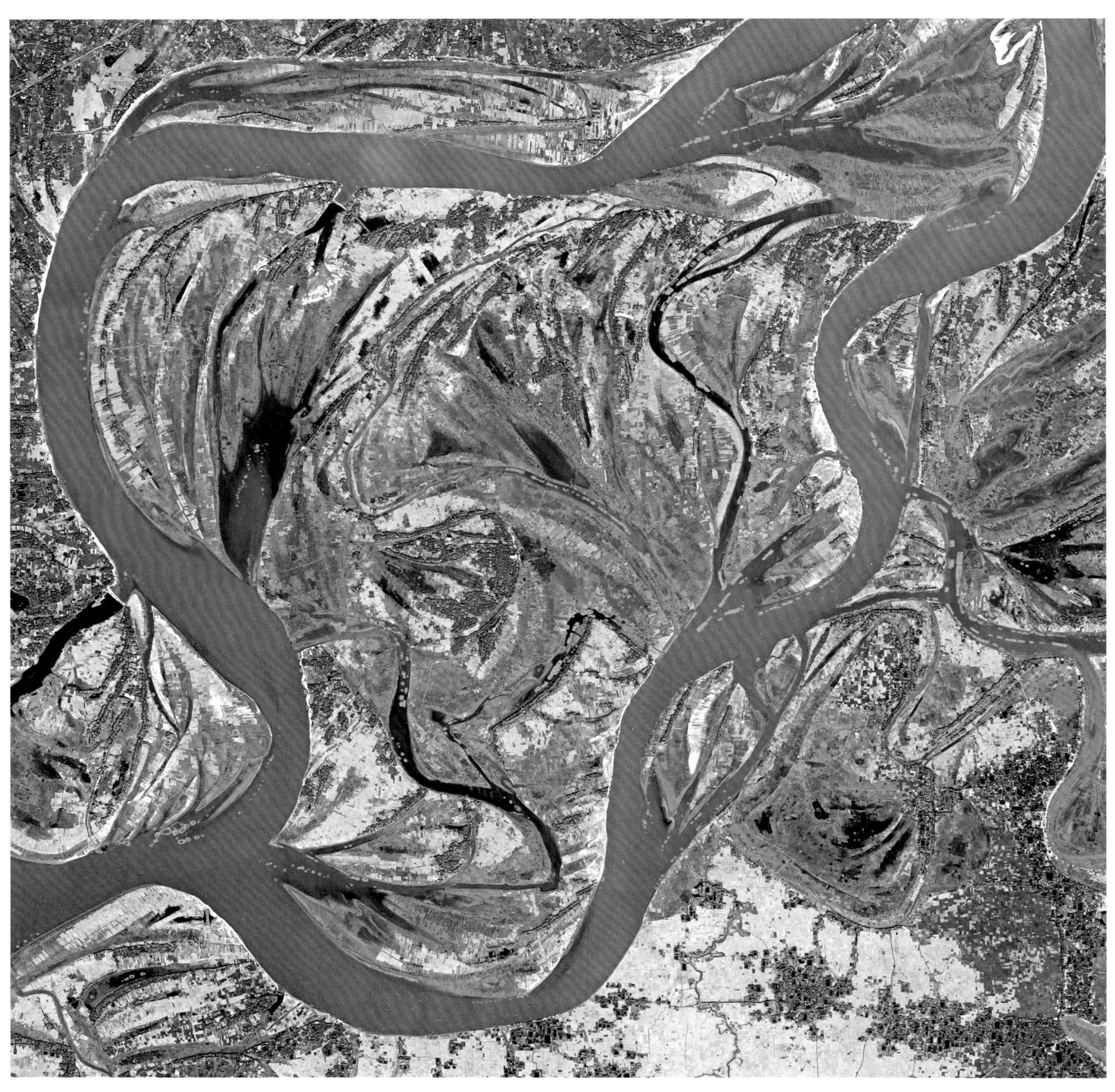

By helping to accurately locate a watercourse's major bed (the area susceptible to flooding), satellite images are precious tools for decision making and planning response to imminent dangers. When the Guadalquivir flooded in February 2012, Andalucia's Agency for Environment and Water used satellite images to determine the zones to evacuate, avoiding significant human loss and economic damages.

UNEQUAL DISTRIBUTION

As with many other resources, the distribution of water across the planet remains very unequal. Nine countries share 60 percent of renewable freshwater resources in the world. There are also significant differences in water consumption levels: on average, a little more than 26 gallons (100 L) per inhabitant per day in Europe and 66 gallons (250 L) in North America versus only 2.5 gallons (10 L) in sub-Saharan Africa.

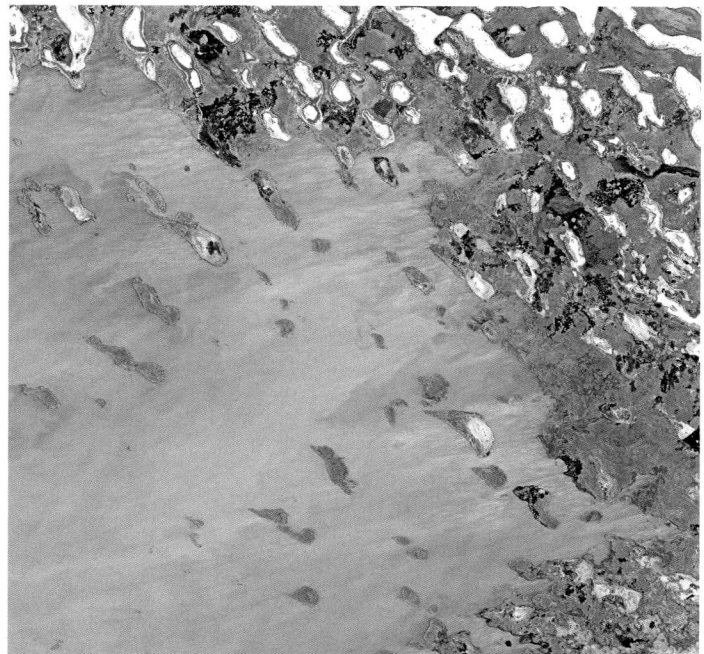

ABOVE, LEFT
EGYPT
THE NILE
24°30' N – 32°54' E

The longest river in the world (4,145 miles/6,671 km), the Nile crosses or skirts nine countries before flowing through Egypt and reaching the Mediterranean. As the principal hydraulic resource in the area, it provides Egyptians with 98 percent of their water. The construction of the Aswan Dam in the 1960s made it possible to regulate the river's flow and to supply water to the country year-round by holding back a volume of water equivalent to two times the Nile's mean annual flow.

ABOVE, RIGHT
CHAD
LAKE CHAD
13°09' N – 14°40' E

Lake Chad supplies water to more than twenty million people. Yet its surface area has been reduced to one-tenth of what it was in the 1960s, partially but not only because of an unsustainable use of the resource. This gradual drying up affects local populations, which have nonetheless been able to adapt. The likelihood that UNESCO will include the lake on its World Heritage List suggests the possibility of a preservation program carried out by a consortium of bordering countries.

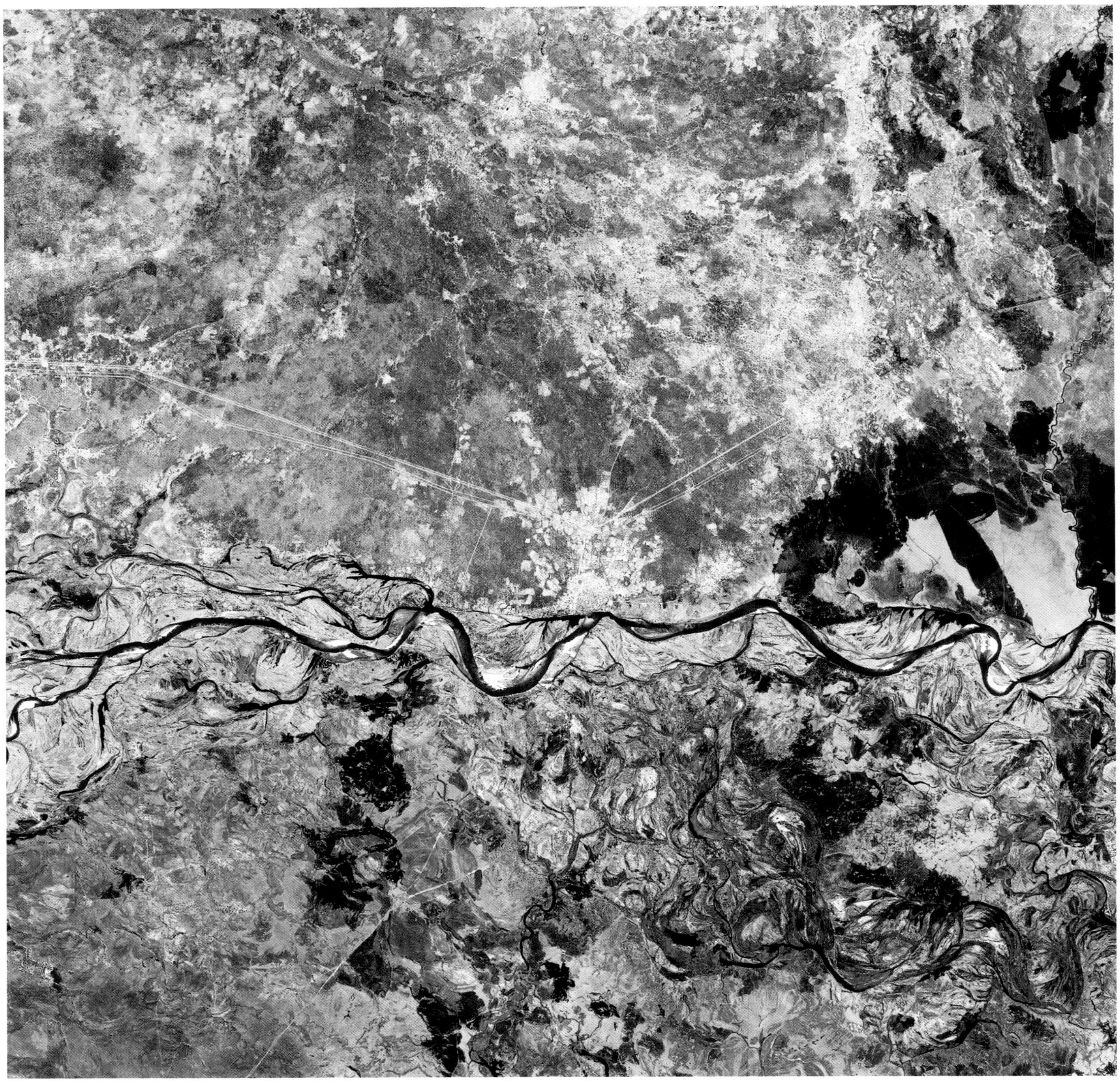

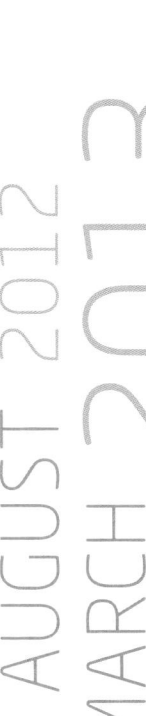

ZAMBIA / NAMIBIA
ZAMBEZI RIVER
17°30' S – 24°50' E

The Zambezi River defines the border of
Zambia and Namibia. Flowing slowly, it
regularly drains into the flood plains during
the wet season from February to April. In
satellite images, taken in the dry season in
August 2012 and the wet season in March
2013, respectively, the Zambezi's level can
be determined by shades of blue, with the
darkest blue indicating the deepest water.
Yet the most striking difference is in the
levels of the plains, with dry and burned
(brown and yellow) expanses giving way to
flowering (green) spaces: The hydrophilic
plants of the river plains depend on these
annual rains to grow.

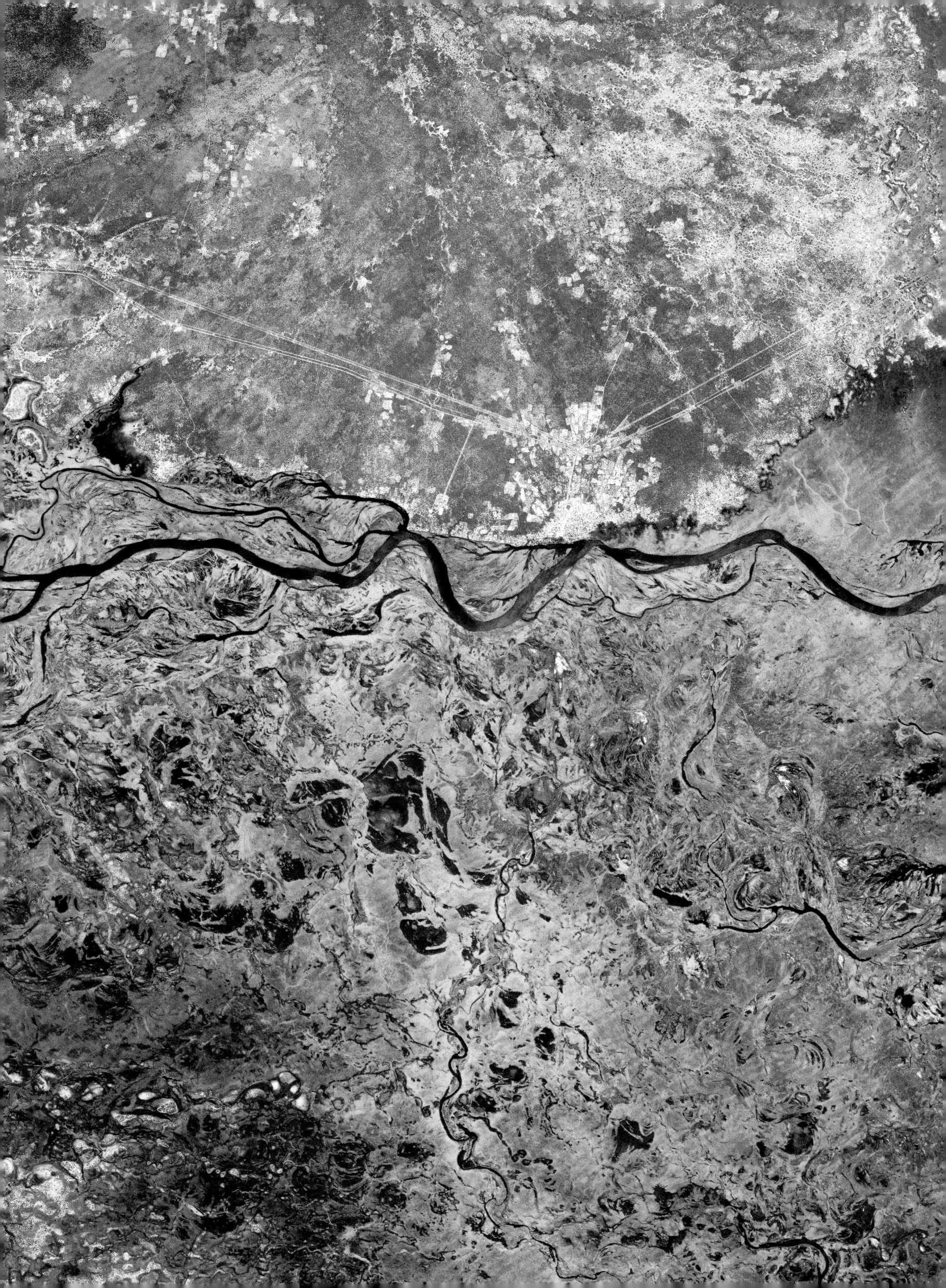

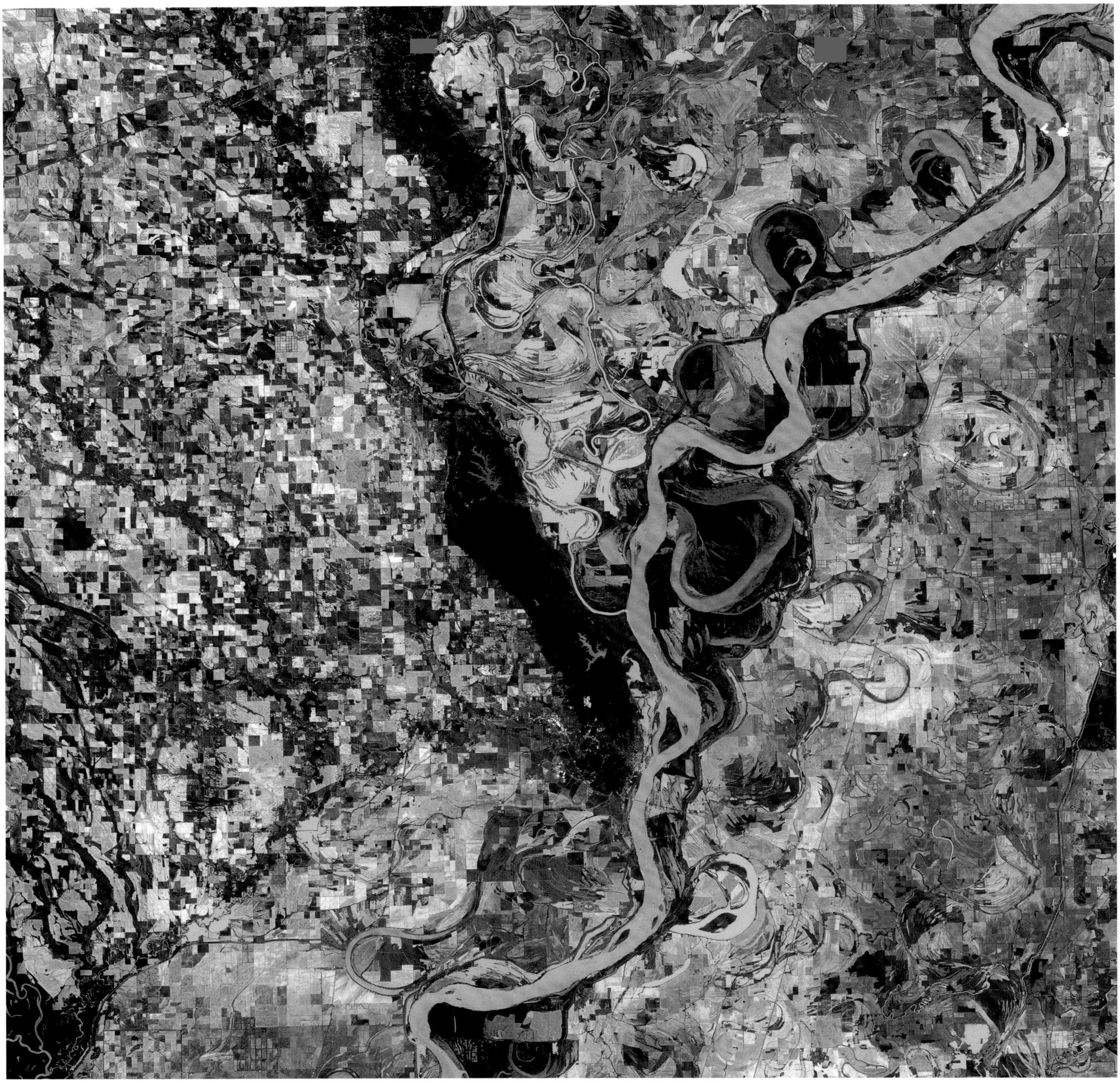

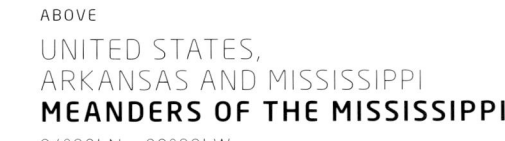

UNITED STATES,
ARKANSAS AND MISSISSIPPI
MEANDERS OF THE MISSISSIPPI
34°39' N – 90°33' W

The grid pattern of fields and pastures borders
the majestic whirlpools of the Mississippi, which
boasts the largest hydrographic basin in North
America. Numerous oxbow lakes underline the
river's meanders south of Memphis, Tennessee,
at the border of Arkansas and Mississippi.

In 2012, the World Health Organization announced that
89 PERCENT OF THE GLOBAL POPULATION
had ACCESS TO DRINKING WATER.

Demand for water will increase over the course of the twenty-first century due to population growth and improving standards of living. Yet, unless we make massive investments, water could easily become the source of conflicts. In many places around the planet, difficulties are not due to the resource being absent, but rather a lack of means to make water drinkable and to store and distribute it. Like hunger, access to water is a problem of poverty. Satellite images alone cannot provide a solution to this crucial issue.

There is one piece of good news, however: Real progress is taking place, thanks notably to the efforts implemented by the Millennium Development Goals adopted by the United Nations in 2000. Access to drinking water was one of the eight objectives to be achieved by 2015. In 2012, the World Health Organization announced that 89 percent of the global population had access to drinking water. This Millennium Development Goal was reached three years ahead of schedule.

In twenty years, the number of people who have accessed improved means of supplying themselves with drinking water—such as pipes or protected wells—has increased by two billion. However, access to quality purification systems remains problematic, with only 63 percent of the global population currently able to enjoy their benefits. The Goal for 2015 was 75 percent. According to projections, it will not be reached.

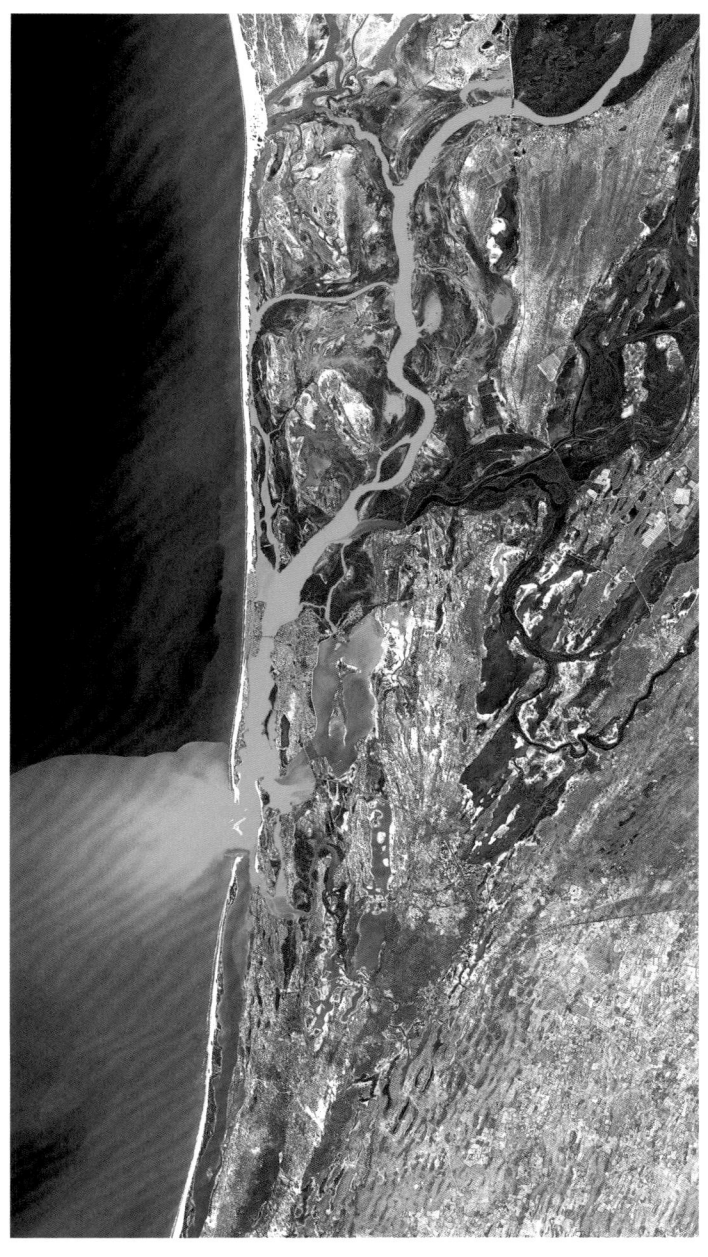

SENEGAL
**MOUTH OF THE
SENEGAL RIVER**
16°01' N – 16°28' W

After running for 1,112 miles (1,790 km), the Senegal River flows into the Atlantic at Saint-Louis. The former capital of Senegal was founded at the mouth of the river by the French in 1659 and was once an important commercial center, particularly for the slave trade. In the delta, the Djoudj National Bird Sanctuary is home to pink flamingos, purple heron, African spoonbills, great egrets, and cormorants. This wet zone is also a stopping point for migratory birds. Saint-Louis and the delta are on UNESCO's World Heritage List.

REPAIRING WELLS IN AFRICA

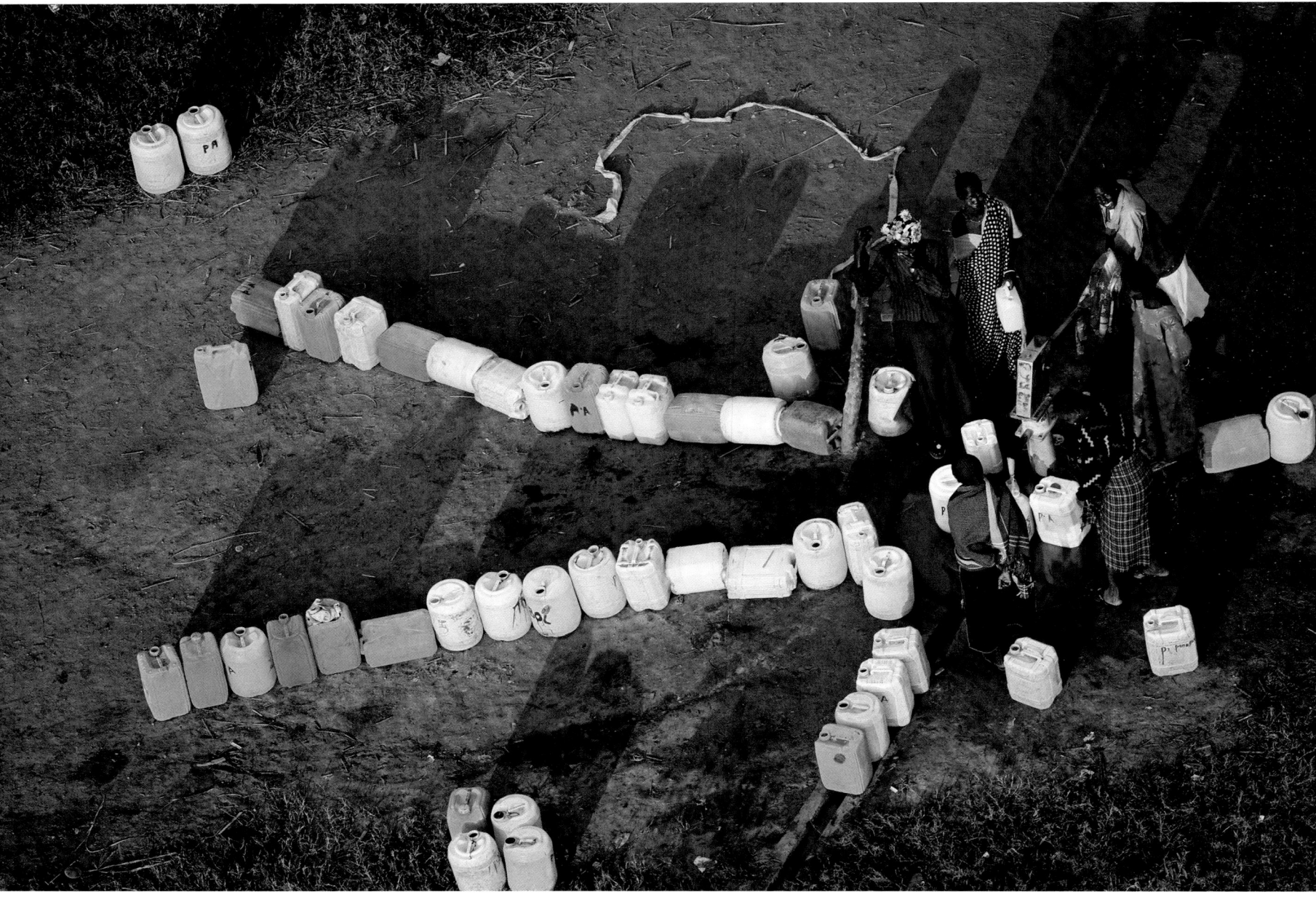

ABOVE

SOUTH SUDAN ,
STATE OF JONGLEI
**WELL AND VILLAGERS
NEAR BOR**
6°13' N – 31°34' E

In many countries, fetching water is traditionally a chore assigned to women. If the
closest source is far away, they must carry
several gallons of water over long distances
for several hours, often on their head. Having
a well nearby radically changes their daily life.

Digging wells is good. Repairing defective wells is even better. Access to drinking water is vital. While generally quite simple in the west, this operation often proves highly complex in many parts of Africa. Living near a well changes a population's life, but close to one billion people in the world still do not have access to drinking water, which represents a major sanitary risk for people who are forced to drink water unfit to be consumed, or for people—especially women—who have to walk miles to get some.

Many wells have been drilled, in both Africa and other continents. Most of the time, they are equipped with hand- or foot-activated pumps. When these pumps break, access to water can be suspended for three months or even a decade. According to Philippe Lacour-Gayet, a former executive with Schlumberger and the cofounder of IDO—a nongovernmental organization under French law that aims to provide underprivileged populations with permanent access to high-quality water—repairs should be carried out every three years.

IDO operates in Logone Oriental in the south of Chad, where, according to Lacour-Gayet's estimate, two-thirds of well pumps are unusable due to lack of maintenance. If you consider the number of broken pumps in the region on a nationwide scale, there are approximately ten thousand unusable wells in Chad.

Given the lack of a reliable inventory of wells, part of IDO's program consists of identifying and listing them. An additional challenge is that maps of the area date from the 1960s and the country has changed considerably since then. IDO's experts, whose backgrounds in the oil industry provide them with a good grasp of drilling techniques, called on Planet Action to obtain satellite photos and locate the wells. They are now able to carry out inspections and, when necessary and possible, have wells repaired by a local artisan. Providing a new well costs about 10,000 euros, while repairing a pump costs about 100 euros on average. But the work to be accomplished is massive. Since 2009, IDO has repaired 167 pumps.

However modest, this success has incited IDO to extend its activities to different areas of Chad and other countries, for the needs are huge. The NGO recently launched an operation in Congo, where the situation appears to be even more disastrous, with nearly all the wells in the country out of service.

PAGES 202-203

RUSSIA
VOLGA RIVER
45°58' N – 47°38' E

The Volga, one of Europe's largest rivers, runs 2,300 miles (3,700 km) through Russia before flowing into the Caspian Sea. With 60 percent of its water supplied by melting snow, the Volga's flow heavily fluctuates from one season to the next, shifting from a mean of 141,000 cubic feet (4,000 cu m) per second in winter to close to 900,000 cubic feet (25,000 cu m) in the months of May and June.

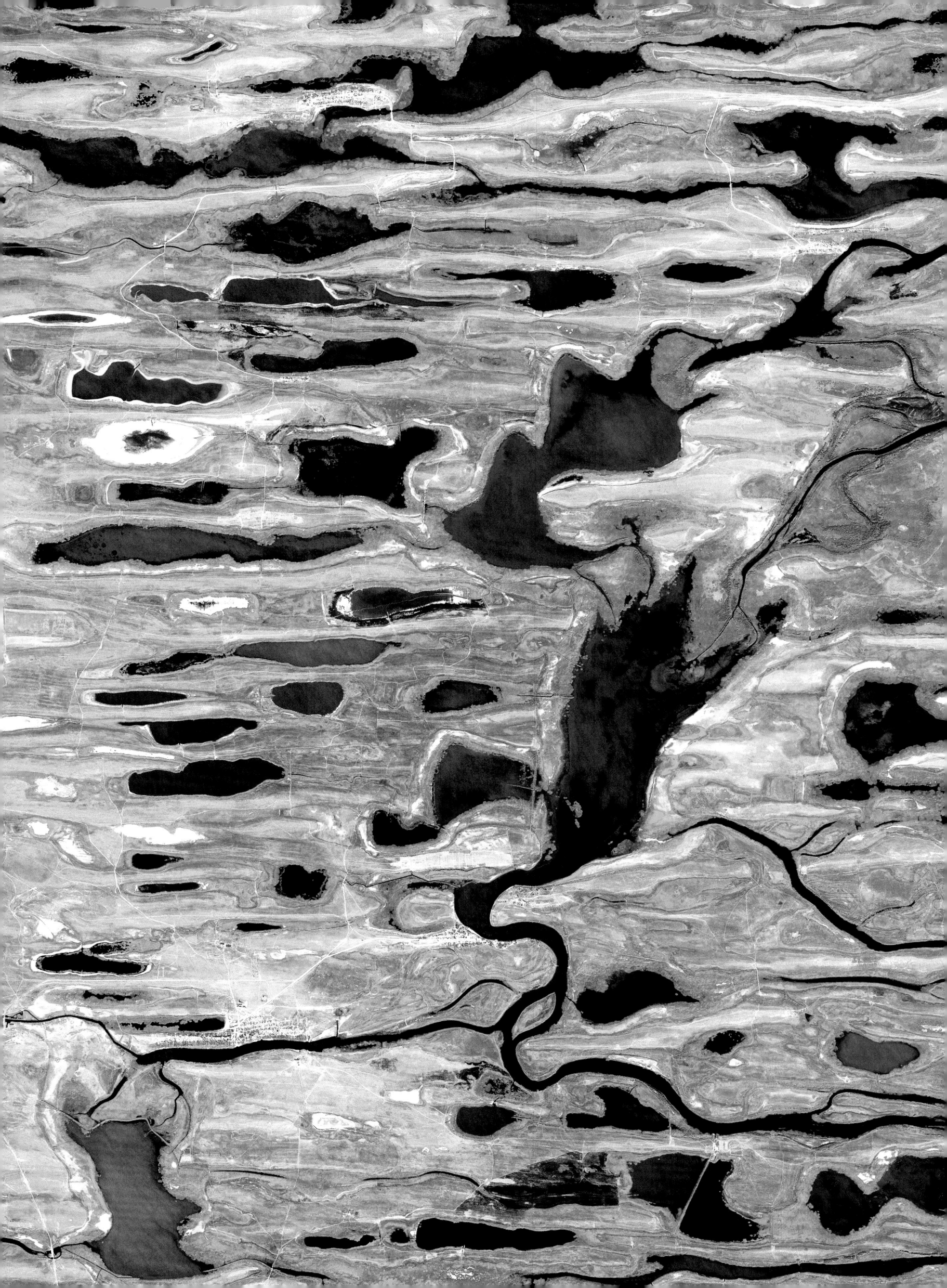

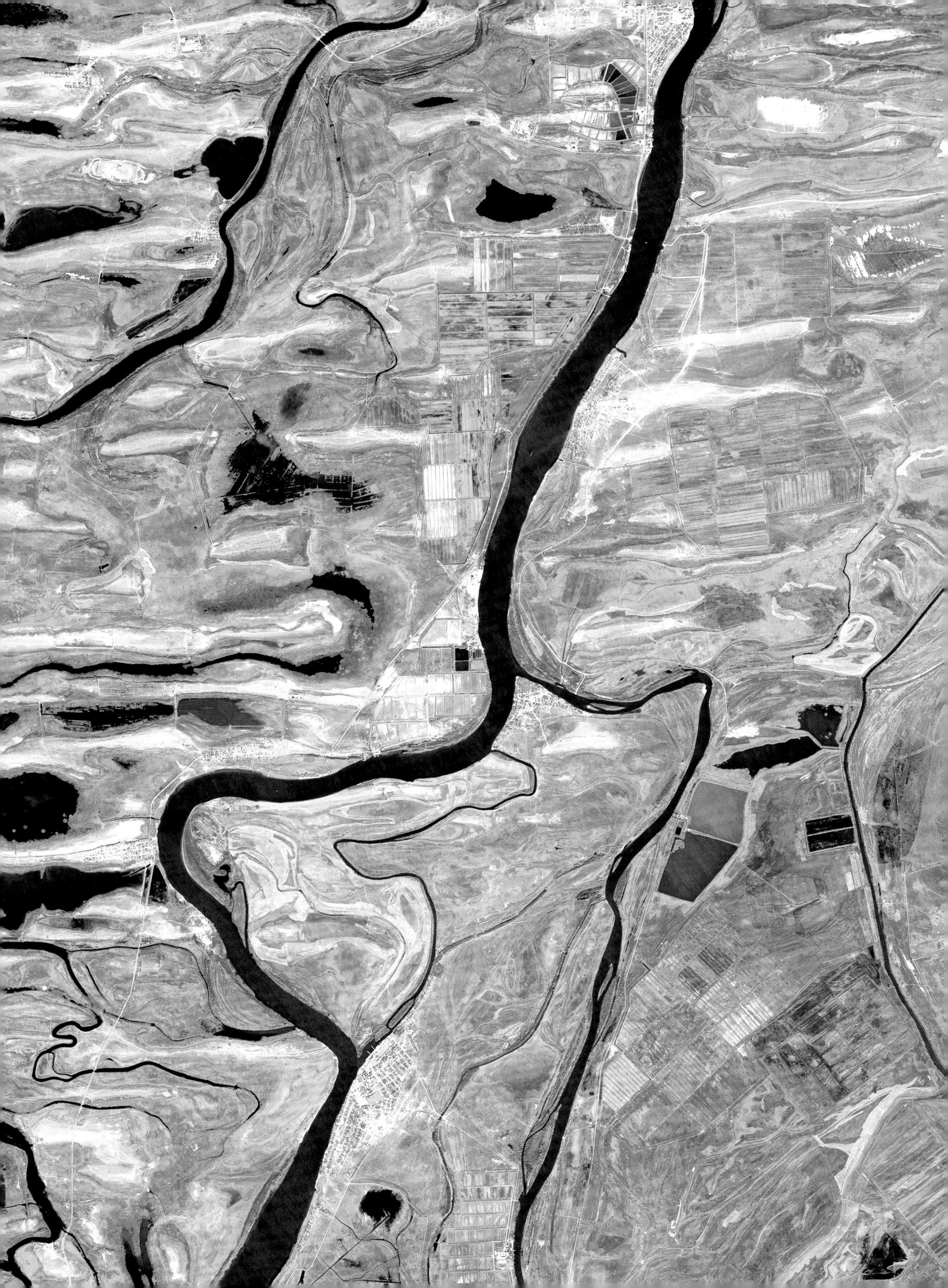

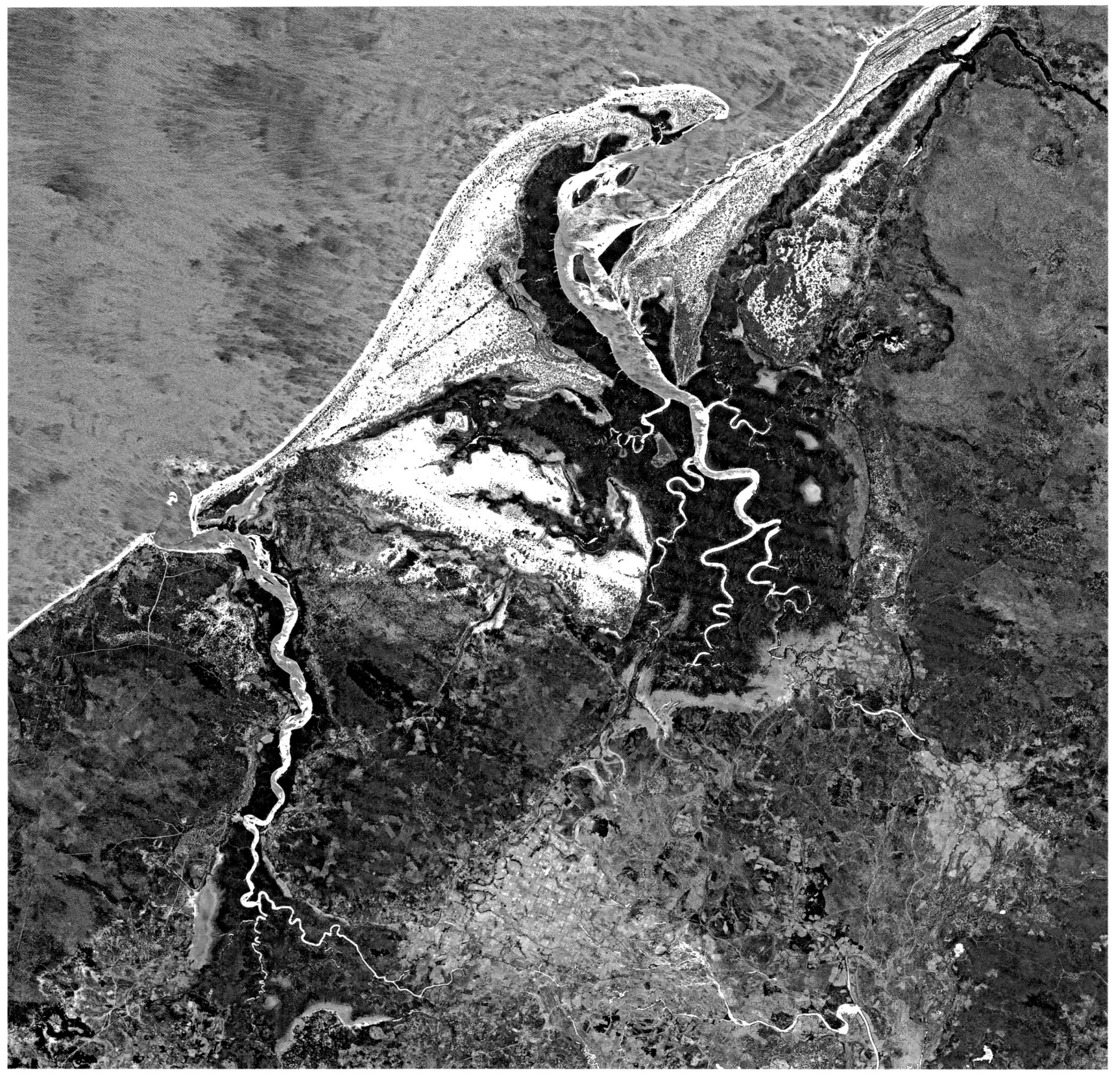

ABOVE
MADAGASCAR
ANTSANITIA
15°31' S – 46°28' E

Some 12 miles (20 km) north of the Betsiboka estuary, two other watercourses shape the coasts of Madagascar, near Antsanitia.

OPPOSITE
AUSTRALIA, NORTHERN TERRITORY
JOSEPH BONAPARTE GULF
15°02' S – 129°10' E

Named in 1803 by the explorer Nicolas Baudin in honor of Joseph Bonaparte, brother of Napoleon and then king of Naples, the gulf is located at the mouths of the Keep and Victoria rivers in the Timor Sea. This sea is a branch of the Indian Ocean between the island of Timor and the Australian coasts of Western Australia and the Northern Territory.

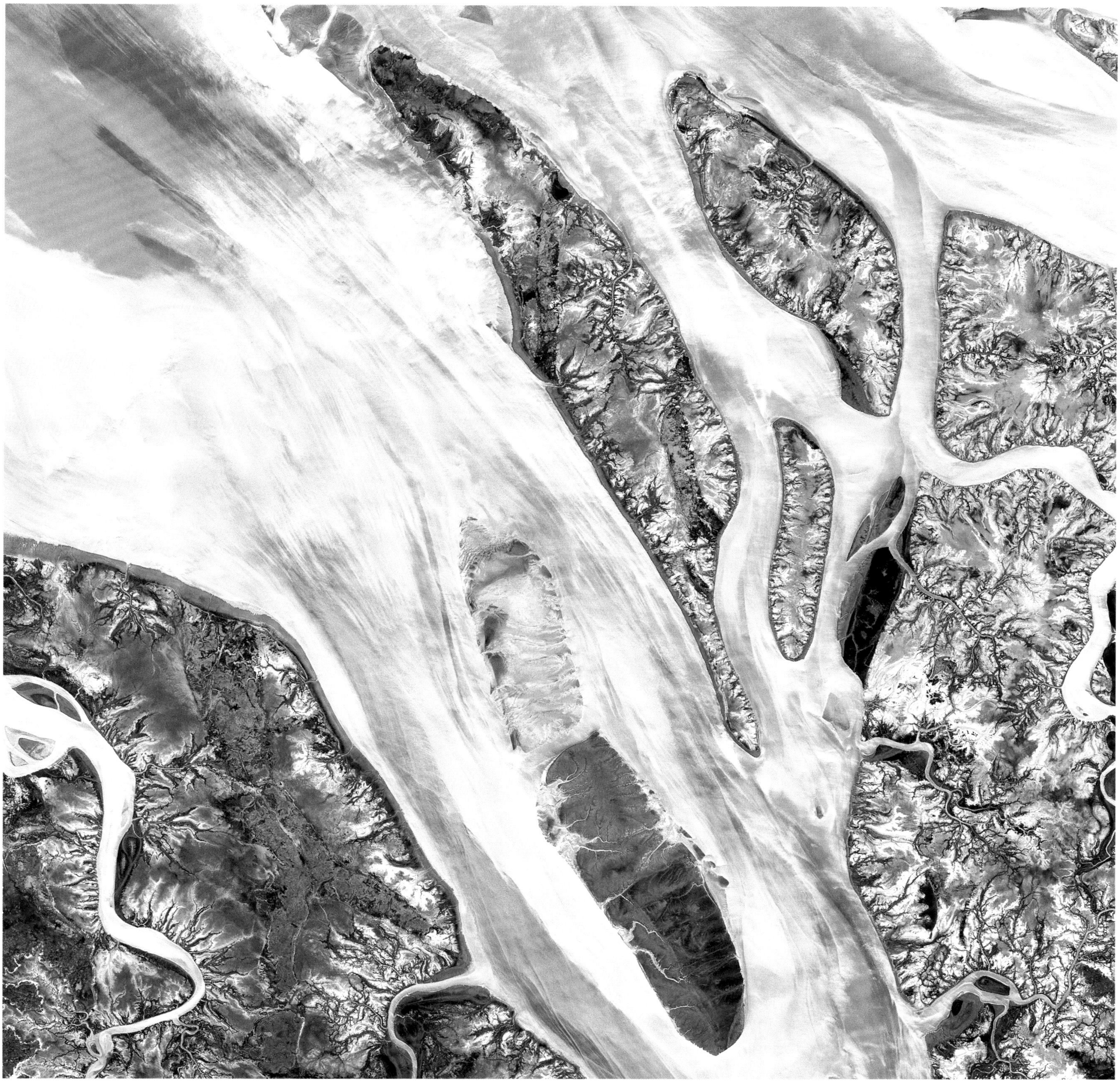

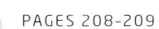

PAGES 206-207

MALI
NIGER RIVER
16°36' N – 03°10' W

The Niger River springs from the foot of Mount Loma in Guinea, close to the Sierra Leone border. Between Macina and Timbuktu, in Mali, the Niger crosses a vast flood plain that spans nearly 15,450 square miles (40,000 sq km) at the height of the floods and is known as the interior delta. Its flow then decreases: The river loses 25–50 percent of its water here, largely by evaporation.

PAGES 208-209

BRAZIL, MATO GROSSO DO SUL, MINAS GERAIS, AND SÃO PAULO
PARANÁ RIVER
20°02' S – 51° W

The Paraná River is formed by the confluence of the Paranaíba (left) and the Rio Grande (right). In this picture, the Paranaíba's waters are lighter in color because they carry less alluvium downstream of the Ilha Solteira Dam. The Paraná runs for 1,864 miles (3,000 km) in Brazil, skirting the borders of Paraguay, Argentina, and Uruguay, before flowing into its estuary and toward the Atlantic at Buenos Aires. The Paraná's hydrographic basin is the third largest in the world after those of the Amazon and the Mississippi.

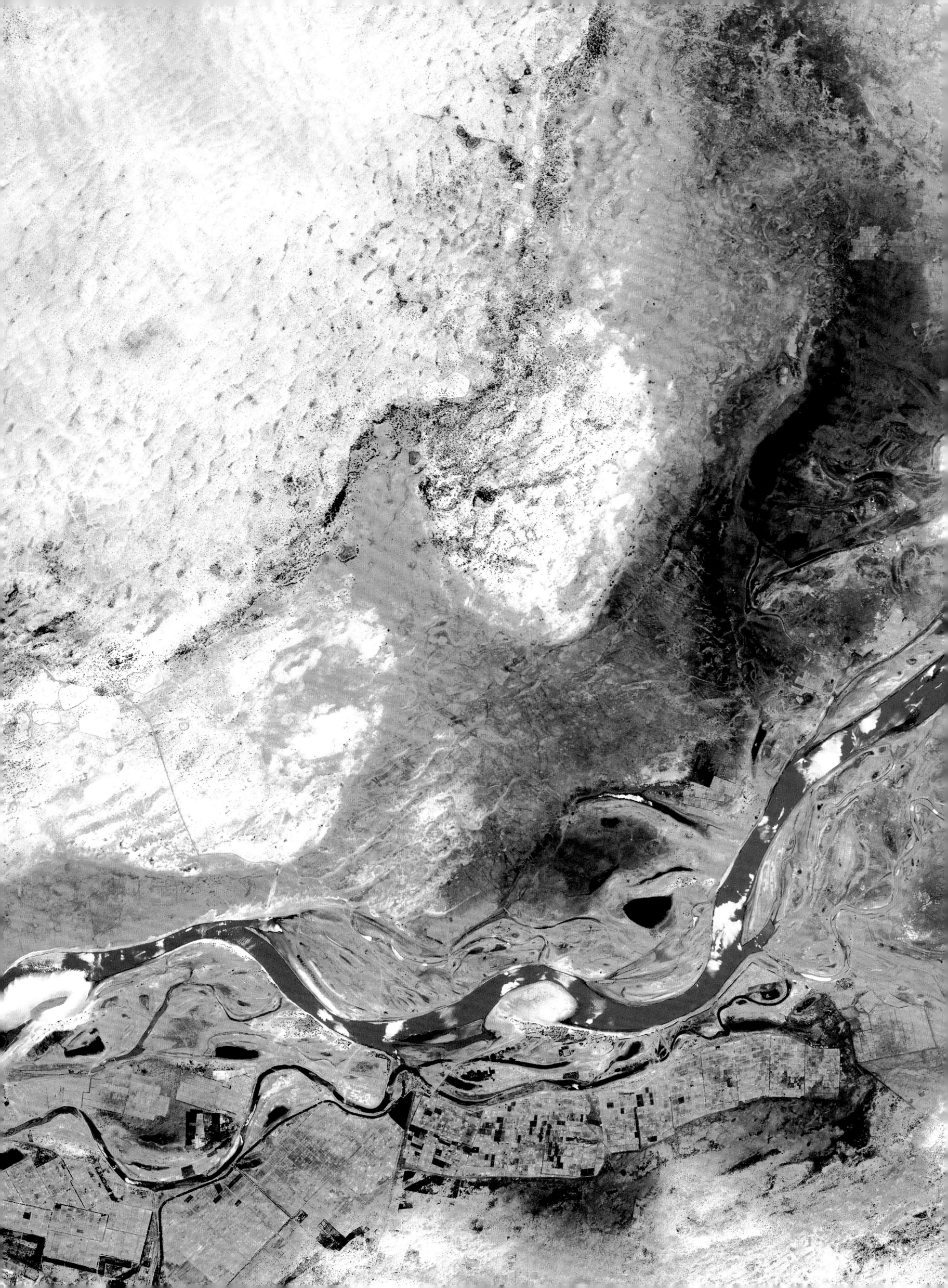

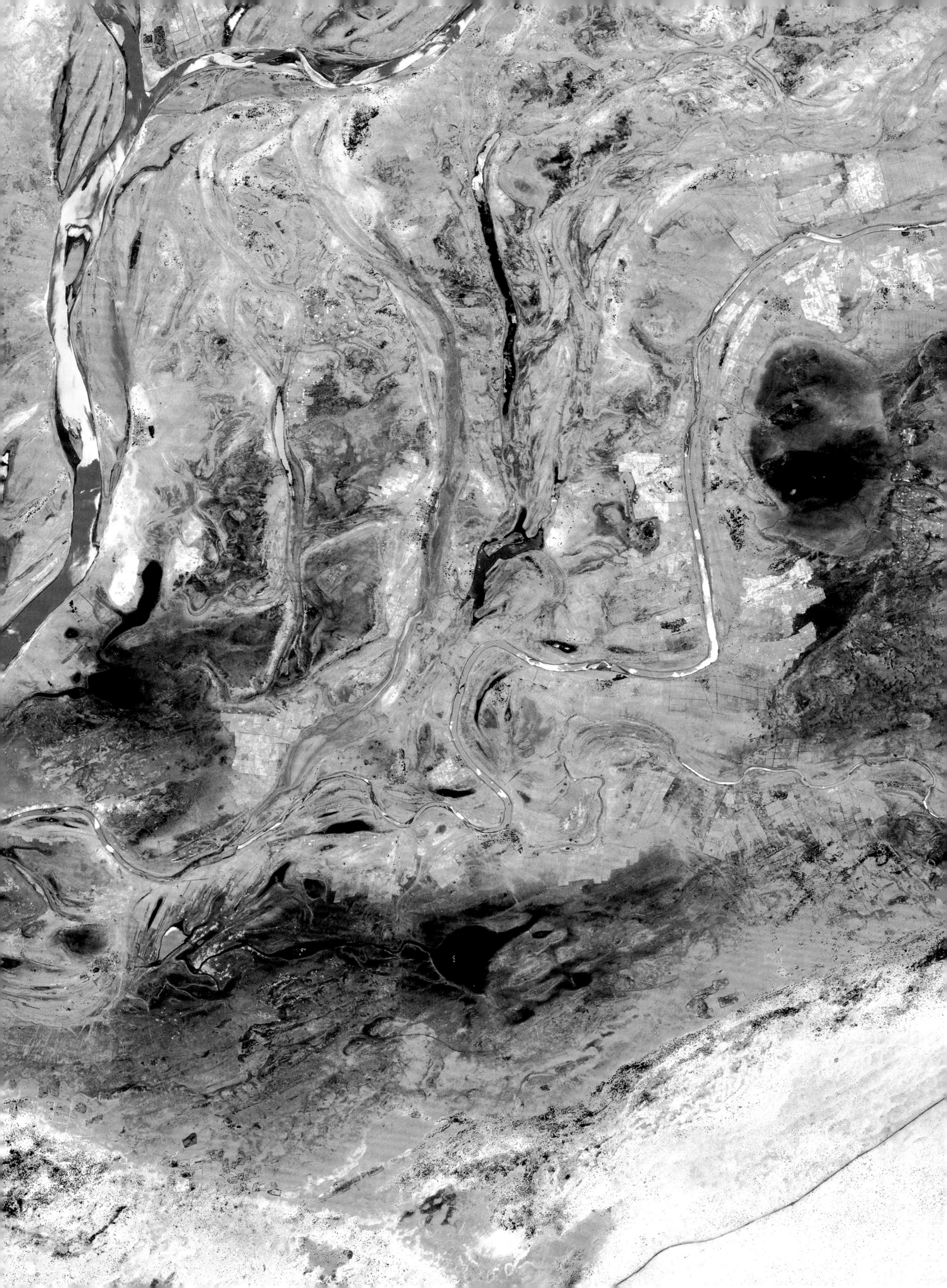

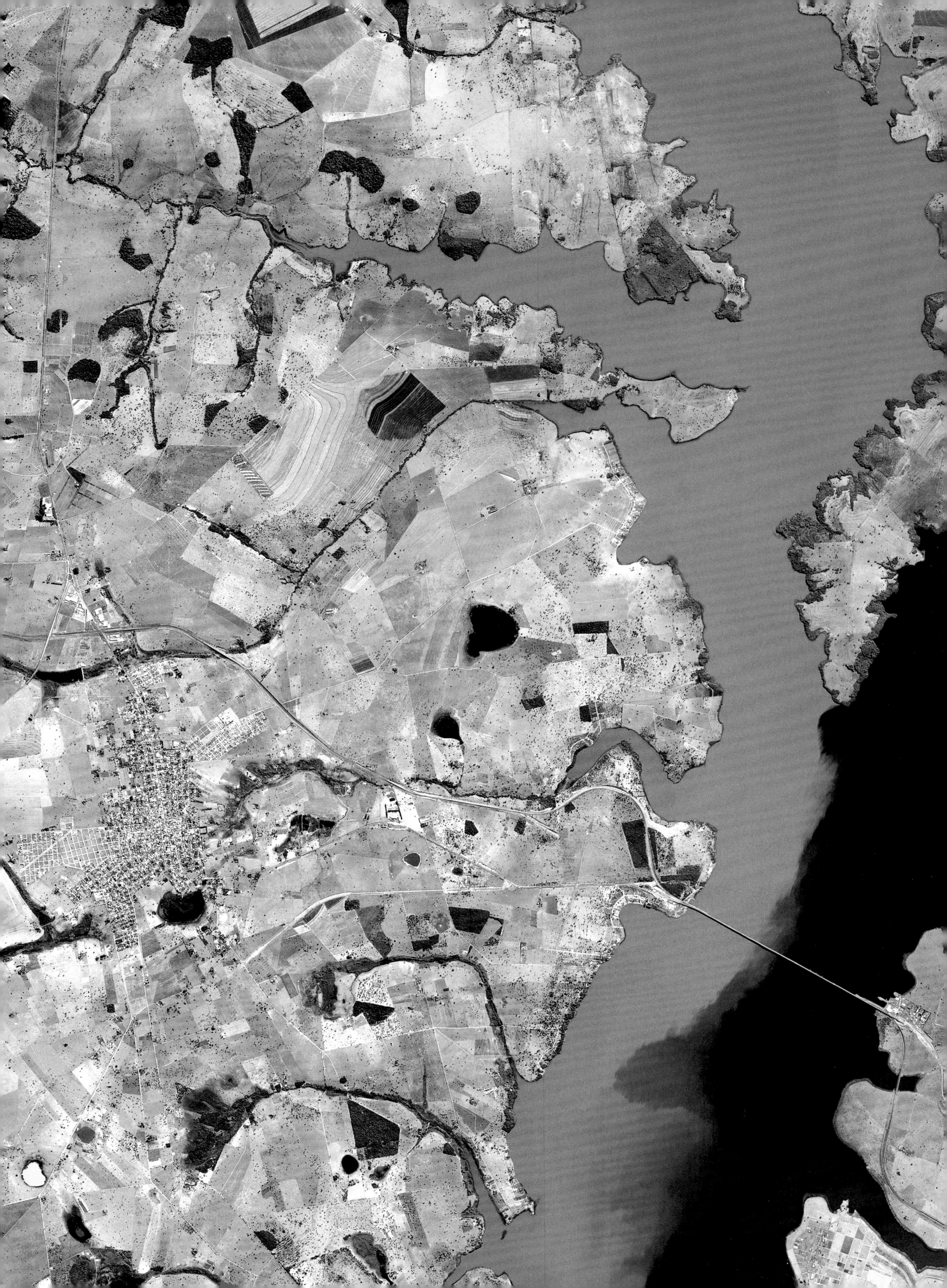

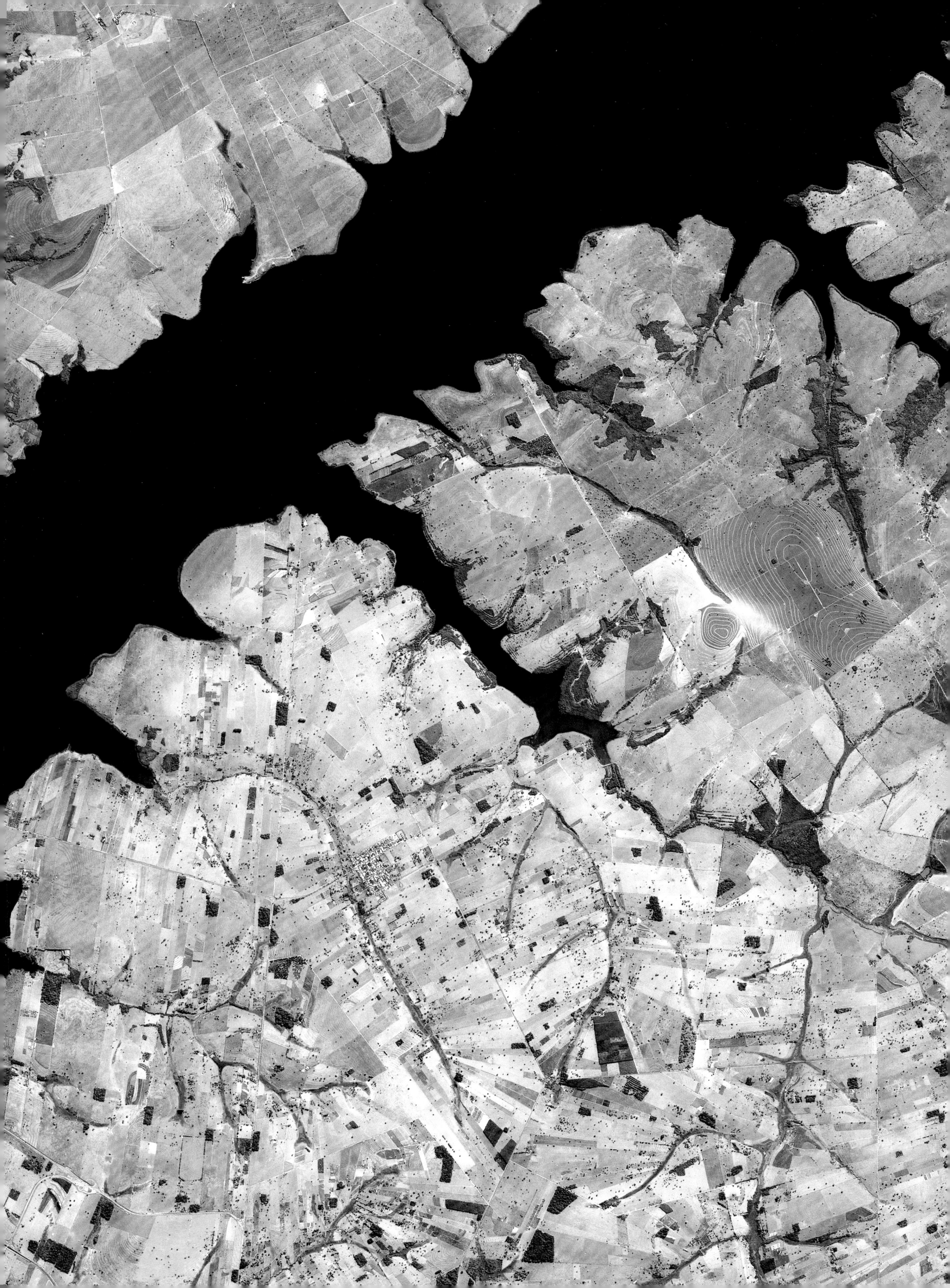

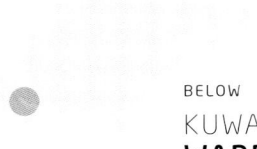

BELOW

KUWAIT
WARBAH ISLAND
30° N – 48°04' E

Warbah Island is at Kuwait's northern
extremity, on the border with Iraq, in the
Euphrates delta on the Persian Gulf. The
island is 9 miles (15 km) long and 3 miles
(5 km) wide, with a total surface area of
14 square miles (37 sq km). It is uninhabited.

OPPOSITE

SUDAN
THE WHITE NILE
14°33' N – 32°09' W

Upstream from Khartoum, in Sudan, the Nile is divided
into two tributaries: the White Nile and the Blue Nile.
The former includes the river's southern network, up
to Lake Victoria. The latter is the western tributary
and drains from Lake Tana, in Ethiopia; it accounts for
close to 90 percent of the river's flow.

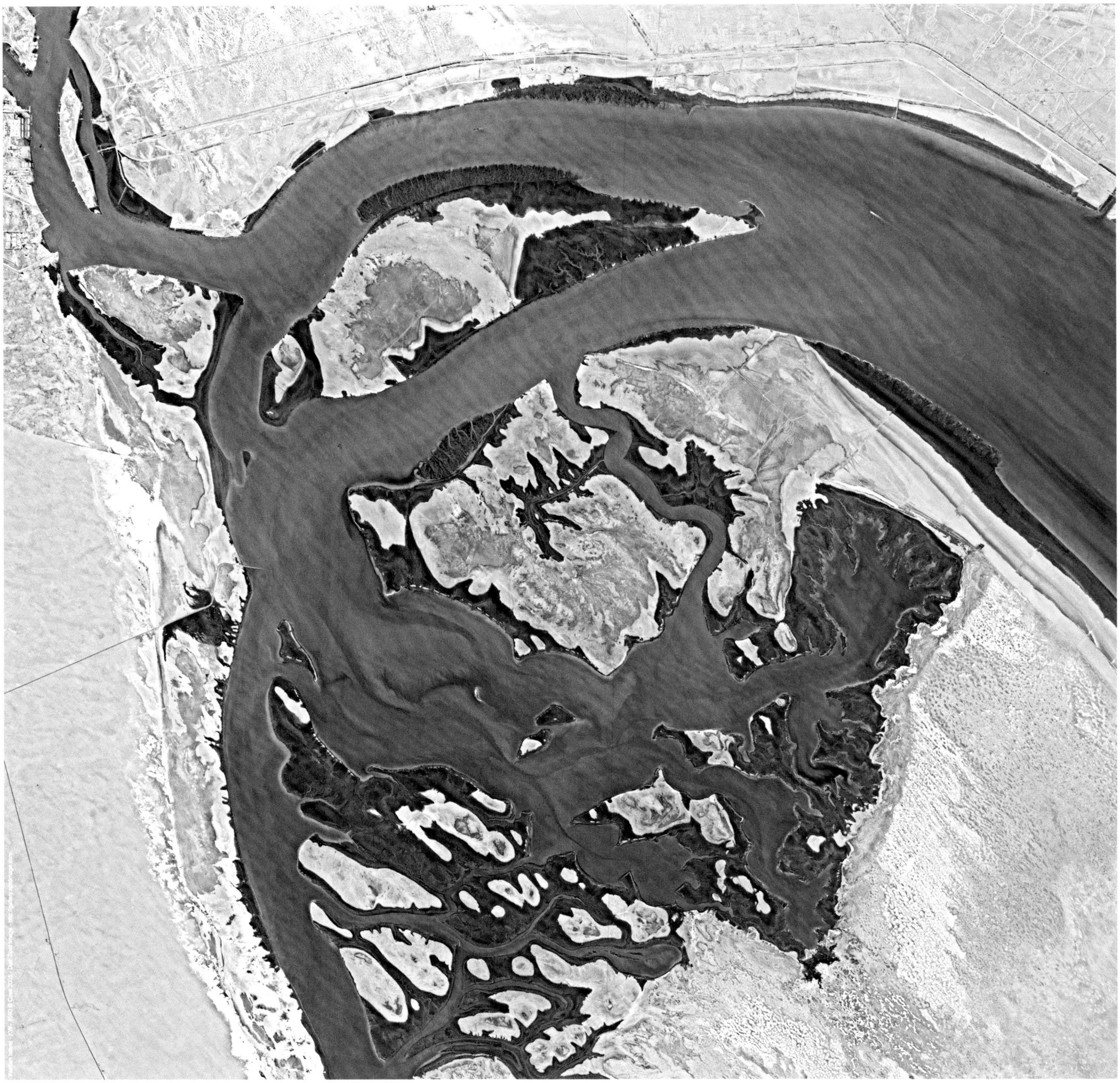

PAGES 212-213

GUINEA-BISSAU
GEBA RIVER
11°38' N – 15°38' W

The Geba River is a major communication channel in Guinea-Bissau. Running only for 341 miles (550 km), this coastal river is navigable over most of its course. It flows into the Atlantic Ocean through a vast estuary, whose right bank is the site of Bissau, the nation's capital.

PAGES 214-215

CHINA, INNER MONGOLIA
ORDOS DESERT
39° N – 109° E

The Ordos Desert is an area of steppes and desert covering more than 34,750 square miles (90,000 sq km) in southern Inner Mongolia. With the exception of the Huang He (the Yellow River) valley, the region is arid and poor, cut off by steep hills in the north and crisscrossed by dunes in the south.

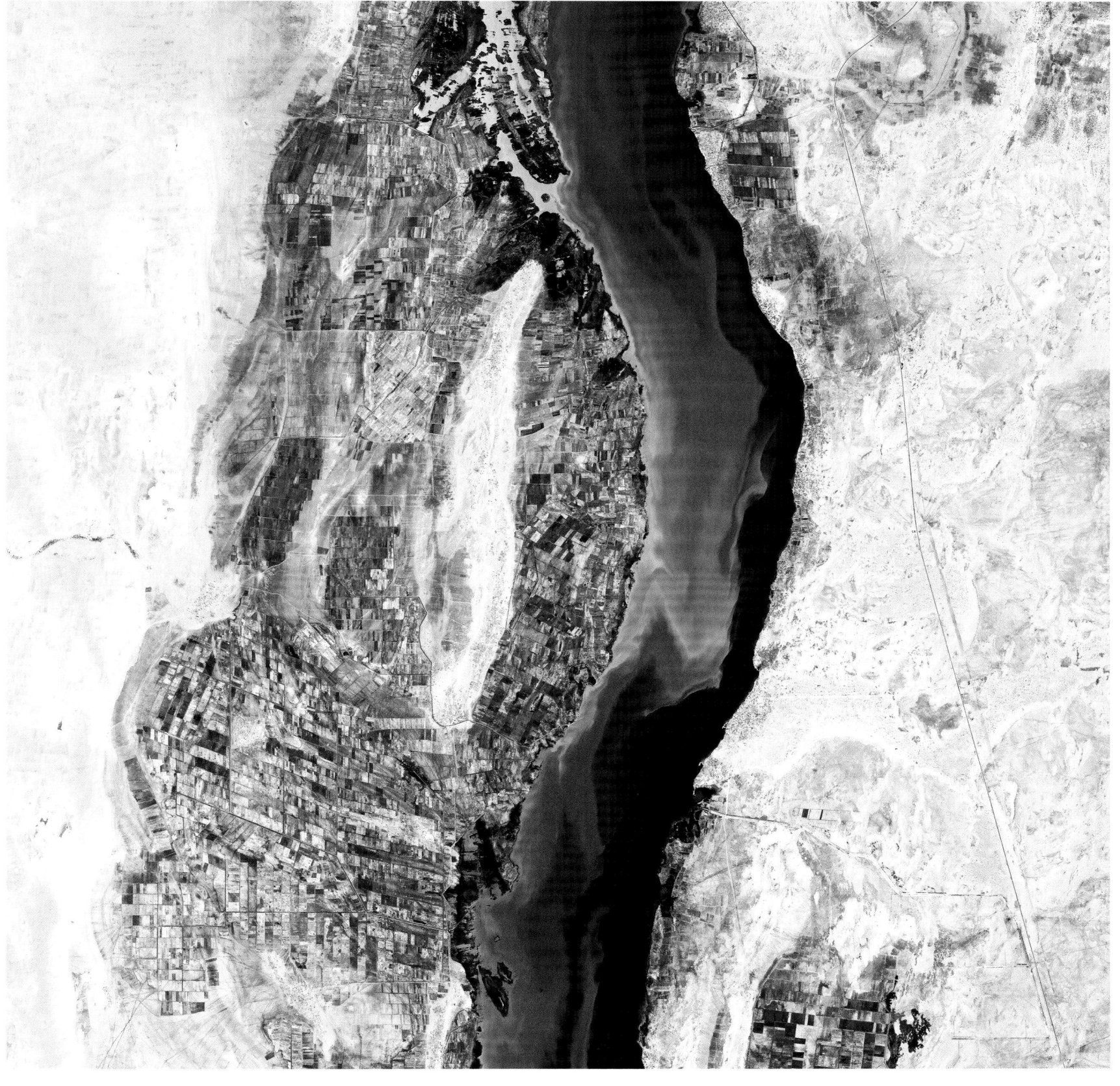

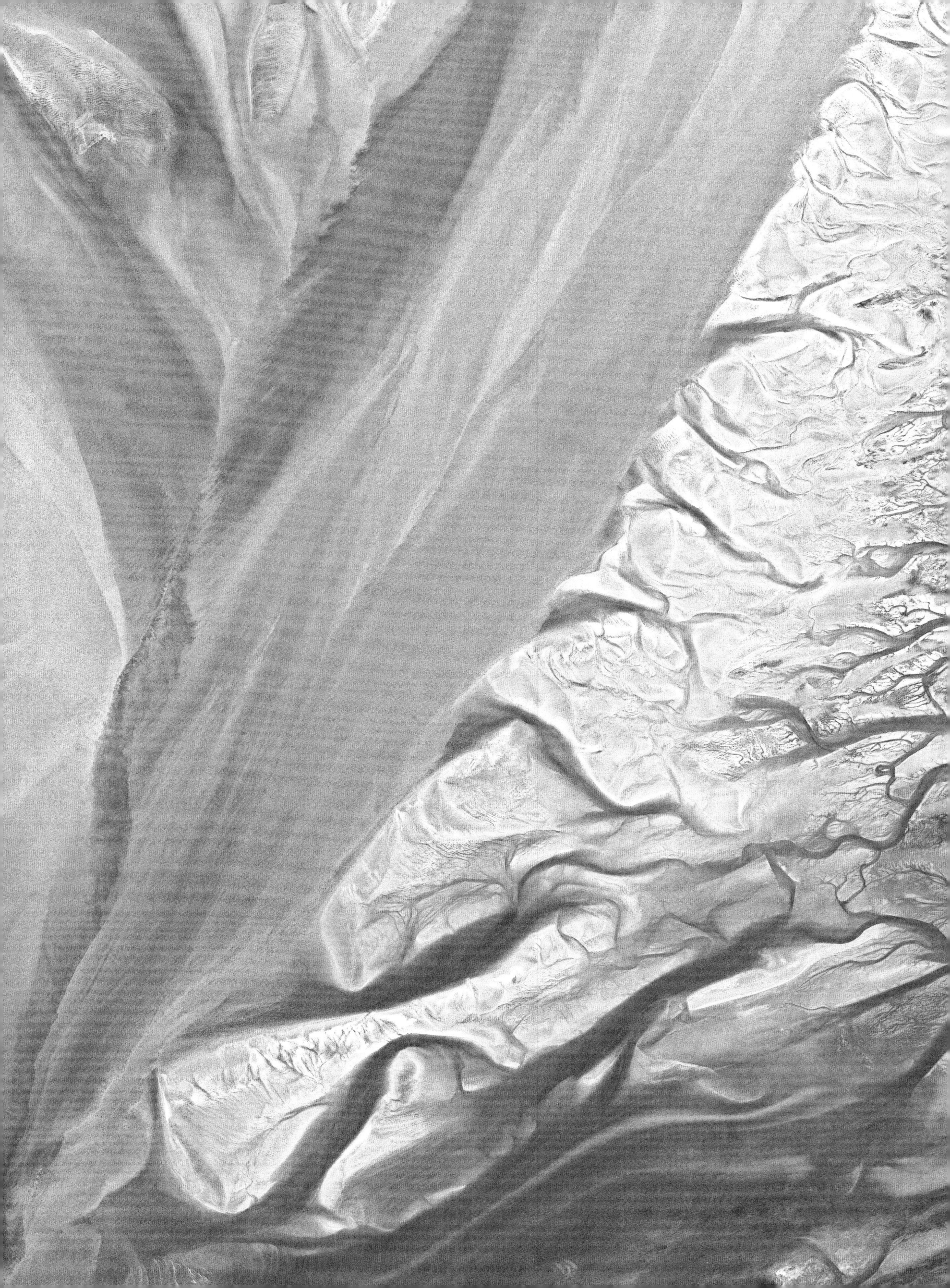

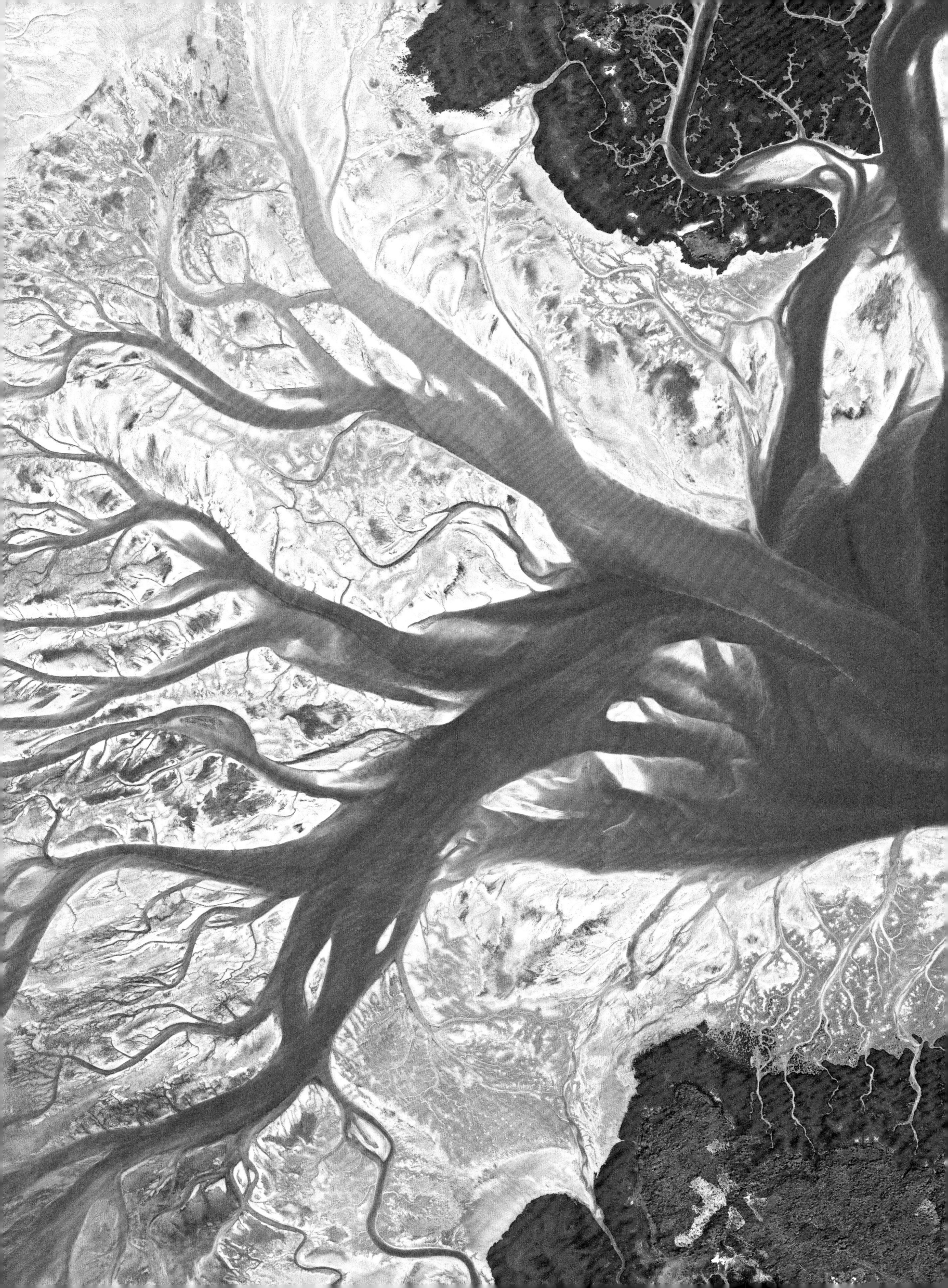

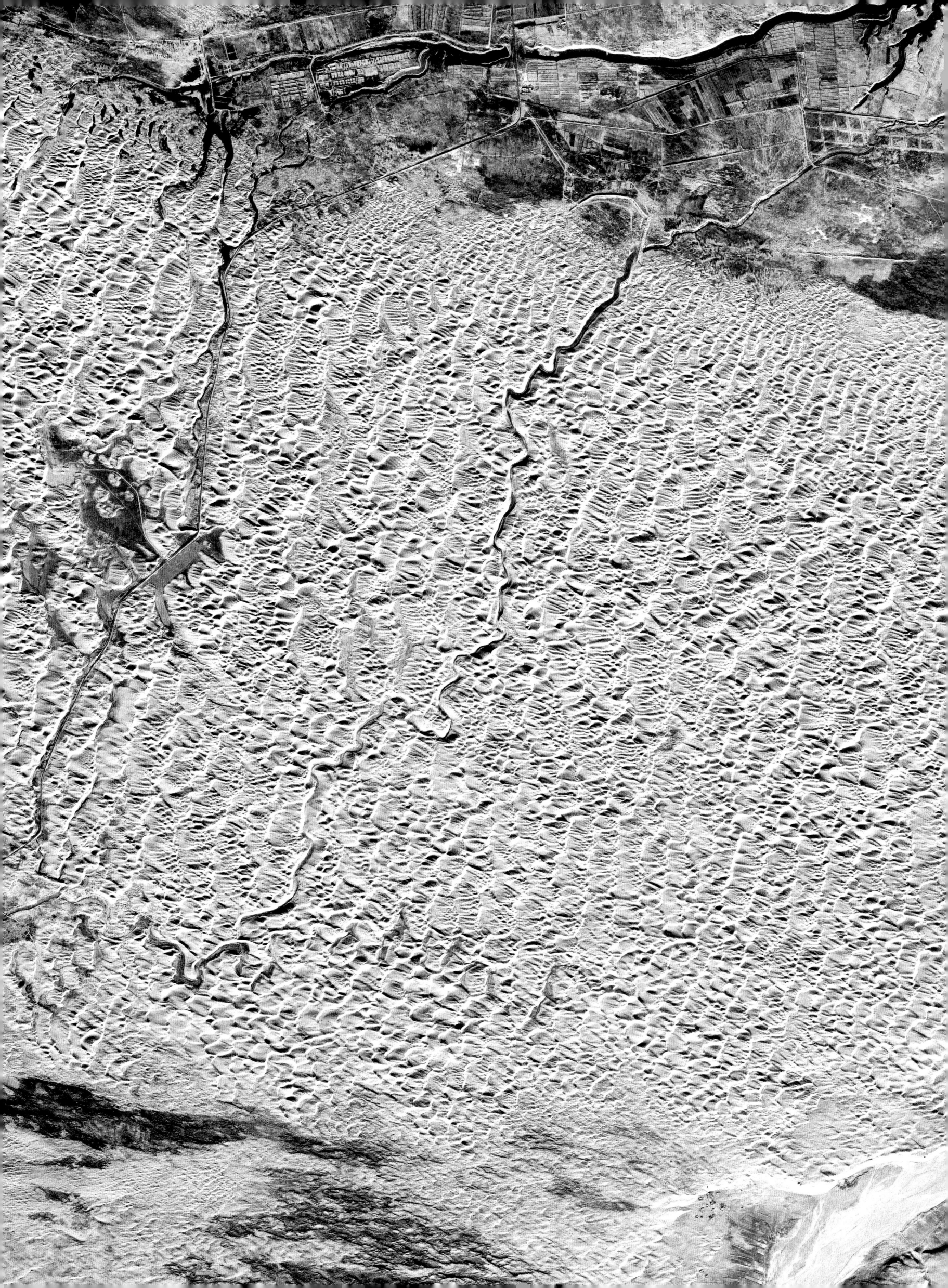

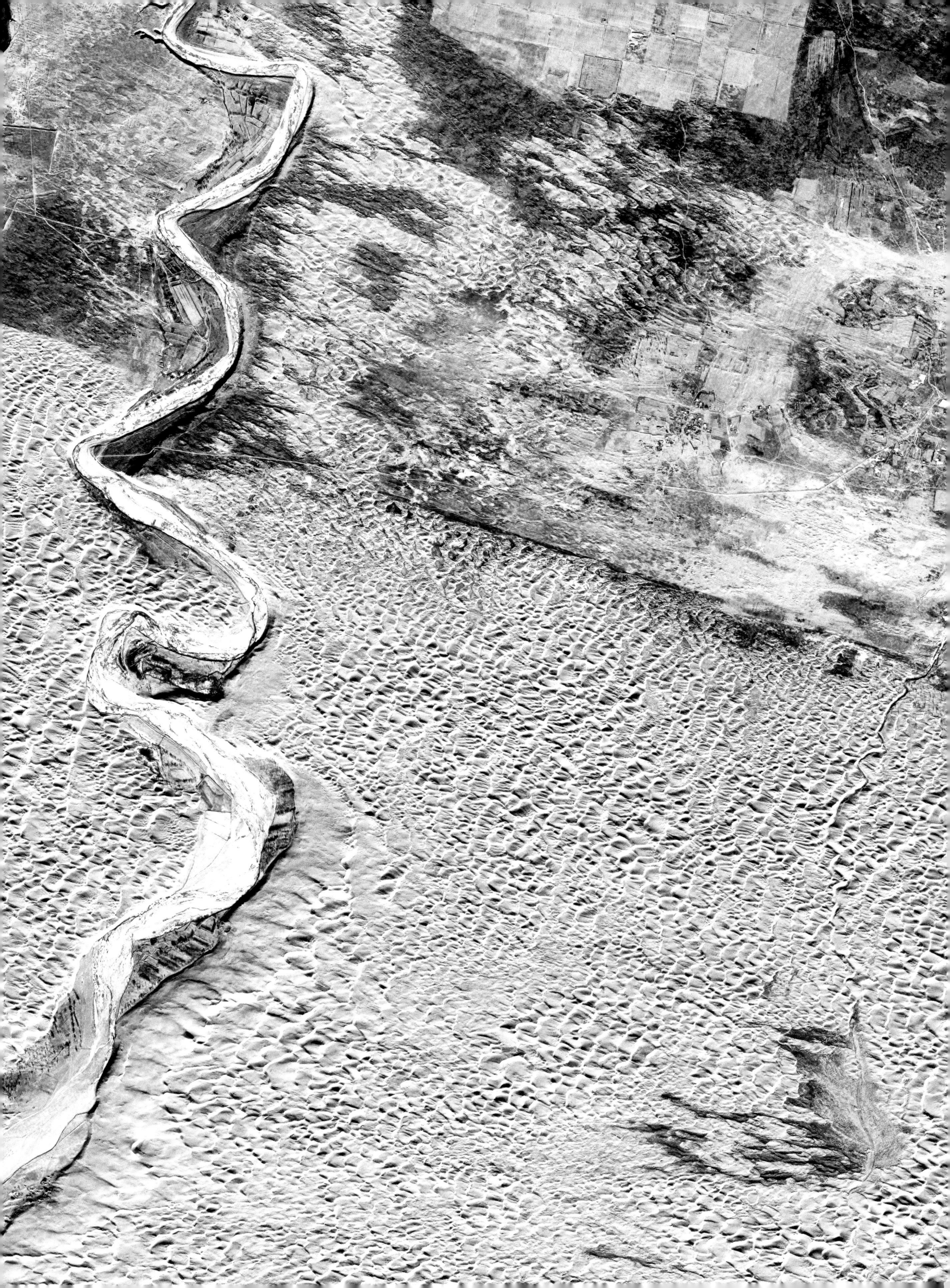

Desertification presents humanity with an unprecedented environmental, economic, and social challenge. But this complex phenomenon remains poorly understood.

___

FIGHTING DESERTIFICATION

___

"**Forests precede civilizations** and deserts follow them." This quote, attributed to Chateaubriand, provides an enlightening summary of the pressure exerted on ecosystems by humans; it points to our species' responsibility.

Deserts are associated with desolation, and this connotation, widely disseminated in our imaginations, colors numerous derived terms such as "desertification"—at least in its modern usage. Yet this is a misleading image. Deserts are ecosystems in their own right, characterized by a unique biodiversity.

"Desertification" commonly evokes dunes, bare and inhospitable ground, crushing heat, and terrible droughts starving and killing millions of African children. Yet that is not what the desert is—or not *only* that.

Desertification cannot be reduced to a natural consequence of the displacement of sand and the expansion of deserts. It refers to the degradation of farmable land in arid zones and the progressive destruction of associated ecosystems. This is a long process, sometimes nearly invisible, which occurs at every latitude and on every continent.

Desertification does not exclusively concern the Sahara, but also the Sahel with its shrubs, farmland, and millions of inhabitants, and the Gobi with its freezing nights. It affects not only Africa, but China, the United States, Australia, Portugal, Greece, and others as well. According to the United Nations, a total of 168 countries are at risk—in other words, nearly the whole planet.

AGRICULTURE

So-called "arid" or "dry" zones are far from isolated cases: They cover nearly half of landmasses and make up 44 percent of cultivated land, produce a significant portion of edible plants, and feed 50 percent of the cattle in the world. Their degradation leads to a drop in agricultural production, a devastating problem at a time when population growth and increased consumption make supplying people with food harder every day.

According to the United Nations' estimates, up to 5 percent of the value of global agricultural production is lost each year due to soil being made unfarmable by desertification and drought. In Africa, the figure is 4–12 percent of the value of agricultural production, but it reaches 24 percent in Guatemala and 20–30 percent in Uzbekistan.

In 1993, an initial study by the United Nations Environment Programme (UNEP) evaluated the annual losses due to desertification at $43 billion. Twenty years later, the figure has grown to $450 billion a year.

POPULATIONS

By causing an agricultural problem, desertification is a highly concrete threat to the poorest populations. For, behind words and figures, there are hunger and exile, suffering, and loss of life.

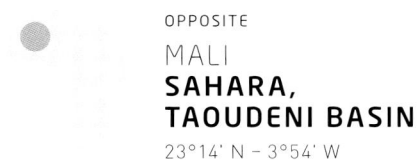

The Sahara is the largest hot desert in the world. Spanning approximately 3.3 million square miles (8.6 million sq km), it extends along the Tropic of Cancer—across the entire African continent, from the Atlantic Ocean to the Red Sea—for a total distance of approximately 2,300 miles (4,800 km). As for the Sahel, it is not a desert but a transition zone to more humid ecosystems. It borders the Sahara to the north and south.

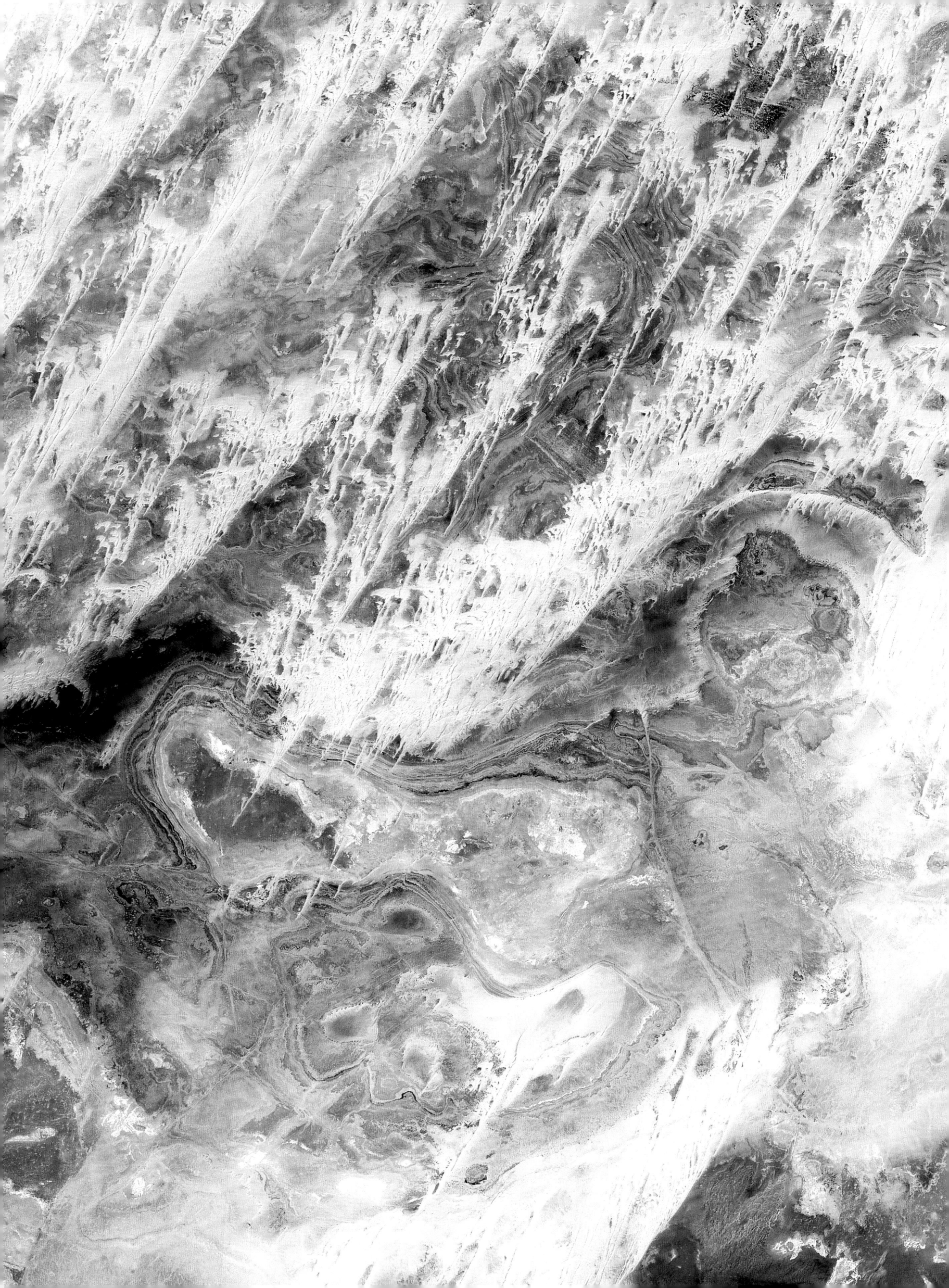

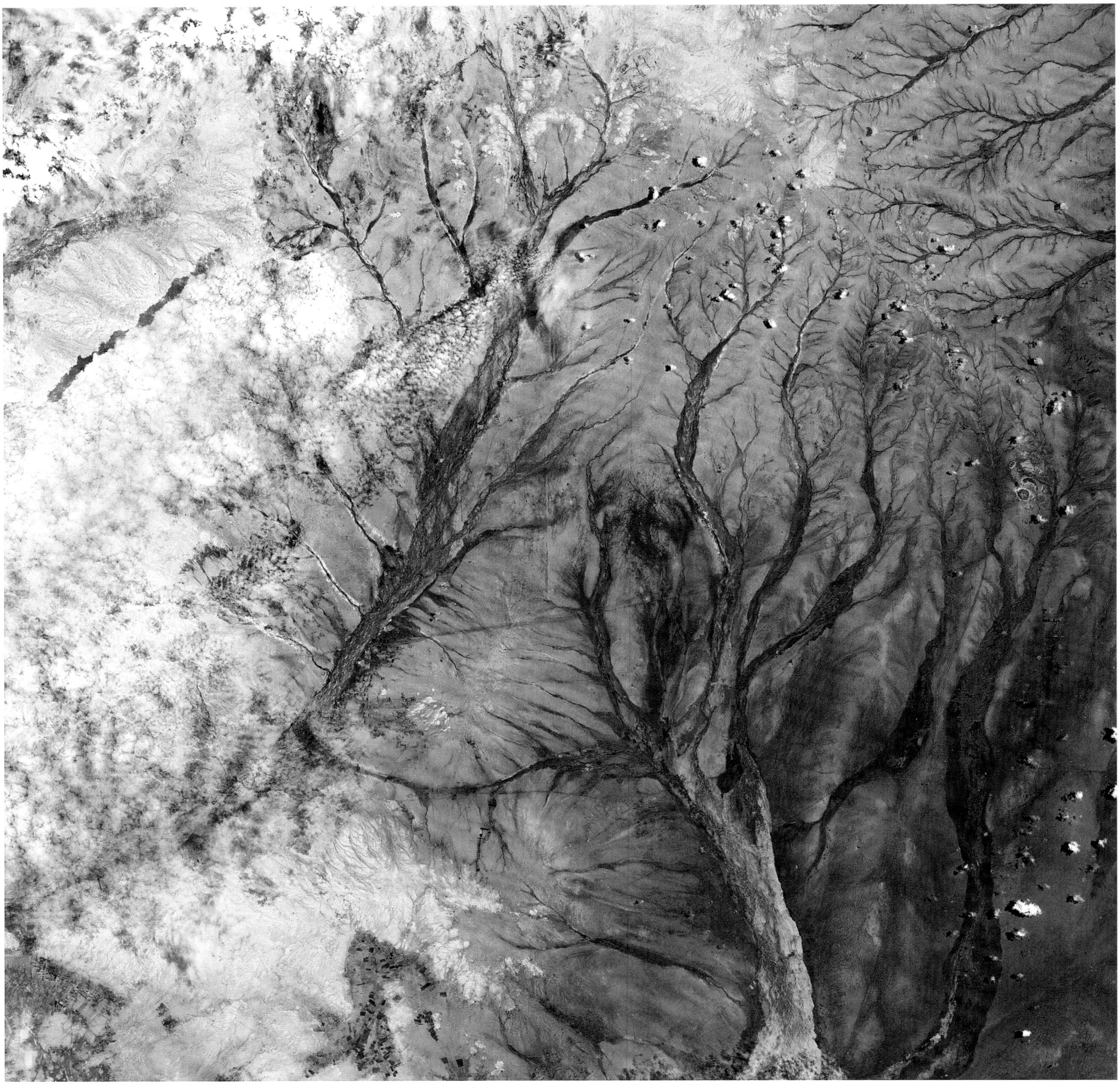

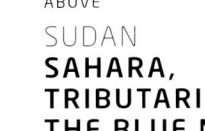

ABOVE

SUDAN
**SAHARA,
TRIBUTARIES OF
THE BLUE NILE**
15°30' N – 33°30' W

Tributaries of the Blue Nile (Bahr el-Azrak) flow into large valleys to the southeast of Khartoum. Contact between the steppe and the beige shades of the sand is partially masked by the mottling of low clouds and their shadows on the ground (in gray), seen particularly clearly in the northwest corner. Despite the proximity to Khartoum, irrigation here is little developed. In this respect, the Blue Nile's right bank is strikingly different from the vast irrigated triangle that separates the White and Blue Niles upstream of their confluence.

BELOW, LEFT
CHAD
SAHARA, TIBESTI
18°40' N – 19°06' E

In northern Chad, near the volcanic relief of the Tibesti Mountains, strips of dunes line up parallel in the direction of the dominant winds: the trade winds, the Harmattan. Slowed by the mountains, the sand-bearing wind forms lines of dunes clearly separated by corridors, only partially covering the rocky base, which is eroded in the same orientation.

BELOW, RIGHT
LIBYA
SAHARA, ERG UBARI
26°50' N – 13° E

This vast range of dunes stretches across the Fezzan in southwest Libya. To the north, large cordons line up in parallel in the direction of the trade winds. The different shades indicate that the dunes are fixed; they range up to 295 feet (90 m) in height. To the south, nearing the Msak Mustafit plateau, the dunes' alignment becomes less regular, before changing and coming to an end. In some places, large tabular surfaces are barely covered by small undulating dunes.

The majority of people directly affected by desertification live below the poverty line. And for poor farmers, a 10–20 percent drop in their crops tips a precarious balance into a disastrous spiral leading to debt, famine, and exodus: When the land becomes impossible to farm, farmers migrate to the cities, often ending up in shantytowns. This exodus, which can cause social tensions and political instability, has played a part in triggering several armed conflicts in Africa.

CAUSES

The causes of desertification are well known. It is not due to dryness so much as it is a combination of anthropogenic factors. Deforestation and overgrazing deprive soil of the plant cover that preserves it from erosion. Without the protection of branches and the reinforcement of roots, the soil dries up and is carried away by wind and rainwater. Plowing dries out and packs the earth. Excessive use of chemical inputs weakens the living organisms that make land fertile. Poorly managed irrigation systems increase soil salinity, which makes the land infertile. All these factors are aggravated by climate change.

SATELLITES

Until relatively recently, it was difficult to measure desertification, particularly if it extended over gigantic surfaces. In China, for example, 1 million square miles (2.6 million sq km), or 27 percent of the country's land area, are affected. Satellites are obviously precious tools to observe the phenomenon on such a vast scale, particularly since the regions concerned are often difficult to access.

This technology has additional advantages. To perceive long-term trends, seasonal and especially interannual

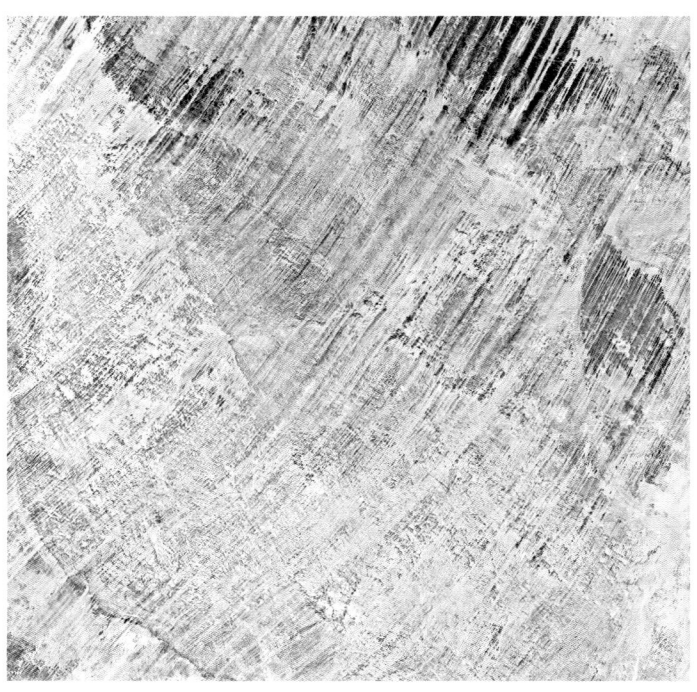

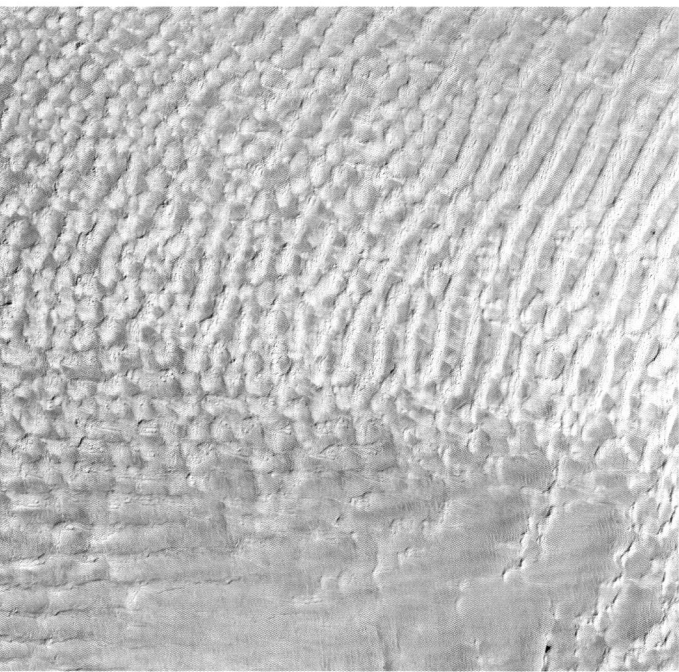

The GREEN WALL OF CHINA calls for the massive planting of trees to slow winds and put roots down in the soil, thus **STOPPING THE EXPANSION OF THE DESERT**—which in the 1990s was advancing at more than **1,100 SQUARE MILES (3,000 SQ KM) PER YEAR.**

variations must be disregarded. These variations are particularly strong in desert zones. In the Sahel, for example, precipitation can double from a very dry year to a humid year. It is therefore necessary to obtain information over long periods of time. The Landsat and SPOT satellites, which have been providing data for more than twenty years, are invaluable in this respect.

Finally, easily measured common parameters must be identified, which is a challenge given that vegetation can vary considerably from one end of the Gobi's 501,932 square miles (1.3 million sq km) to the other. Criteria, such as plant cover, photosynthesis intensity, and rate of soil humidity, are used to estimate the level of desertification and its evolution with increasing accuracy.

SOLUTIONS

Once the phenomenon is identified, it is possible to take action. For nothing is irreversible. The most important initiative, launched in China in 1978 and known as the Green Wall of China, calls for the massive planting of trees to slow winds and put roots down in the soil, thus stopping the expansion of the desert—which in the 1990s was advancing at more than 1,100 square miles (3,000 sq km) per year. In China, 400 million people's means of subsistence were threatened or affected by desertification, soil degradation, and the expansion of the Gobi, Taklamakan, and other deserts. This desertification caused by overgrazing and inappropriate agricultural practices could be felt all the way

to Beijing and sometimes throughout the entire subregion, notably in the form of increasingly severe sandstorms.

Planned to be completed in 2050, the Chinese initiative calls for close to 90 million acres (36 million hectares) of forest in a 2,800-mile (4,500-km) strip. While some doubts have been expressed regarding the project's actual effectiveness, the situation does seem to be improving since the middle of the 2000s: The rate of desertification has slowed to less than 800 square miles (2,000 sq km) per year, while, according to the State Forestry Administration, more than 800 square miles (2,000 sq km) of land have been rehabilitated annually since 2004. The trend of desertification could therefore be reversed.

However, the Green Wall of China remains very fragile. The trees planted are often species that are not native to the region. Adapting poorly, they have a low survival rate, and many were destroyed by the winter storms of 2009.

Another major Chinese project is dedicated to combating the desertification of the Loess Plateau in the north of the country. Launched in 1994, this program was even more successful than anticipated. In this case, the government encouraged a change in agricultural practices by limiting the land used for grazing and offering multiple incentives for agroecology, as well as—but not limited to—the planting of trees. An area the size of Belgium, whose bare land had become sterile, was brought back to life. Thanks to the restoration of ecosystems, agriculture is now prospering on the plateau and local populations have considerably improved their standard of living.

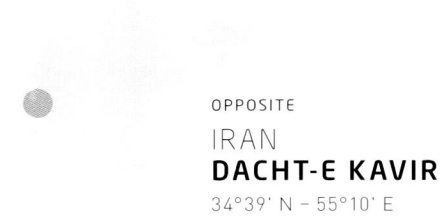

OPPOSITE

IRAN
DACHT-E KAVIR
34°39' N – 55°10' E

The Dacht-e Kavir is the larger of the two deserts occupying most of Iran's central plateau. This sparsely populated salt desert in the country's center north is close to 500 miles (800 km) long and 200 miles (300 km) wide. Located at the site of a former inland sea, this arid region covered in salts is known for its kavir, "salt swamps," which sometimes form areas of quicksand.

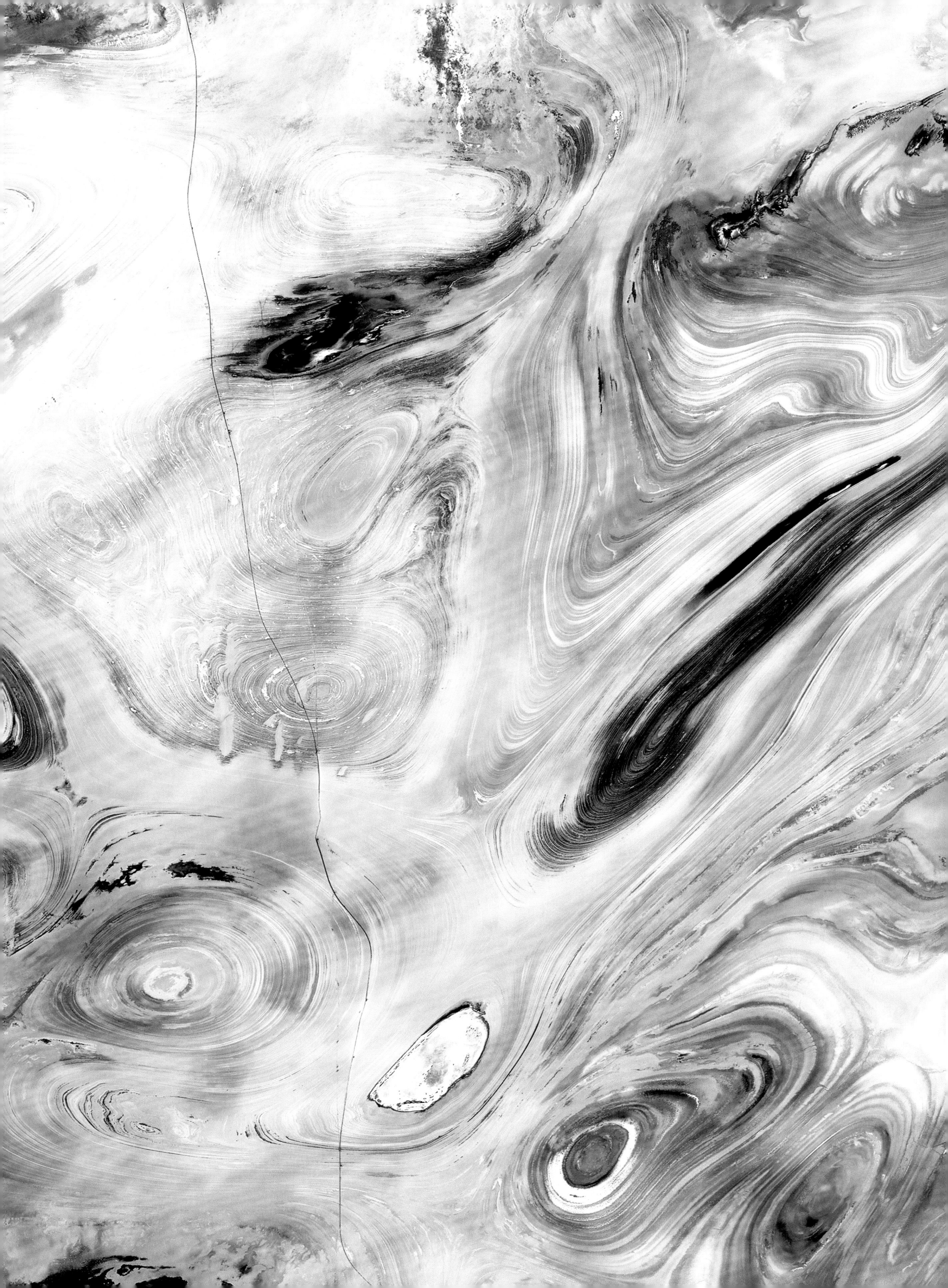

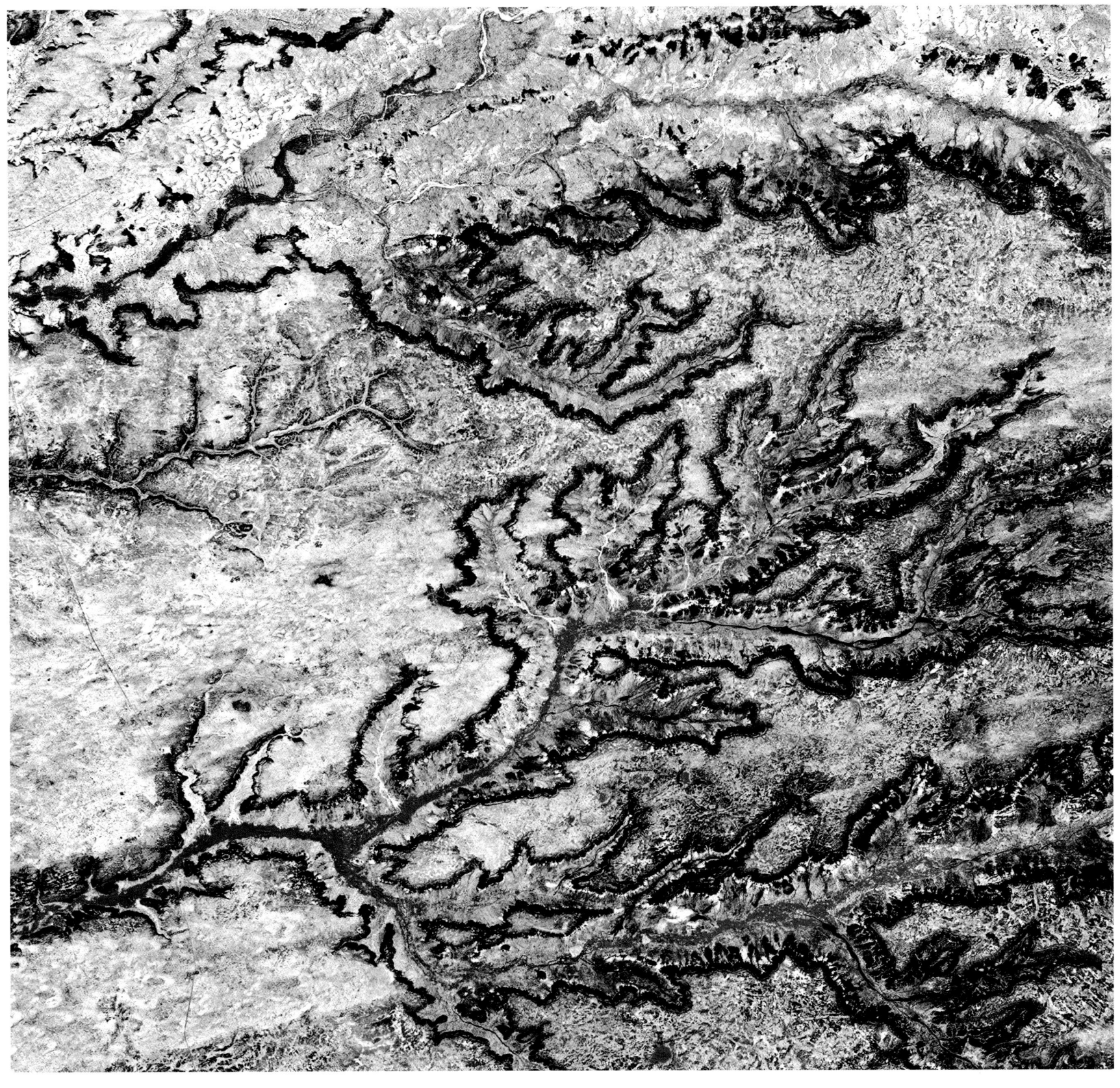

ABOVE

NIGER
ADER

14°51' N – 5°21' E

In the south of Niger, near the city of Tahoua in the Sahel, the hydrographic network has carved into plateaus topped by a ferralitic crust (rich in iron oxides). The picture shows a cross-section of the crust (green) and the underlying rock (beige) on each side of the valley. The width and homogeneity of the dark lines provide data about the thickness and extent of uncultivated land.

OPPOSITE

LIBYA
SAHARA, KUFRA

24°15' N – 23°20' E

In the Kufra oasis in the southeast, Libya launched an extensive project aimed at developing agriculture in desert environments. Fossil waters are pumped from the aquiferous system beneath the scorching dunes of the Rebiana Sand Sea. The water is then directed toward crops irrigated by a honeycomb network of pivots with a 0.6-mile (1-km) diameter. Since 1970, satellites have observed the spectacular development of irrigated land in the heart of the Sahara.

Thanks to satellite imagery, indications of
REGREENING HAVE RECENTLY BEEN OBSERVED
NEARLY EVERYWHERE in the Sahelian strip.

SUCCESS STORIES

These two iniatives have been imitated throughout the world. The Green Wall of China inspired the pan-African project for another Great Green Wall that will eventually span about 4,660 miles (7,500 km) through eleven countries, from Senegal to Djibouti. The program is being implemented in a far more delocalized manner and moving along modestly. The model for regreening China's Loess Plateau has been applied in Rwanda with tremendous success.

Today, fighting desertification rests on a wide range of initiatives. There are inexpensive agricultural techniques suited to the poorest populations that are able to preserve biodiversity, optimize water usage, and improve the yield from crops without damaging soil. These include no-till farming, letting land lie fallow, and agroforestry.

Improving agricultural practices by listening to and involving local populations leads to other advances, such as returning to the planting of crops native to particular environments, which are too often abandoned to satisfy international markets, or, particularly, the promotion of the emancipation of women, who provide 70 percent of the food production in Africa but rarely hold any rights to the land they work.

And it is working. Thanks to satellite imagery, indications of regreening have recently been observed nearly everywhere in the Sahelian strip.

Satellite studies of desertification are therefore doubly precious. They allow us not only to show the extent of the problem and to raise decision makers' awareness of the issue, but also to attest to the effectiveness of the measures taken on a local and governmental scale. Seeing successful results is a major boon to encouraging future initiatives.

Desertification is not unavoidable, and stopping or neutralizing it is not beyond our reach. In 2012, the United Nations set an objective of zero land degradation by 2030. It is not completely illusory.

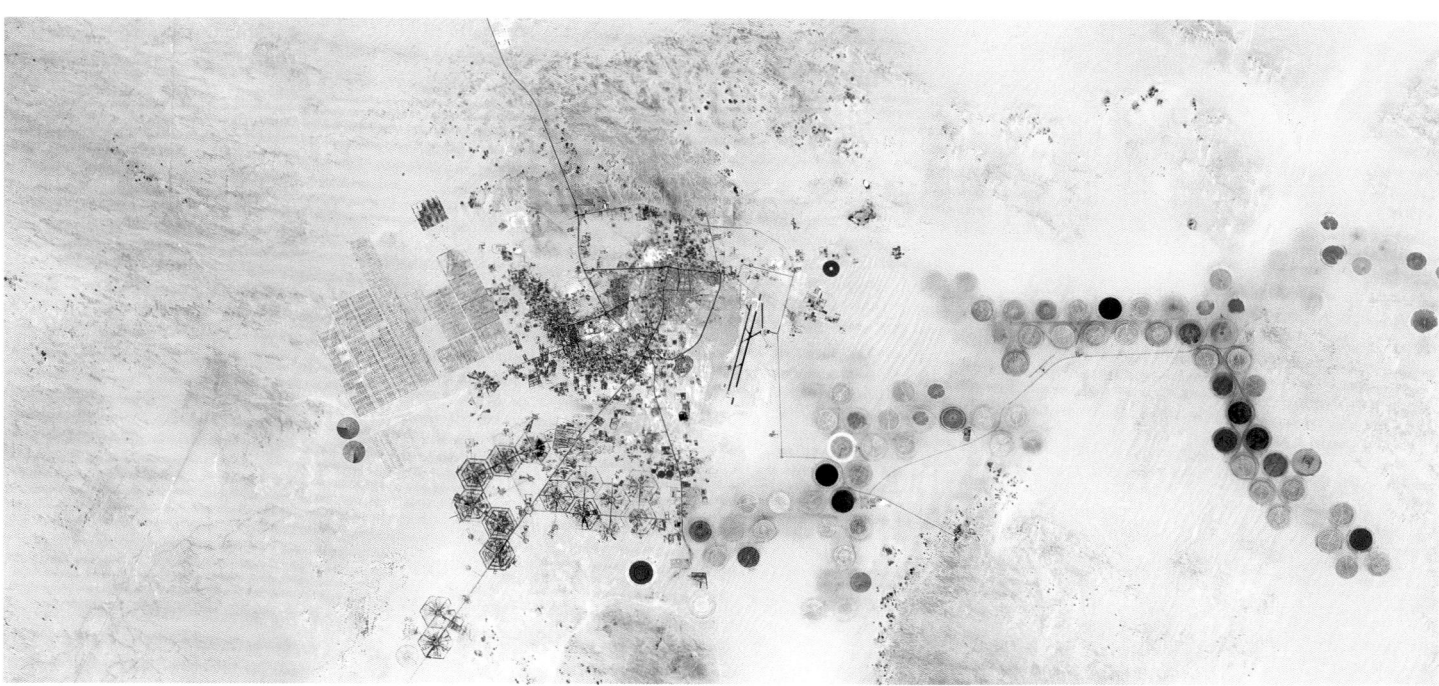

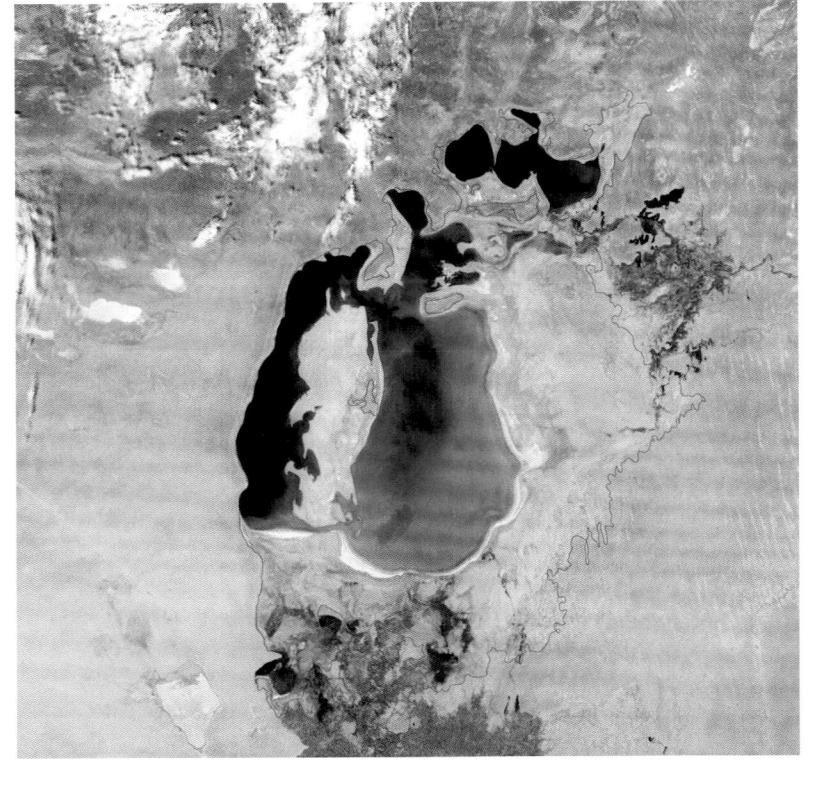
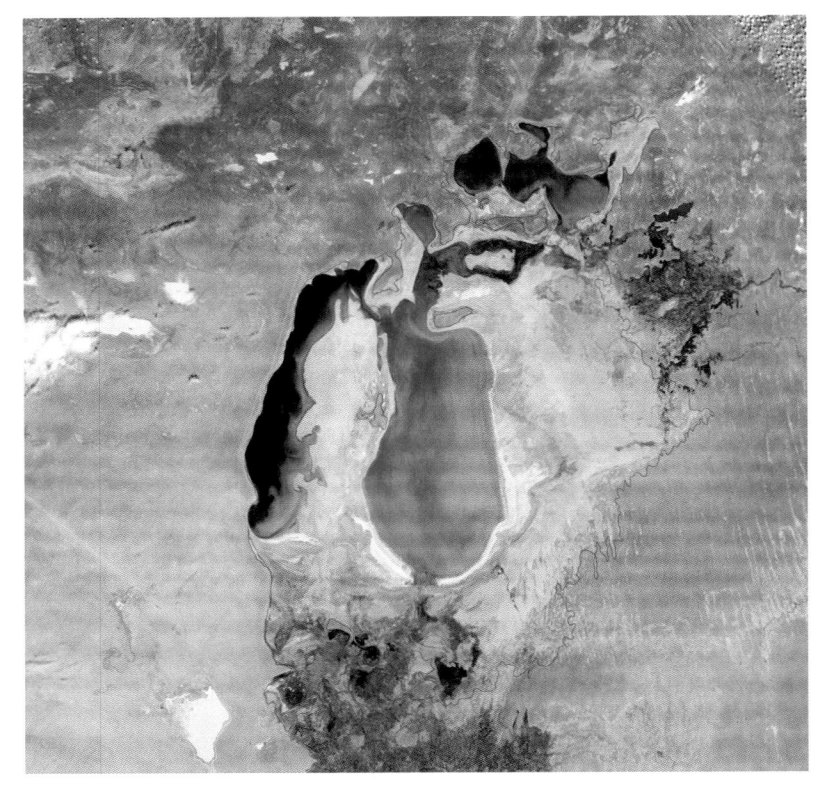
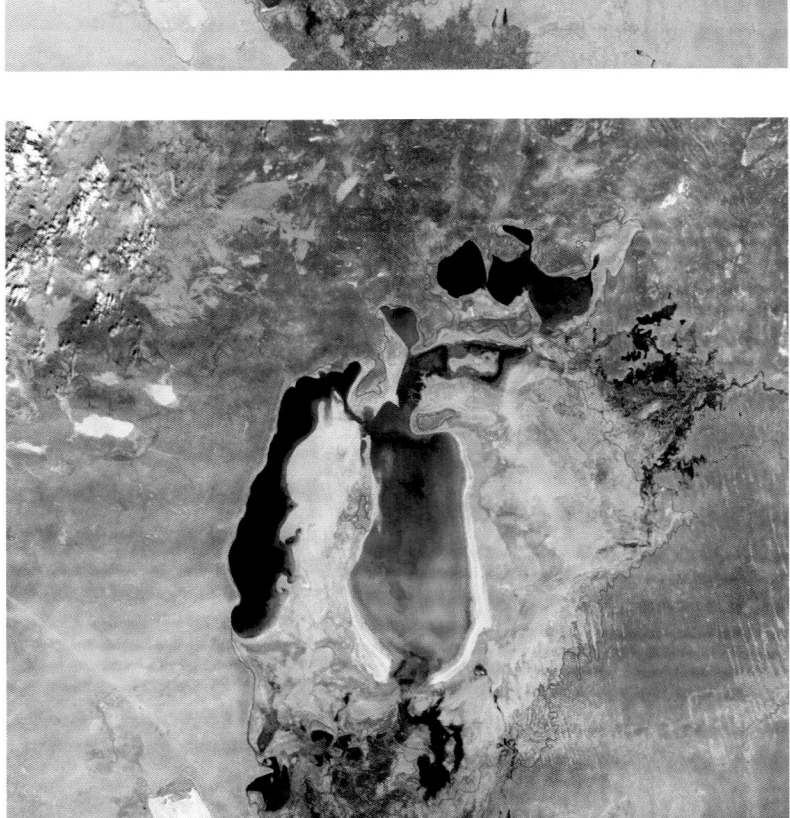
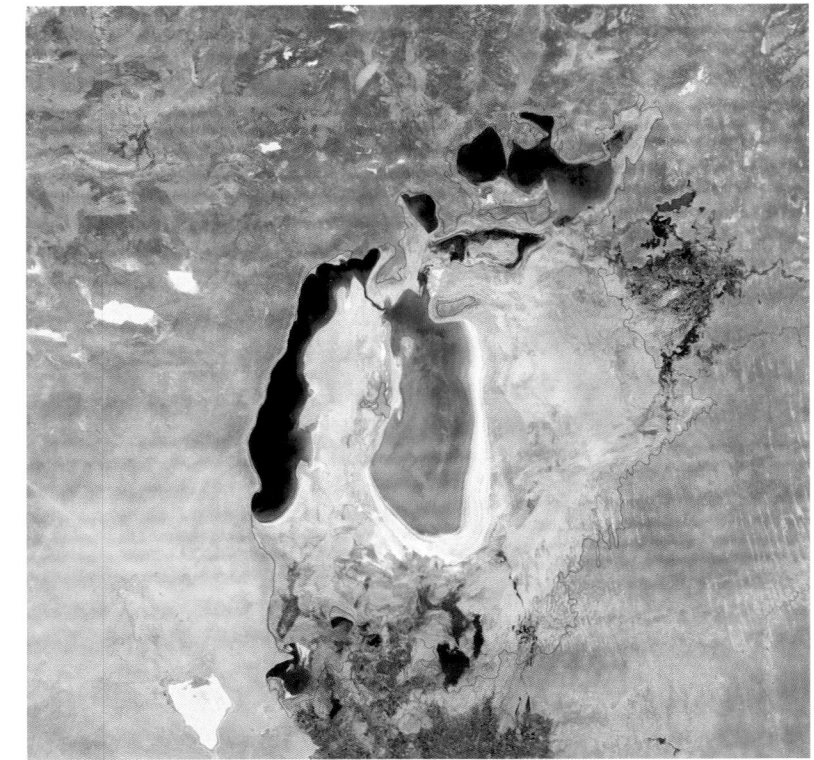
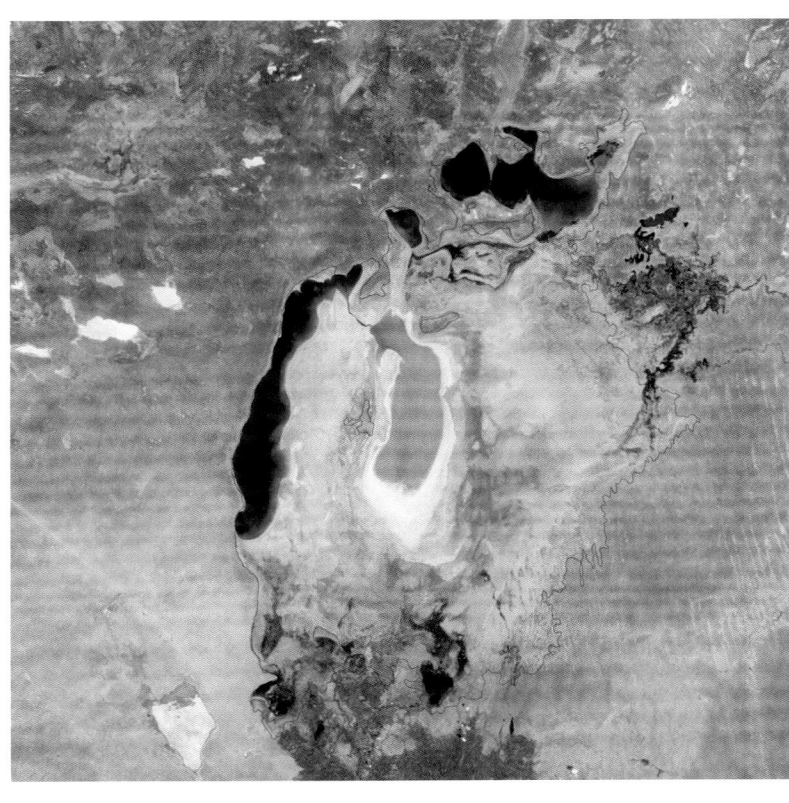
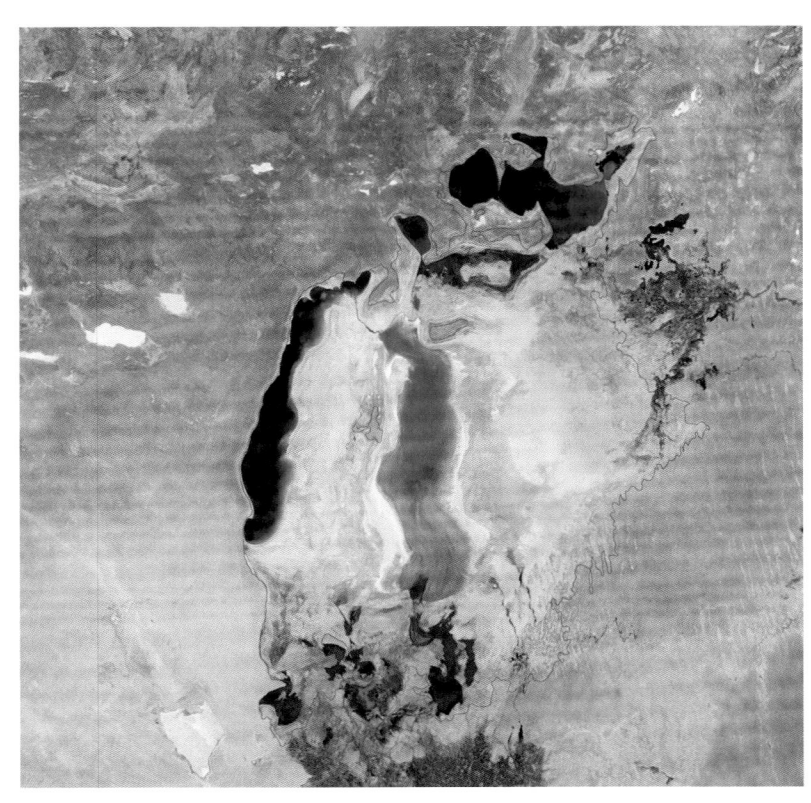

2000 - 2002

2004 - 2006

2008 - 2010

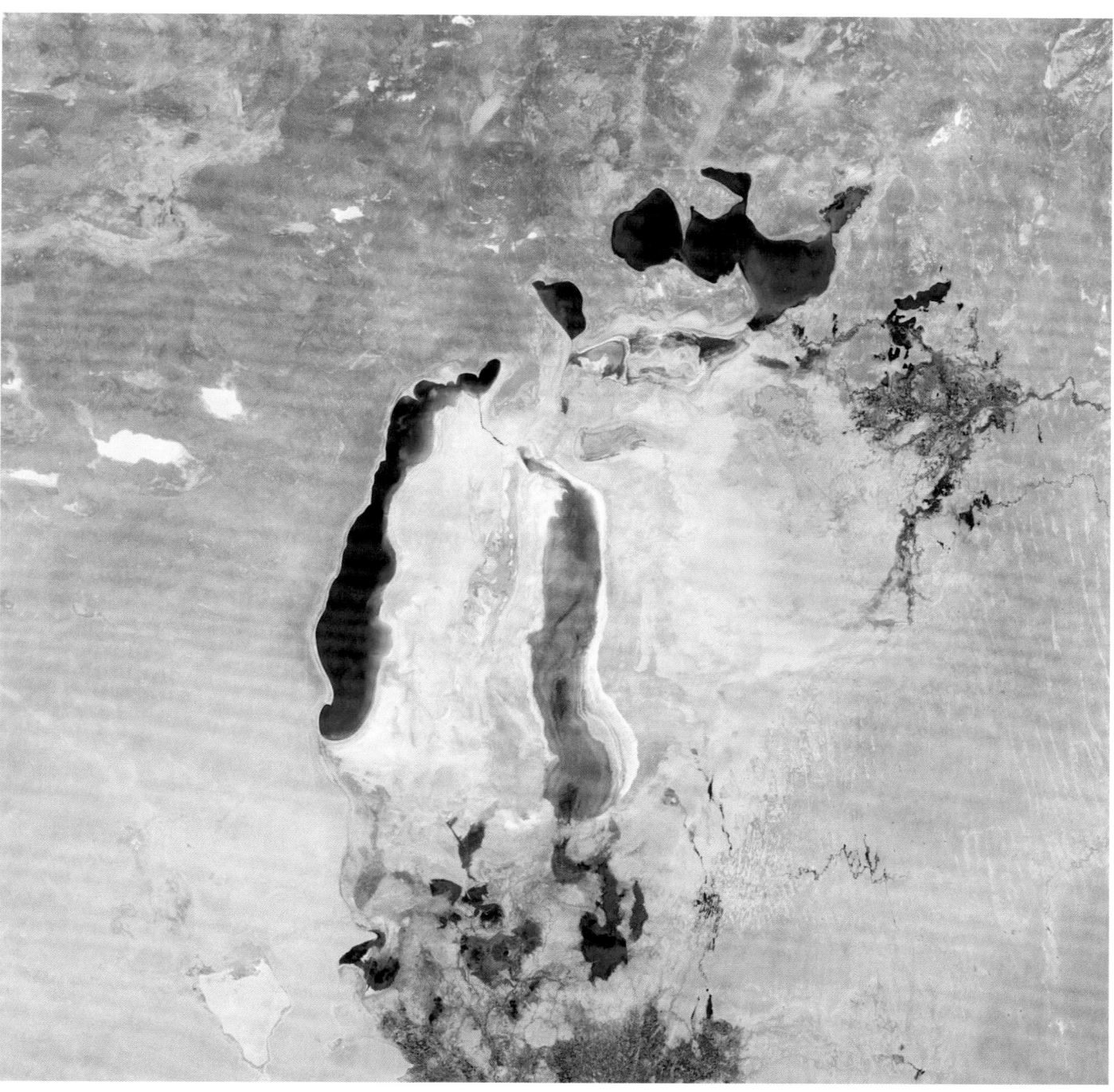

2012

KAZAKHSTAN / UZBEKISTAN
ARAL SEA
45°30' N – 60°30' E

In Central Asia, the massive use of irrigation to extend cultivated areas— primarily for cotton—relied on the diversion of the Amu Darya and Syr Darya rivers, which dried out the Aral Sea; it has lost 60 percent of its surface area and 80 percent of its water volume since the 1960s. Not only has fishing collapsed, but the entire ecosystem has been wiped out. Salt and pesticides accumulate on the dried-up seafloor and are dispersed by the wind, contaminating drinking water and having a significant impact on locals' sanitary conditions.

OPPOSITE
CHINA
GOBI DESERT
44° N – 110° E

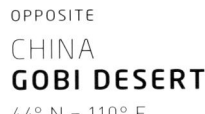

The fight against desertification is a global battle coordinated by the United Nations.

Talking with Luc Gnacadja

Toward a Global Atlas of Desertification

What are the principal aspects of land degradation and desertification that we need to consider?

Desertification and drought are a threat to humanity's long-term security. Each year, they are responsible for the loss of about 30 million acres (12 million hectares) of land, which becomes sterile, depriving us of the possibility of growing twenty million tons of grain. As for land degradation, it has an annual cost of up to 5 percent of the international community's gross domestic agricultural product.

What role does satellite observation play in the fight against desertification? Can you give us some examples?

Mapping the state of land degradation and its progression remains difficult on an international scale because it is a phenomenon that manifests itself on a local level through the interaction of human and biophysical factors. Satellite imaging and particularly the use of remote sensing is a rapid and efficient means to identify changes relating to land use and productivity, since it consists in measuring variables that indicate if land degradation or restoration is occurring and where.

Do you have any specific examples of satellite imaging allowing us to combat desertification?

In June 2012, several international leaders requested global monitoring of the degradation and restoration of land in arid, semi-arid, and sub-humid dry land. Given the absence of more accessible means of gathering comparable date, satellite imagery remains the best solution to evaluate global trends in these three types of zones.

Thanks to satellite imagery, the World Atlas on Desertification (WAD), a new reference tool particularly useful for locating desertification and land degradation, is currently being developed by the European Commission's Joint Research Centre, in partnership with the United Nations Environment Programme and a large network of international experts.

Luc Gnacadja is the executive secretary of the Convention des Nations Unies sur la lutte contre la désertification (CNULCD). He coordinates international initiatives on desertification according to the guidelines set by the major convention signed at the Rio Earth Summit in 1992.

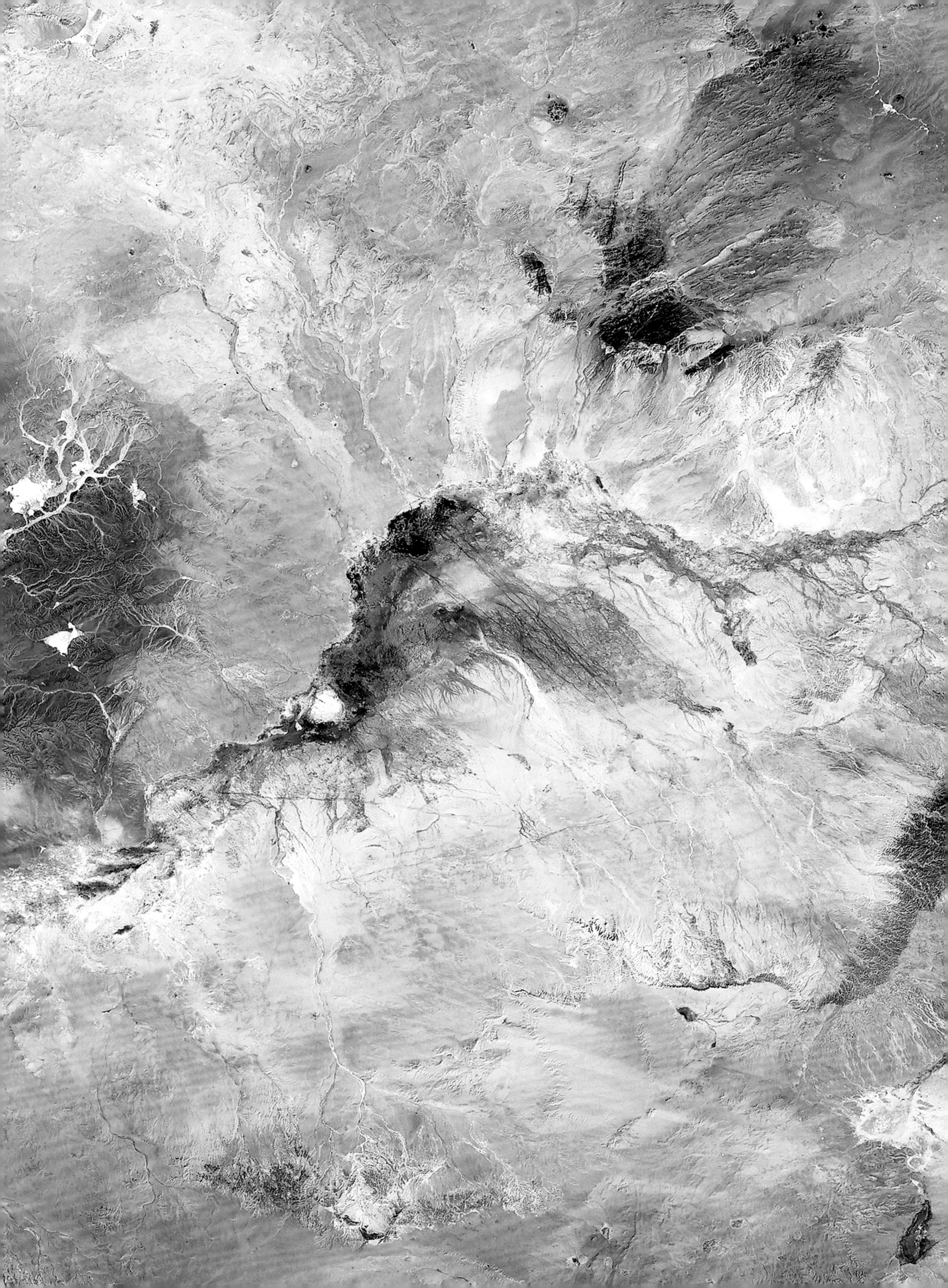

DESERTWATCH

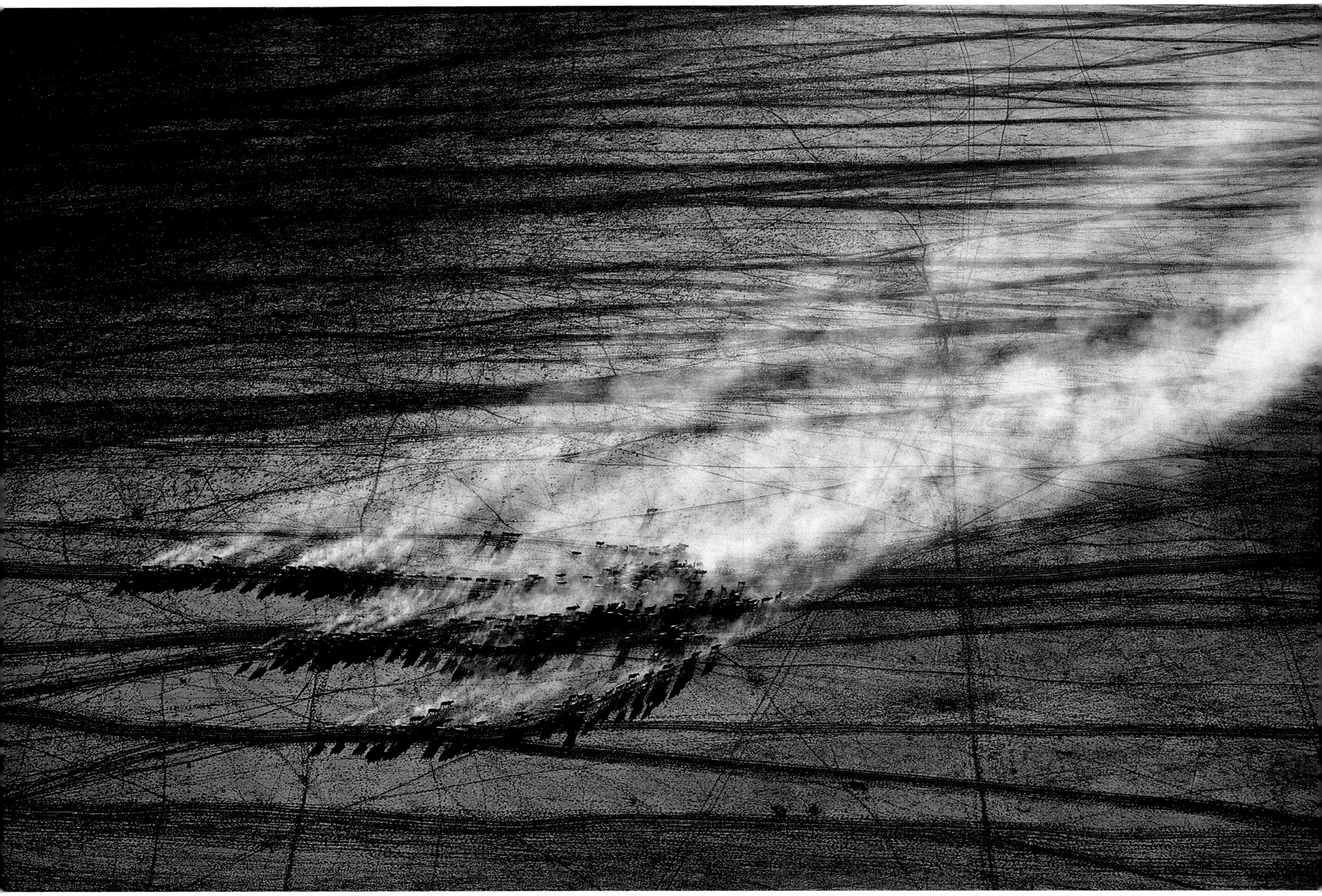

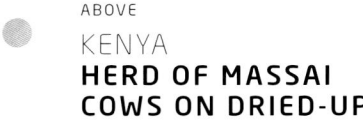

ABOVE

KENYA
**HERD OF MASSAI
COWS ON DRIED-UP
LAKE AMBOSELI**
2°37' S – 37°8' E

One of the causes of desertification is overgrazing. Additionally, periods of drought put major pressure on cattle and the populations that depend on them: The severe drought in the Horn of Africa in recent years threatens 12 million people. Yet even arid and semi-arid land can be put to use by farmers.

Recurring droughts and fires in Europe have shown the continent's vulnerability to desertification: 115,830 square miles (300,000 sq km) are believed to be currently affected by desertification, which could impact the living conditions of sixteen million Europeans. Portugal is one of the most vulnerable countries.

How can areas affected by desertification be identified from outer space? Launched by the European Space Agency in 2004, the DesertWatch project aims to provide a practical, manageable system and to create maps that can be easily interpreted on a national, regional, and local scale—respectively covering approximately 38,600 square miles (100,000 sq km), 3,860 square miles (10,000 sq km), and 100 square miles (250 sq km).

The project has been implemented in three "control" countries—including Portugal—and will subsequently be adapted to other areas such as Brazil and Mozambique and, eventually, the rest of the world.

The DesertWatch system is based on two indicators: the Desertification Sensitivity Index (DSI) and the Land Degradation Index (LDI), which take into account multiple factors including precipitation, plant cover, soil characteristics, and the like. Maps are established for each index. DesertWatch's concept is like a geographic information system (GIS) provided to researchers free of charge.

What are the results? According to DesertWatch experts, approximately 33 percent of Portugal's land is degraded, principally in the country's southern Alentejo and Algarve regions. Observation reveals that the phenomenon has been going on for a very long time and that the land surface still in the process of being degraded is limited. This recent degradation actually accounts for only 1.5 percent of the total, which implies that by concentrating the necessary efforts on these areas, the aggravation of the problem could be more directly prevented. Experts have even identified specific causes—particularly agricultural practices in the Alentejo—and one culprit, Europe's Common Agricultural Policy (CAP): "European subsidies are generally the most important factor in farmers' decision-making: Wheat continues to be cultivated though its commercial value no longer justifies it; cattle is maintained at artificially large numbers by subsidies." CAP's reform has clearly proven to be a crucially important project, though no one would deny such an initiative is highly complex.

—

PAGES 230-231
OMAN, RUB' AL-KHALI
UMM AL SAMIM SALT MARSH
22°00' N – 58°50' E

Along the eastern strip of the Rub' al-Khali desert, the Umm al Samim salt marsh receives what little rainwater trickles down from the surrounding hills. The Umm al Samim then becomes a very dangerous quicksand area.

PAGES 232-233
CHAD
SAHARA, LAKES OF OUNIANGA
19°05' N – 20°30' E

The Lakes of Ounianga are two large groups of approximately fifty stretches of water—the only remains of a vast lake that occupied the basin several millennia ago.

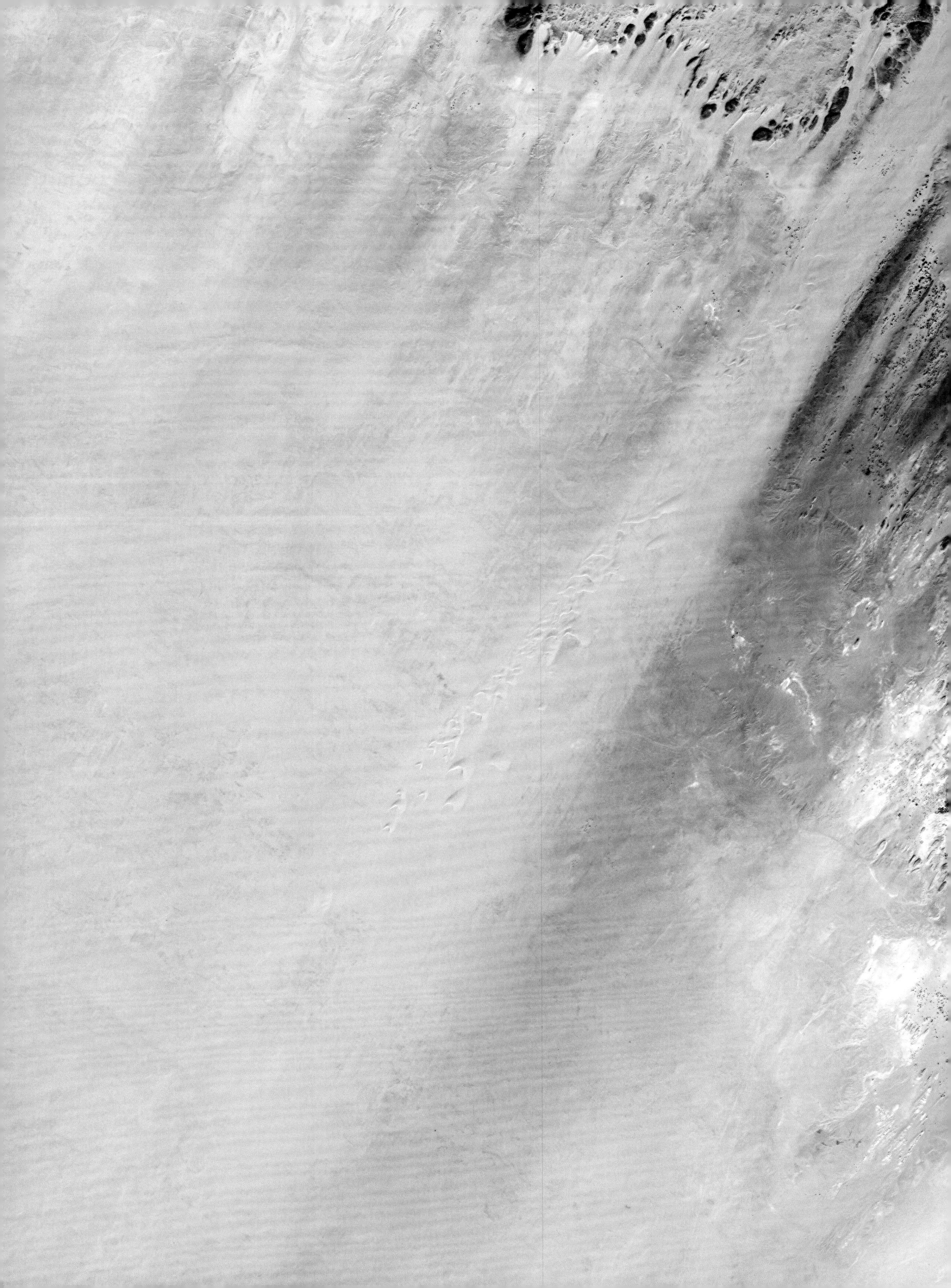

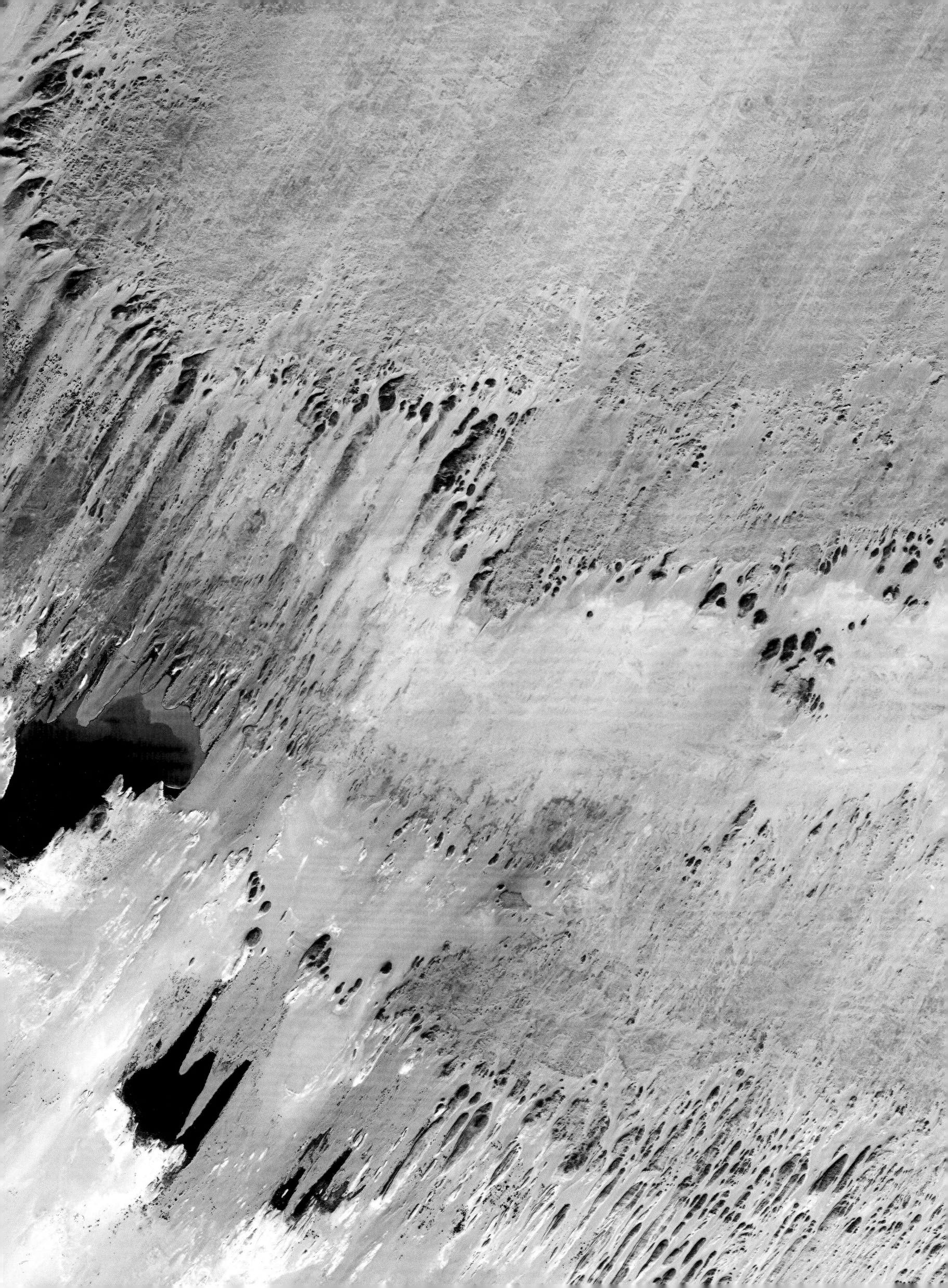

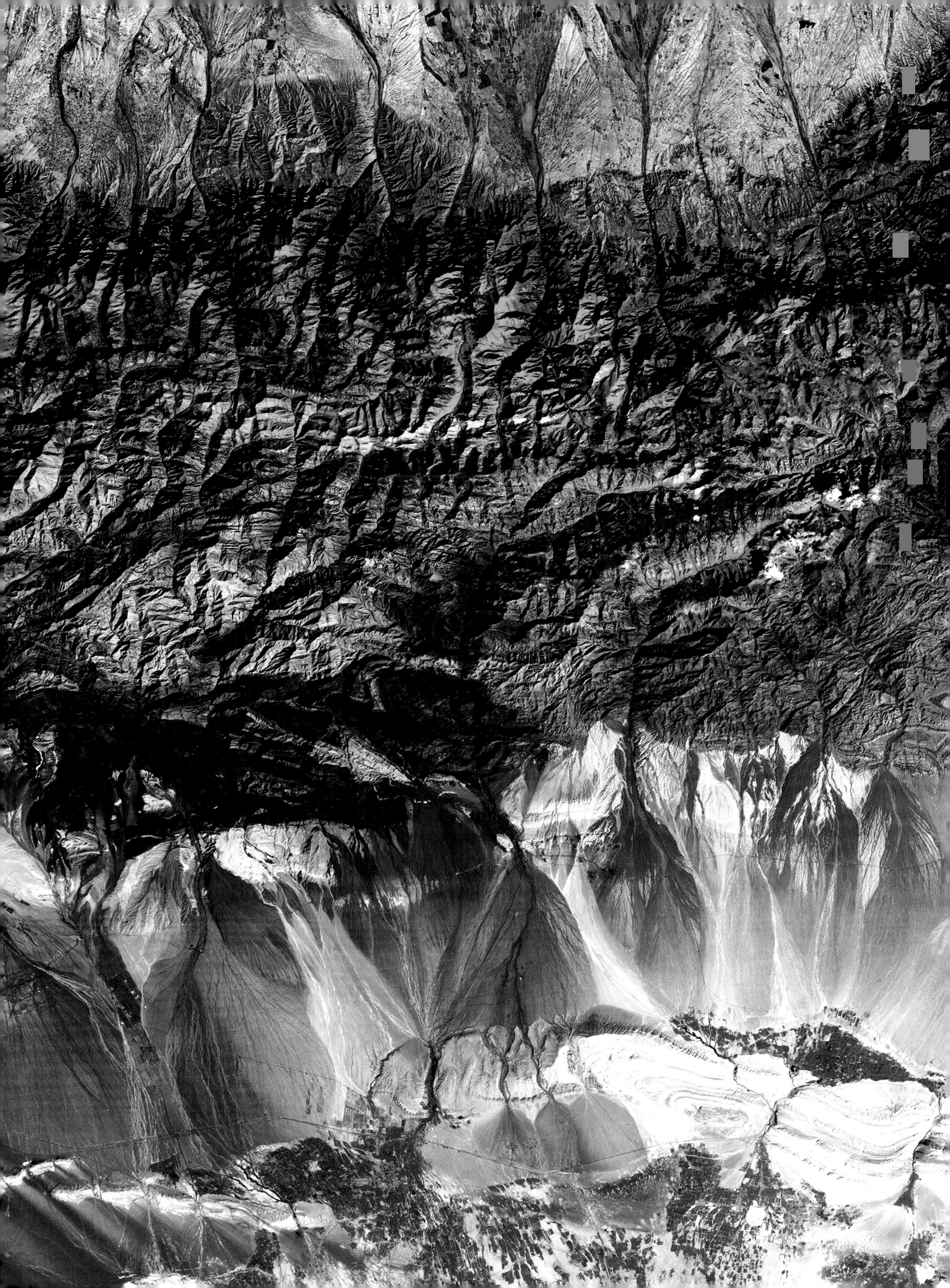

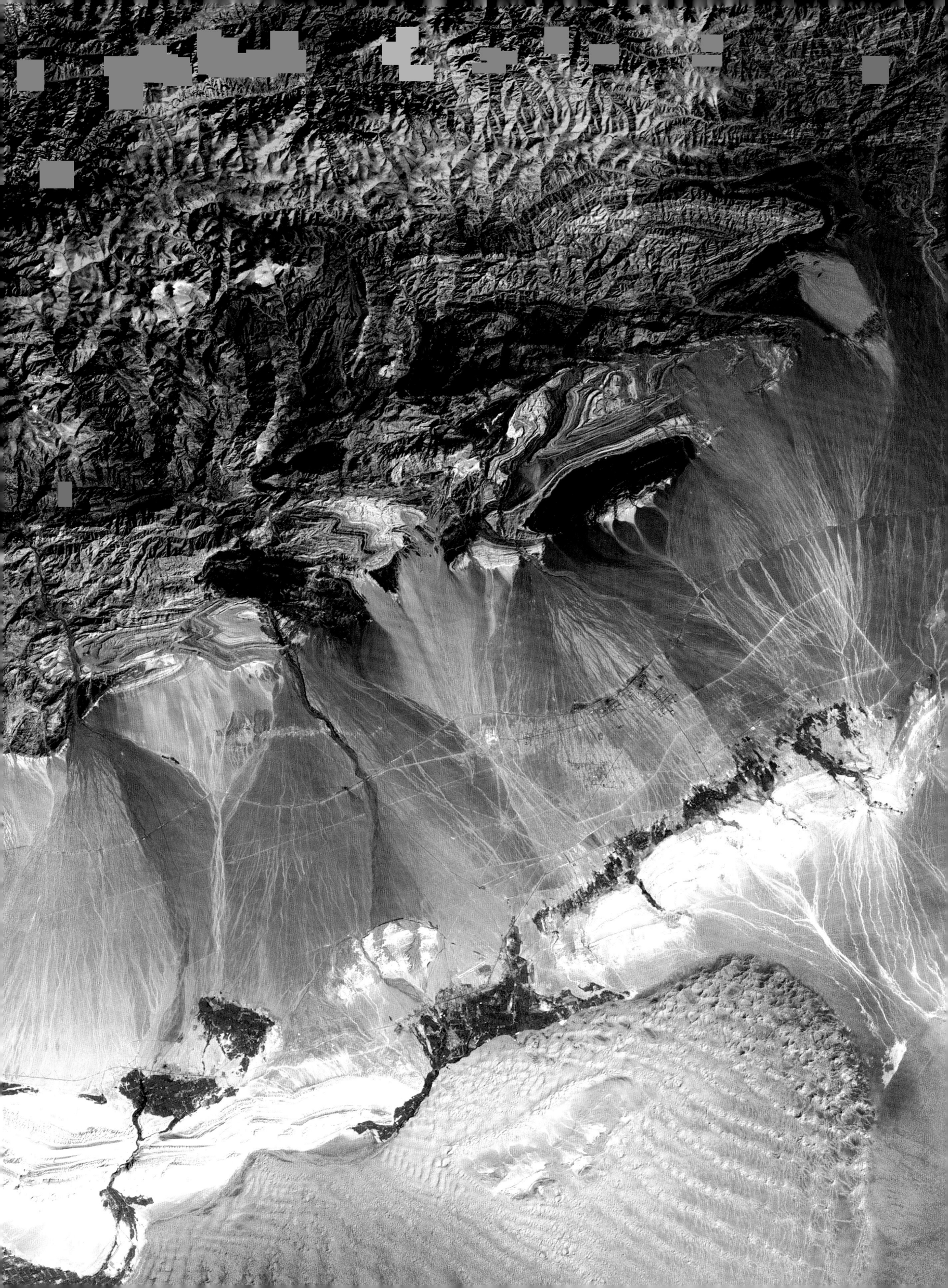

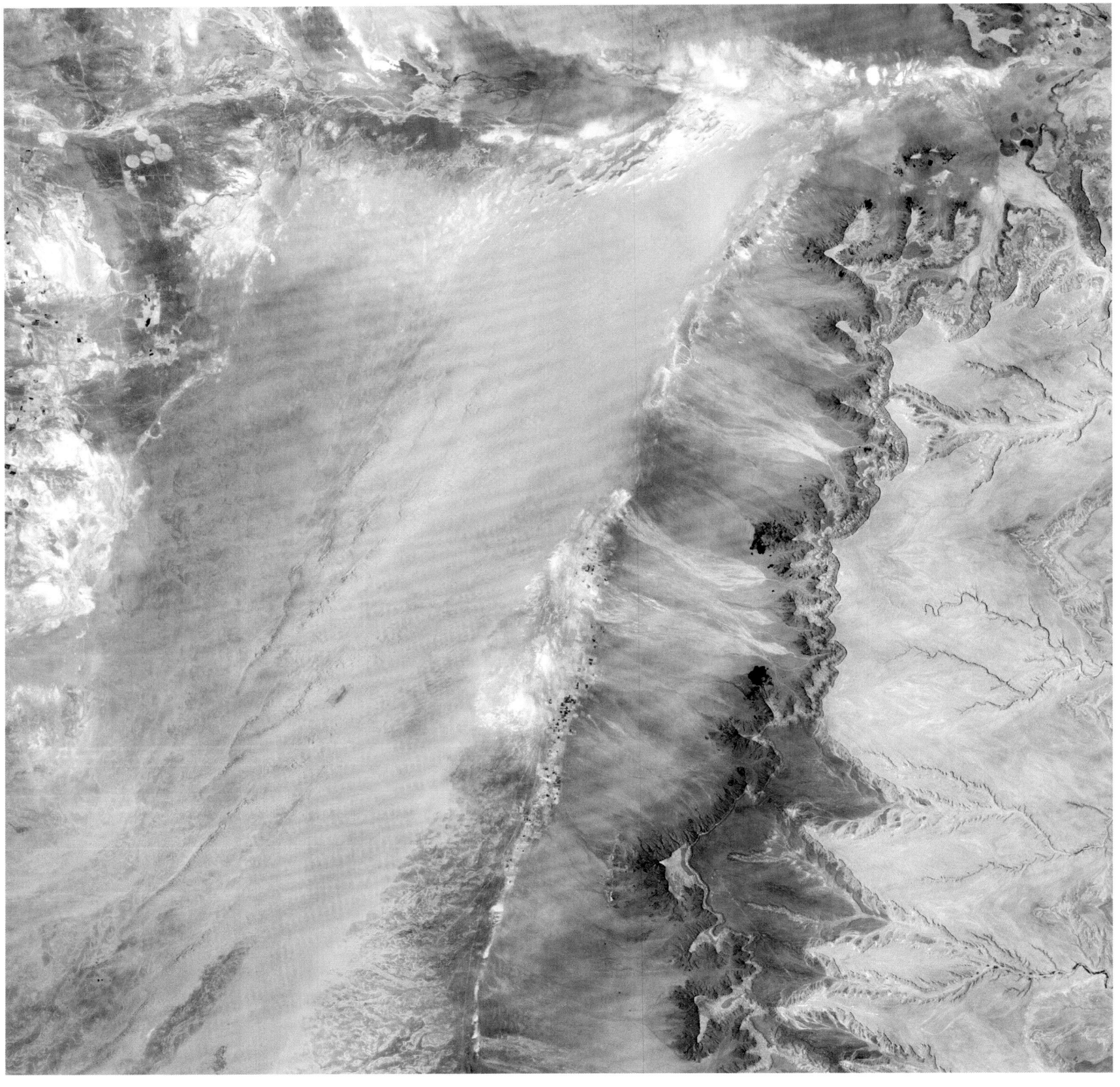

PAGES 234-235

CHINA, XINJIANG
TURFAN DEPRESSION
42°46' N – 89°20' E

The Turfan Depression in northwestern China is one
of the world's lowest points. Located at about 505 feet
(154 m) below sea level, this windy pass at the foot of
Mount Bogda saw the caravans of the Silk Road pass
between its dunes and salt lakes.

ABOVE

SAUDI ARABIA
DJEBEL TOUEIK
23° N – 46° E

Located some 125 miles (200 km) south of Riyadh, Djebel Toueik,
seen forming a barrier in the right of the image, is a cuesta, or the
ridge of a plateau. Eroded and scored by deep valleys, it is part of a
sedimentary basin that rises gradually to the west. A large erosional
glacis (in gray) prolongs the formation to the side, connecting it to a
second plateau. The difference between the slopes of the face and
the glacis can be seen here through the play of shadows.

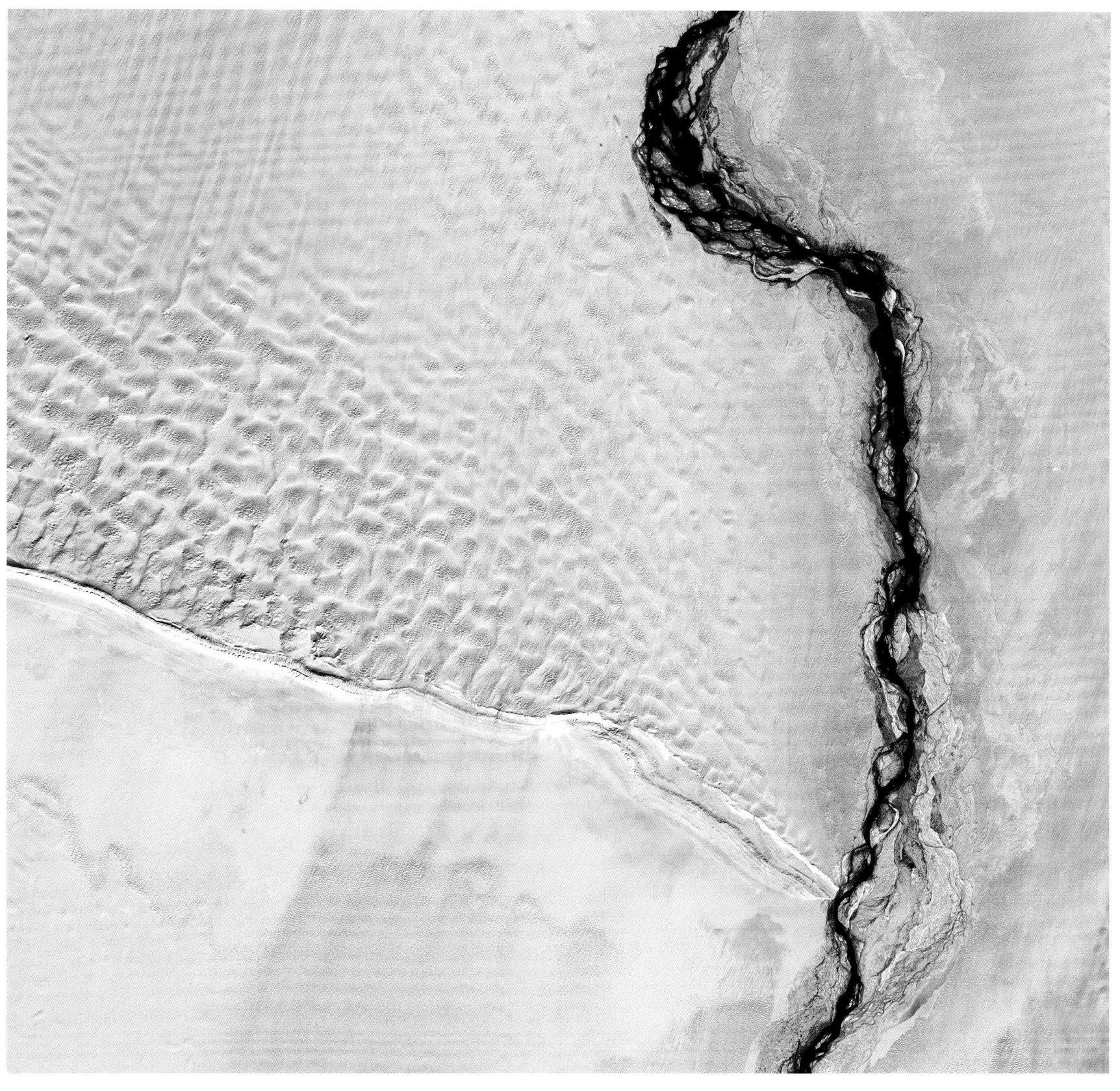

ABOVE

CHINA, XINJIANG
TAKLAMAKAN
38°27' N – 80°54' E

The Taklamakan is a sandy desert stretching across
western China to the foot of Tibet, in the autonomous
region of Xinjiang, populated by the Uighur minority.
Here, we see the Hotan River, the only river to cross
this desert from north to south.

PAGES 238-239

CHINA, XINJIANG
TAKLAMAKAN UNDER THE SNOW
39° N – 83° E

The Taklamakan Desert occupies the basin of the Tarim,
a river that runs across its northern section from west
to east. The Tarim River is unusual in that it does not
flow into the sea or another watercourse, but drains into
the soil of the endorheic Lop Nur basin to the east of the
desert. Another example of this type of watercourse is the
Okavango, which ends in the Kalahari Desert.

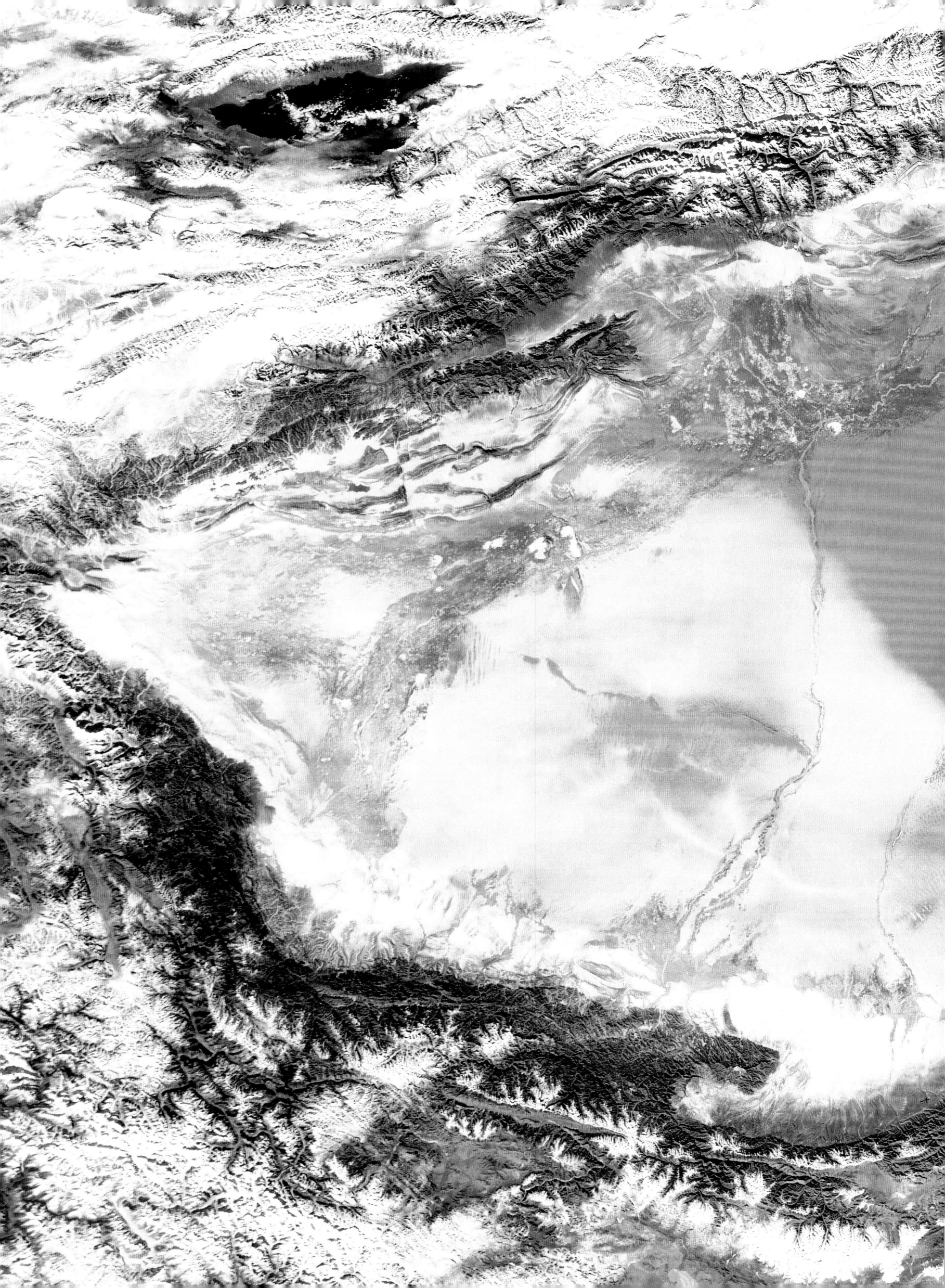

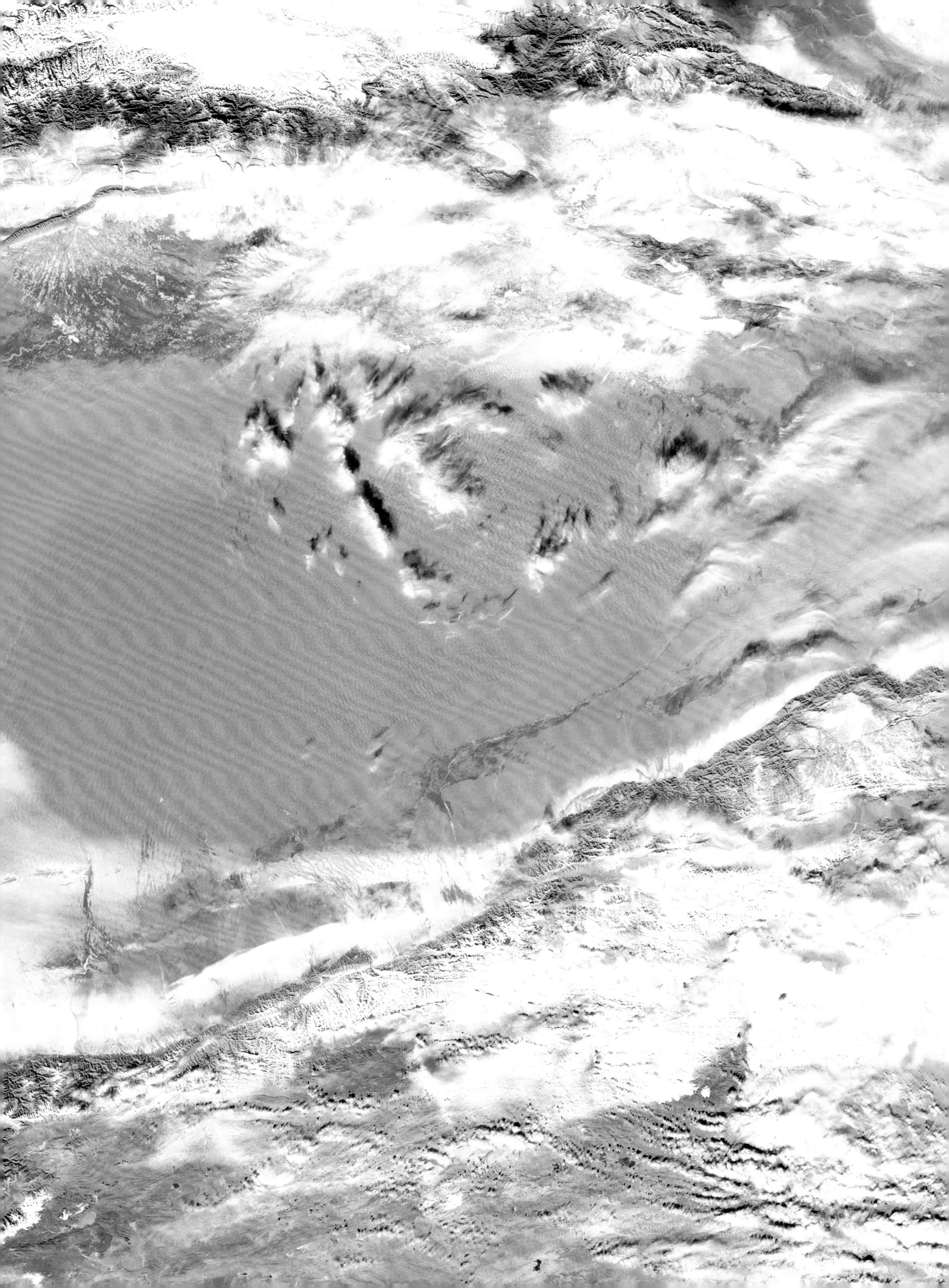

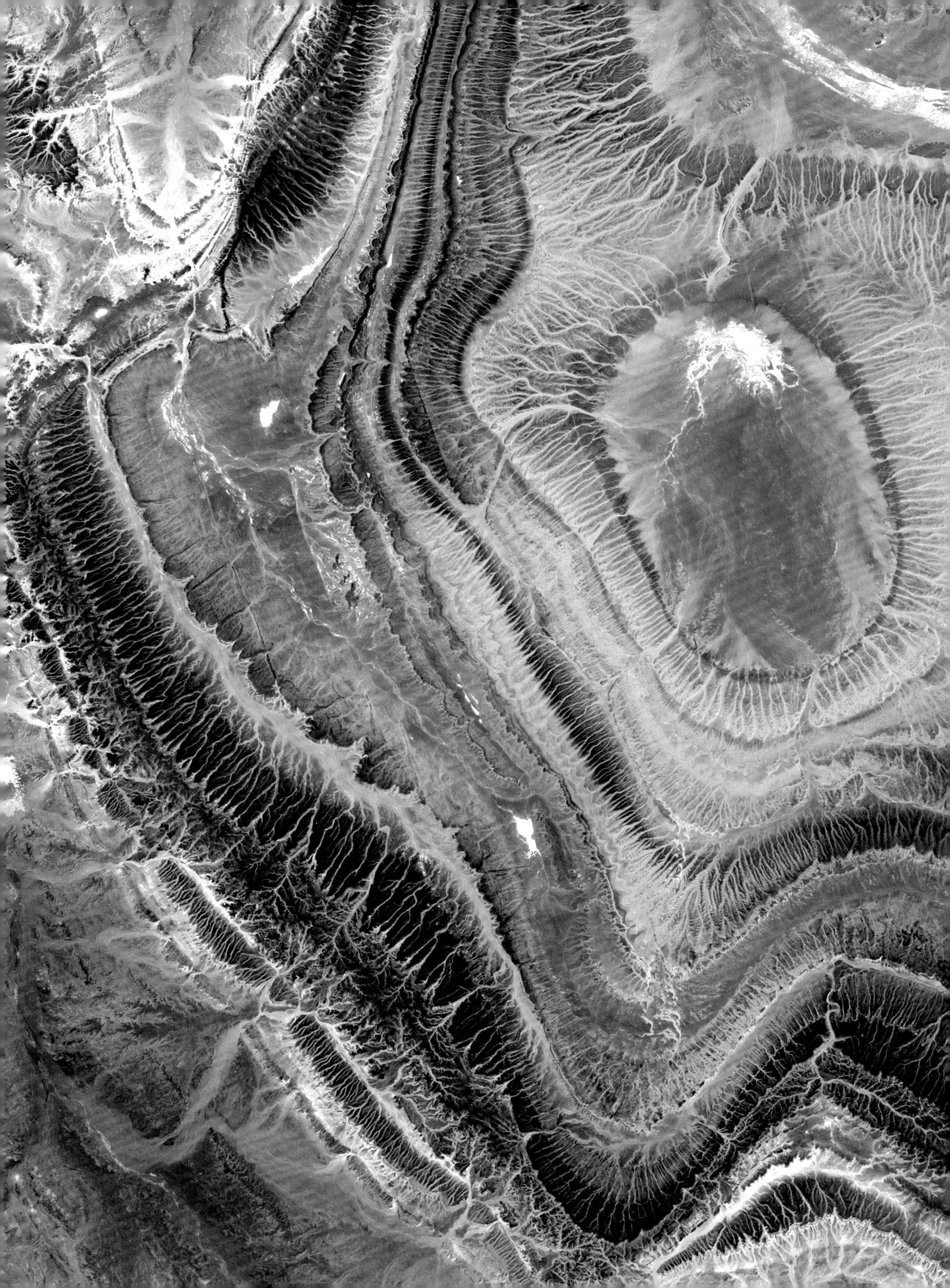

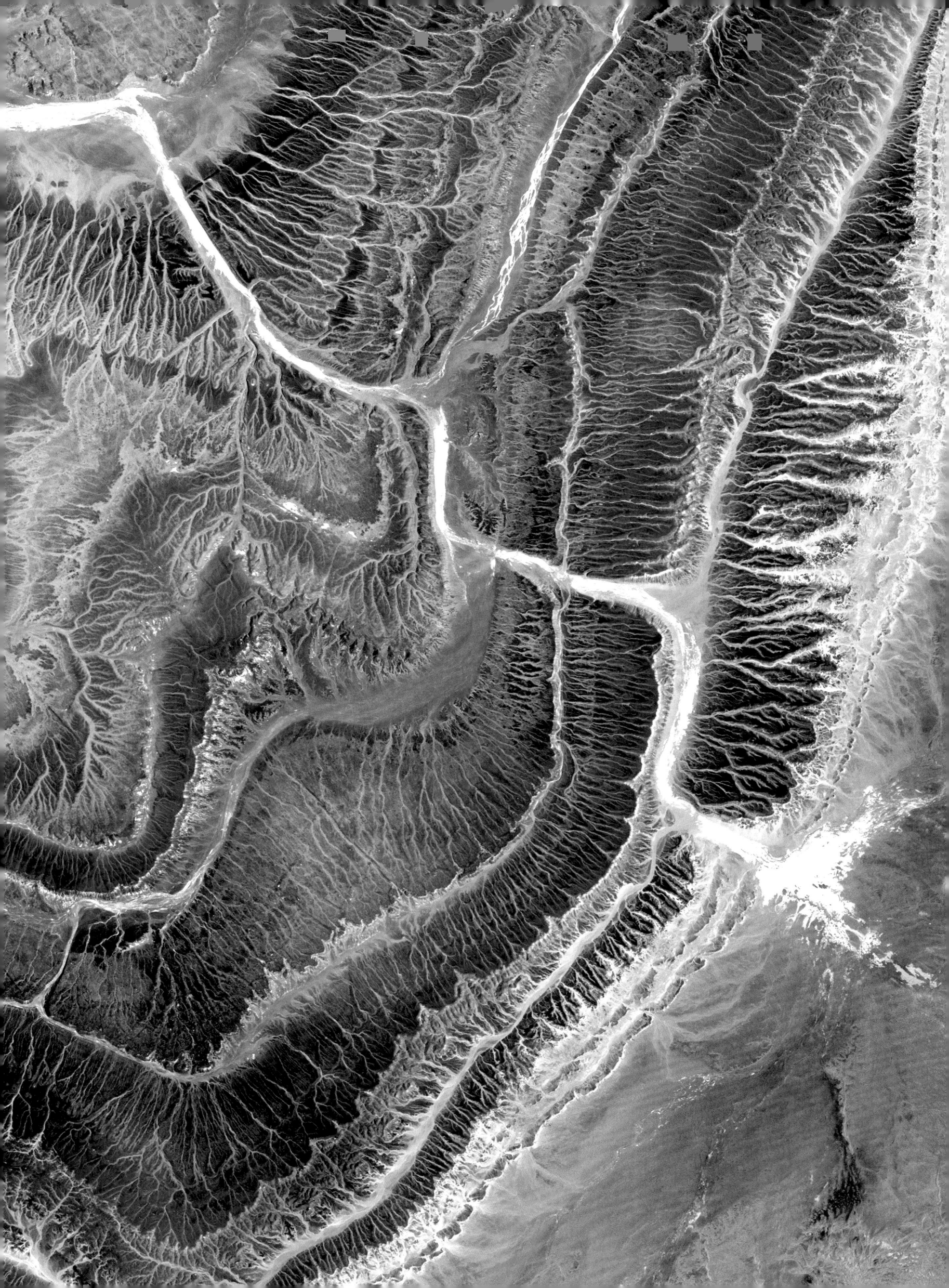

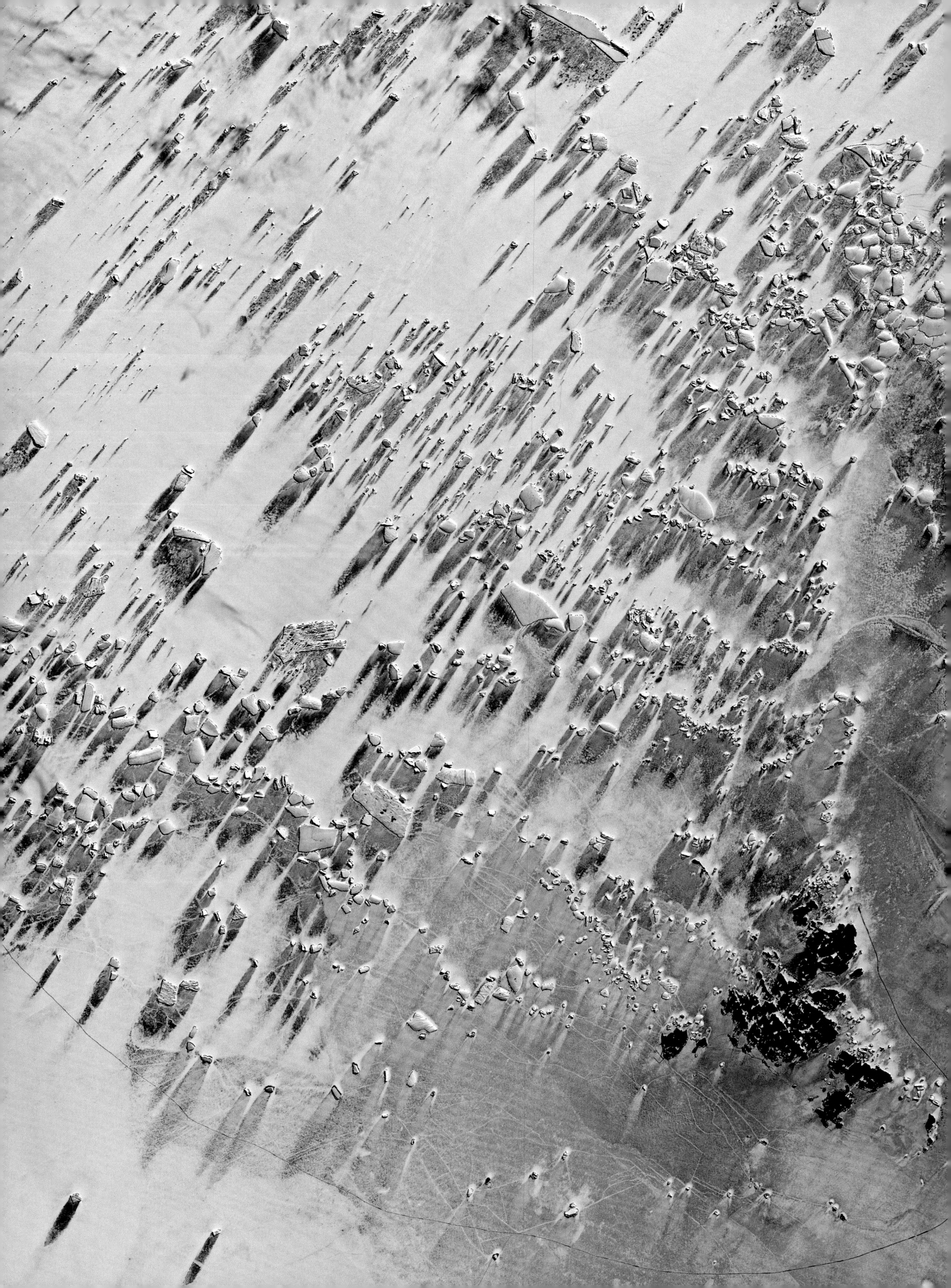

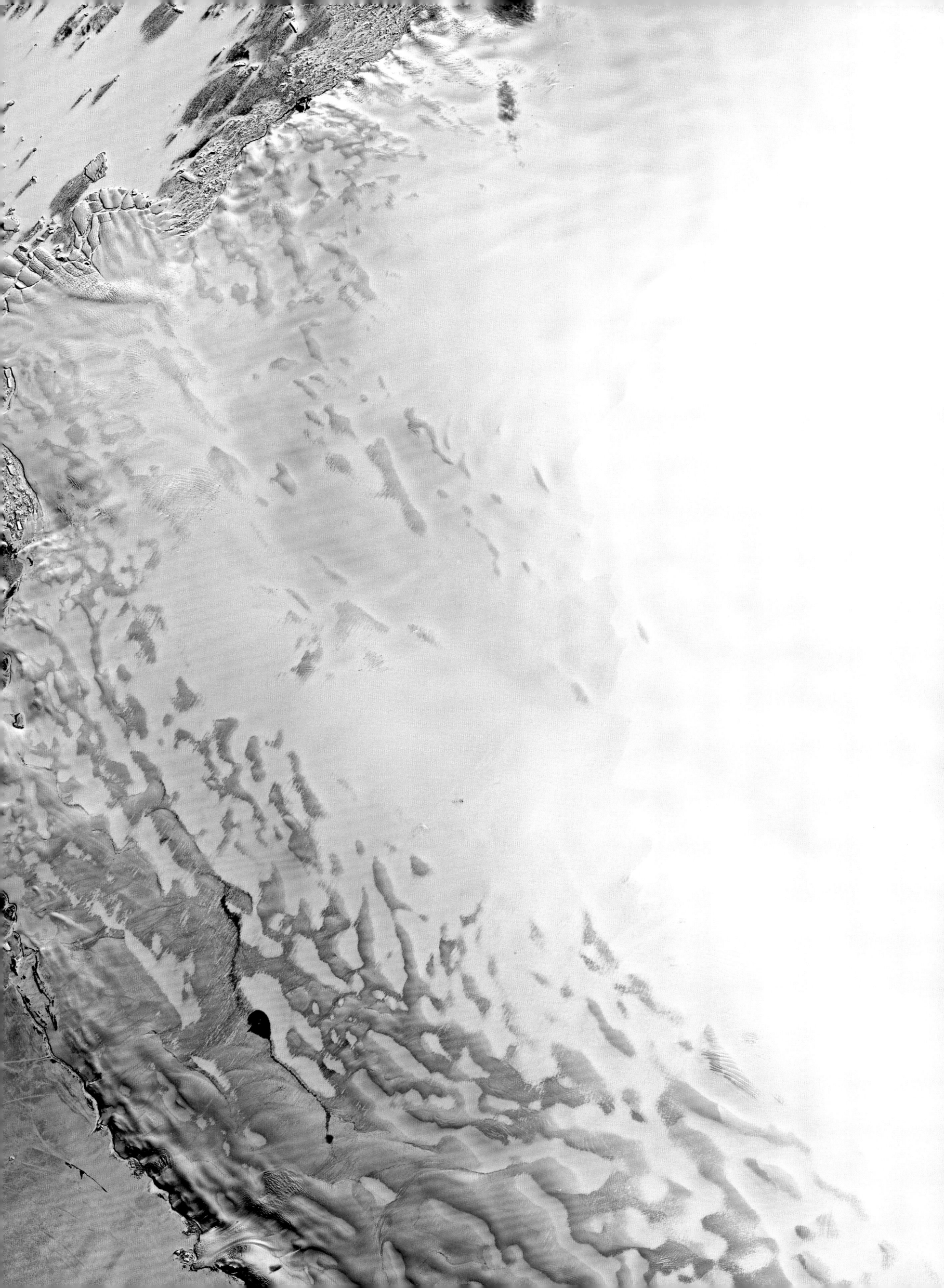

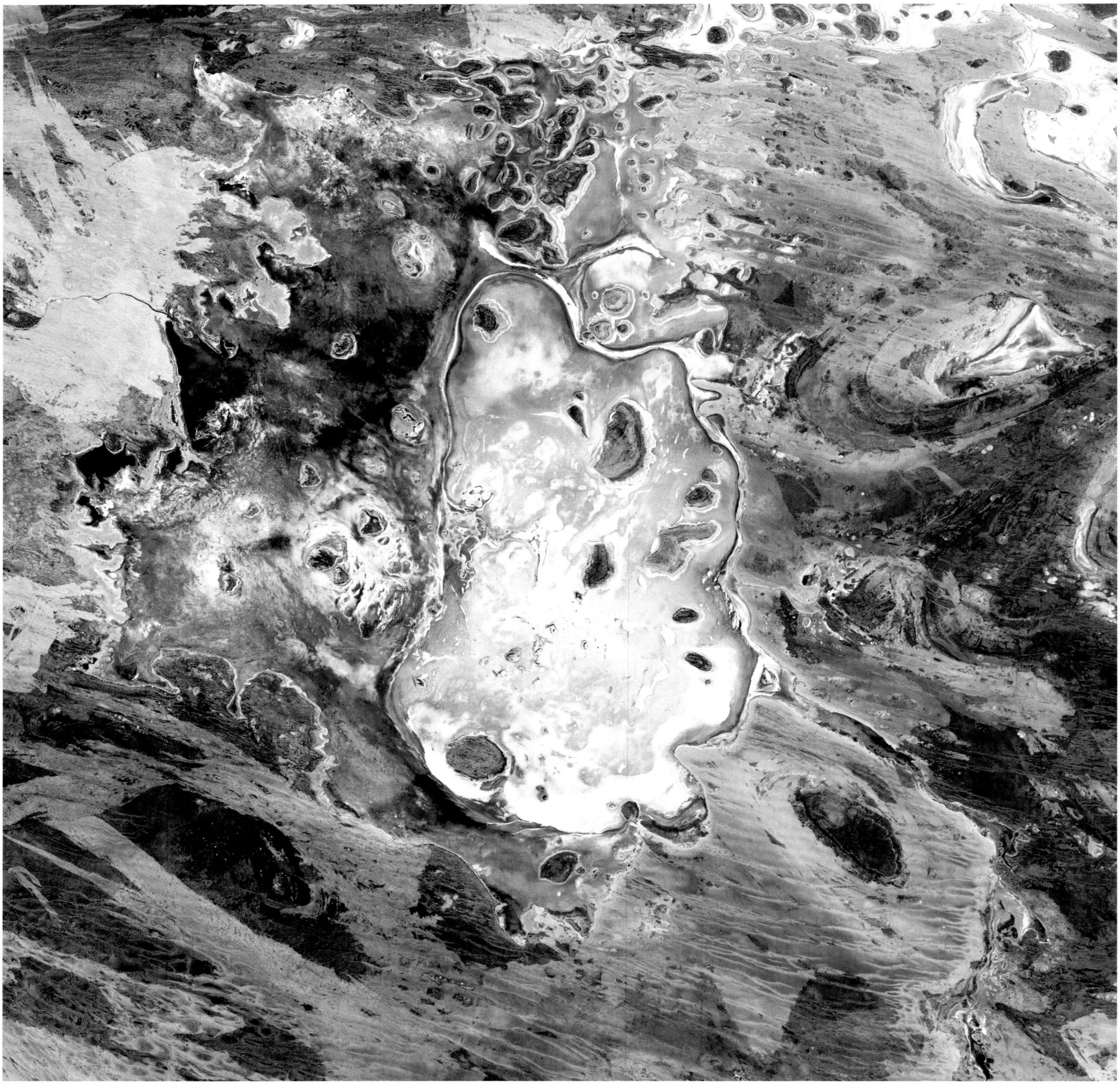

OPPOSITE
AUSTRALIA, NORTHERN TERRITORY
LAKE DISAPPOINTMENT
22°26' S – 122°48' E

This is one of the largest salt lakes in the north of Australia. Covered with a thick crust of salt for most of the year—it is believed to hold 600 million tons of salt—it is sometimes partially watery during the wet season. However, agricultural activity has not developed in the surrounding area, due to its remote location in the heart of the Australian desert, north of Uluru.

BELOW
CHINA, XINJIANG
TAKLAMAKAN
37°20' N – 84°20' E

Here we see a vast alluvial cone between the Altun and Kunlun mountains on the southern edge of the Taklamakan Desert. Formed by the draining of water from the two neighboring mountains, this cone is 37 miles (60 km) long and 34 miles (55 km) wide. Its western part, in light blue, is considered active, because it is continuously irrigated. Flora, seen in red in the top left of the image, can therefore develop; local farmers cultivate small plots of land there.

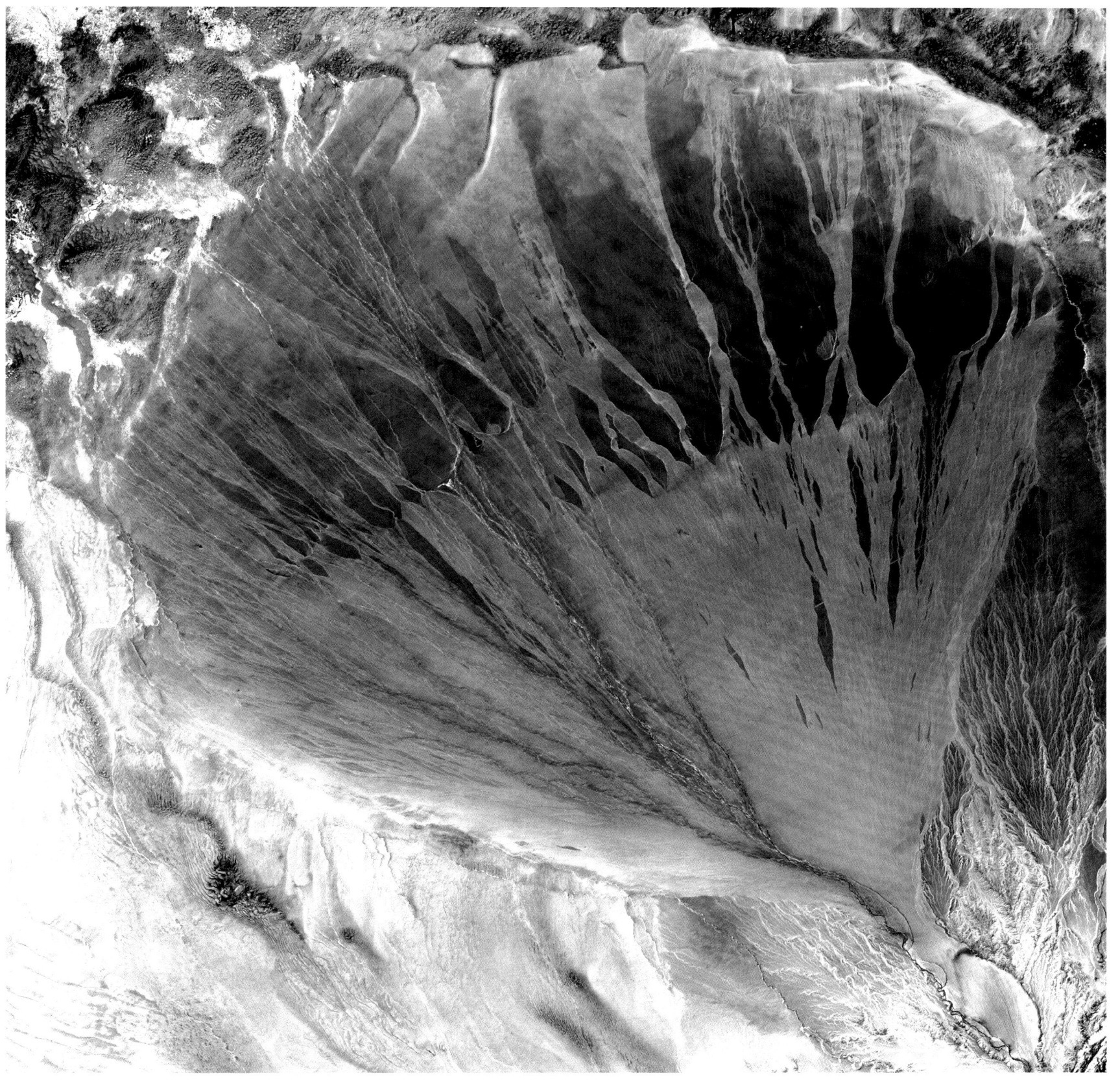

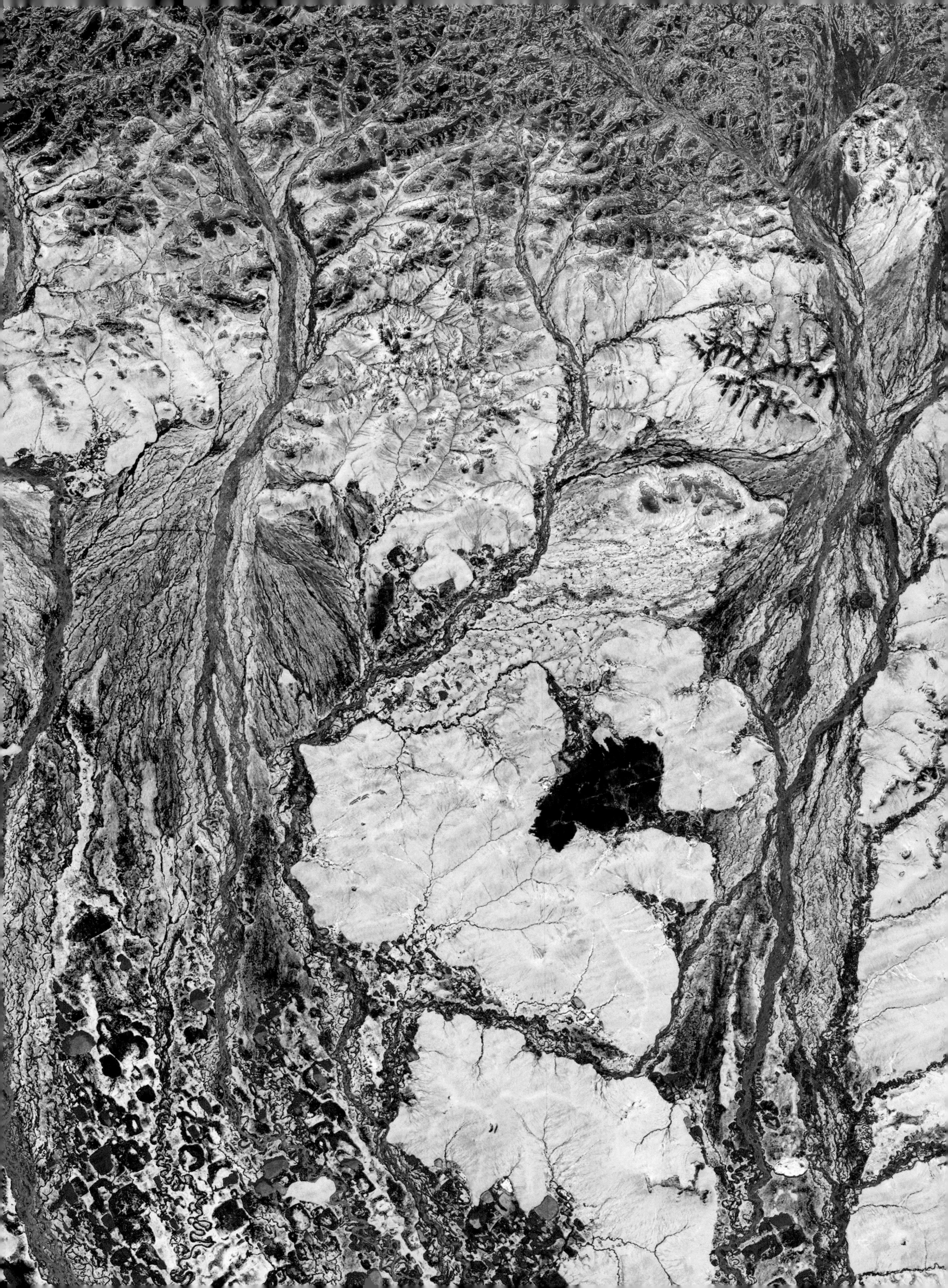

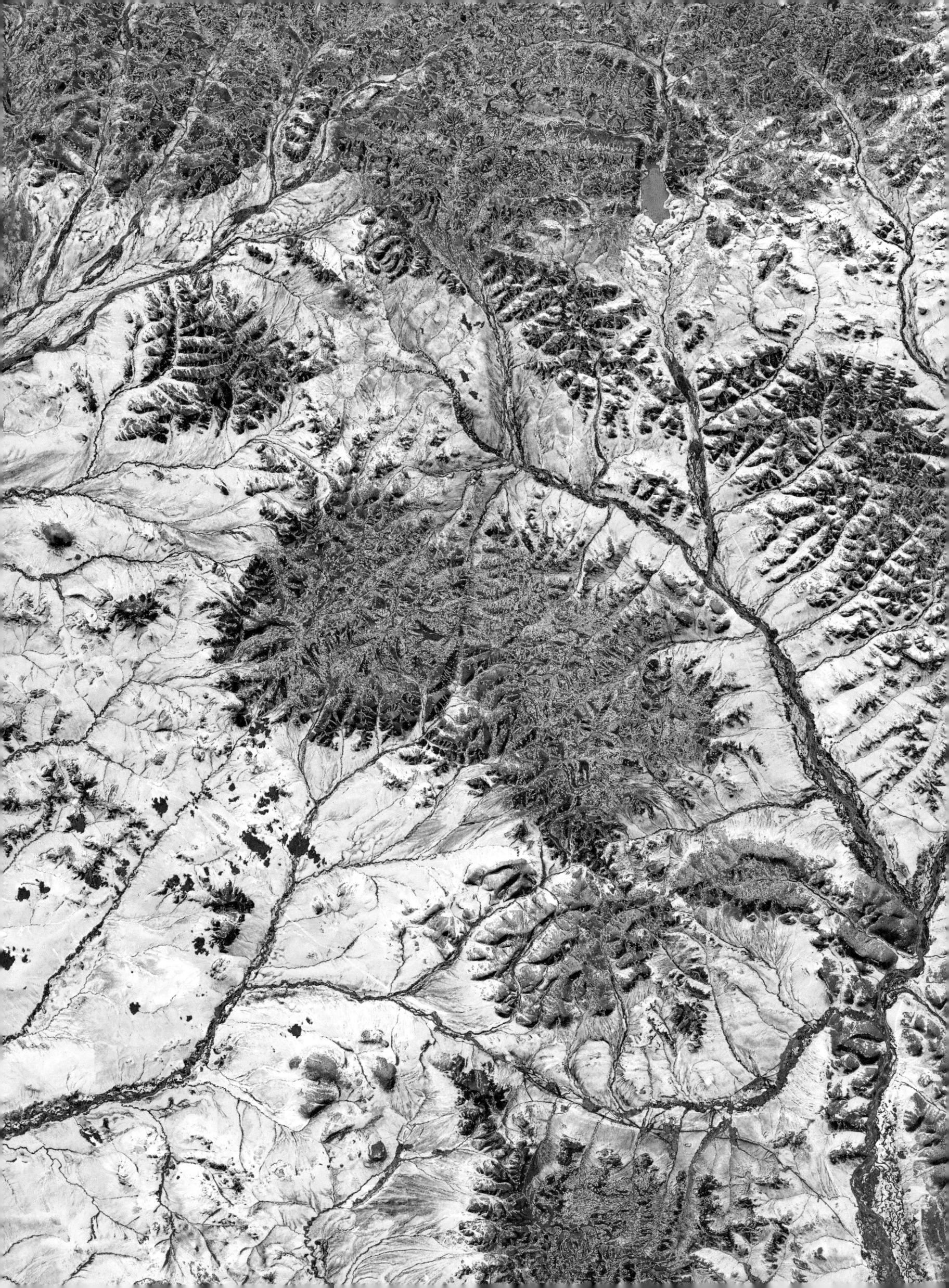

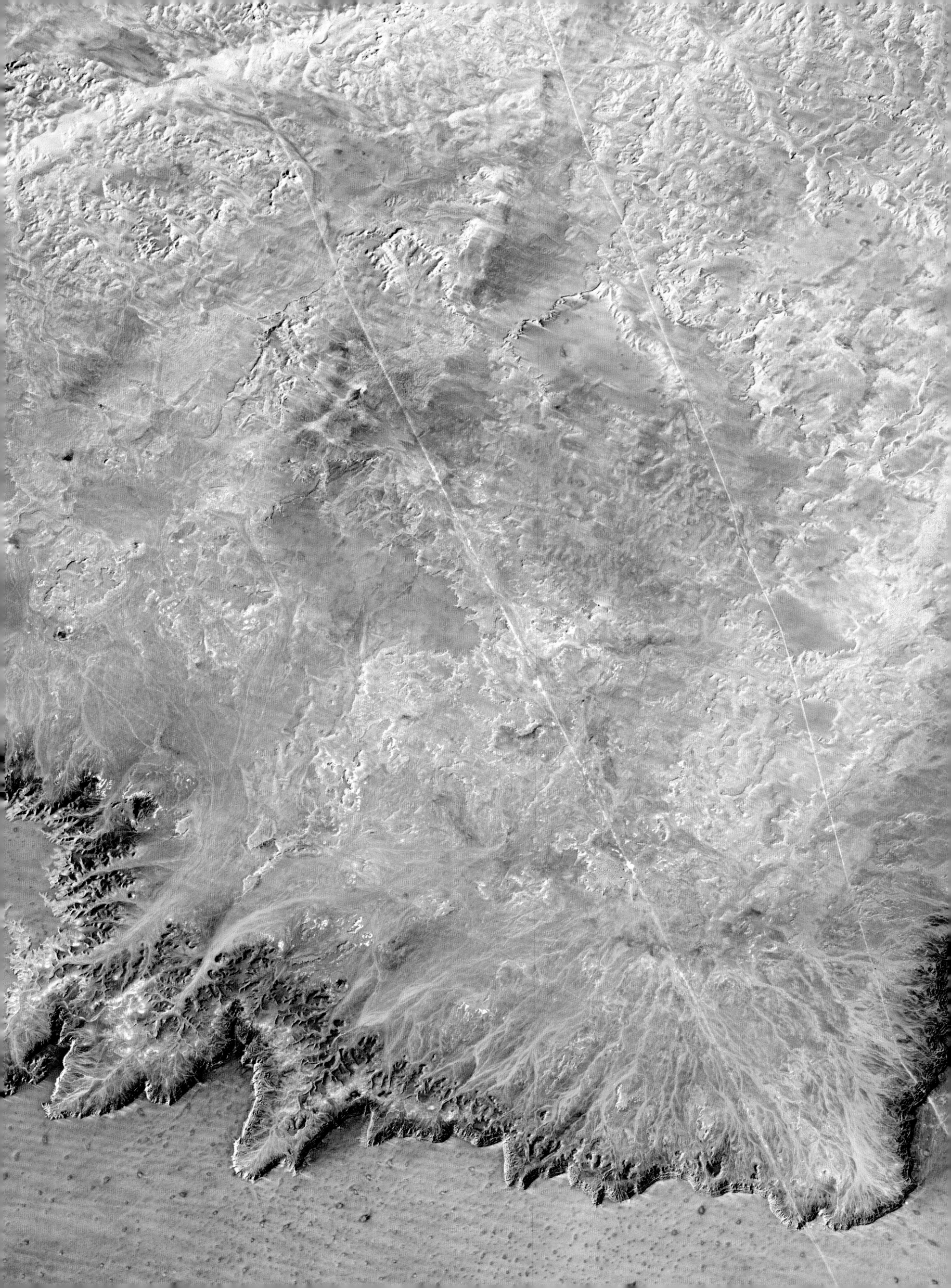

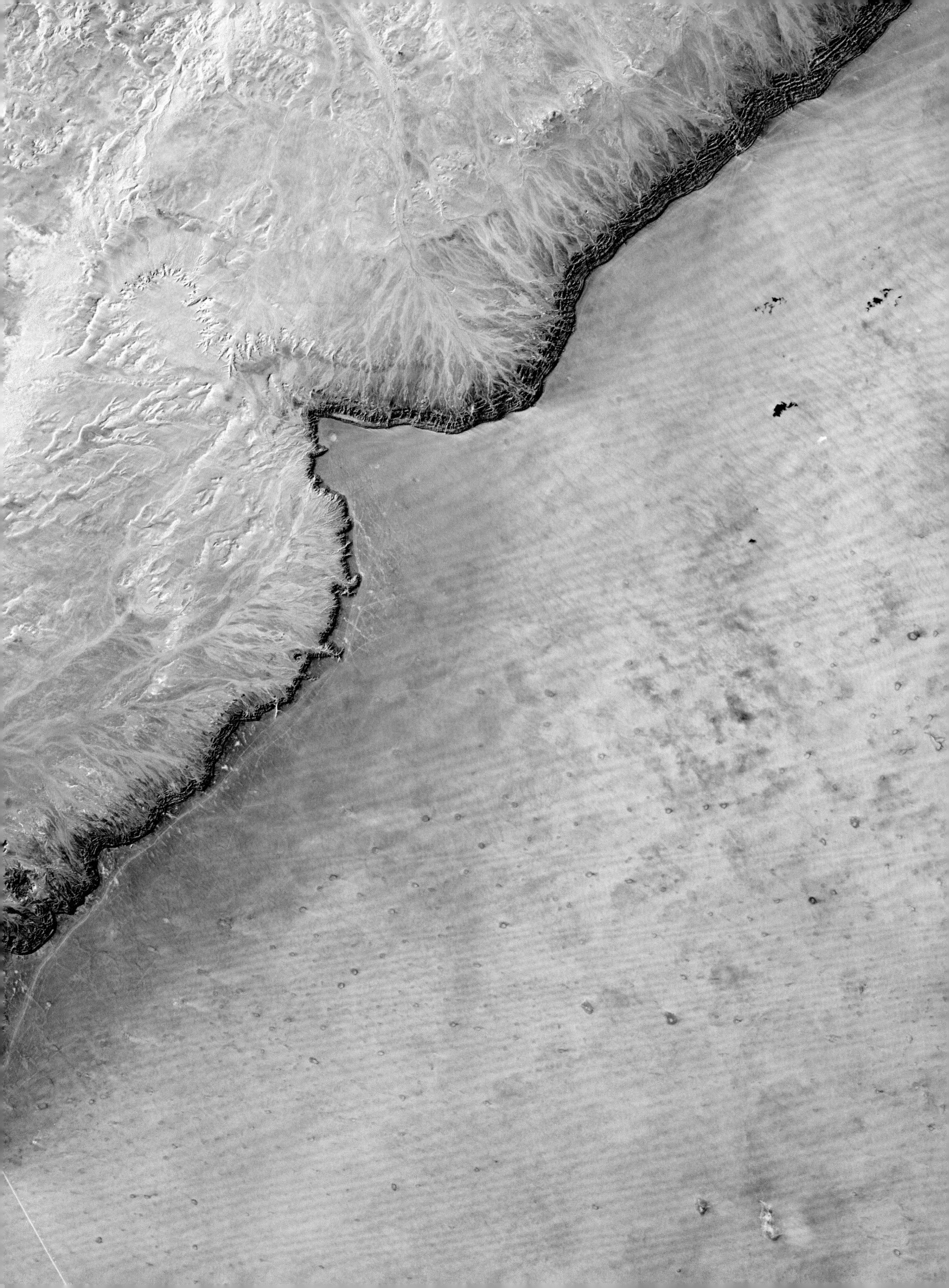

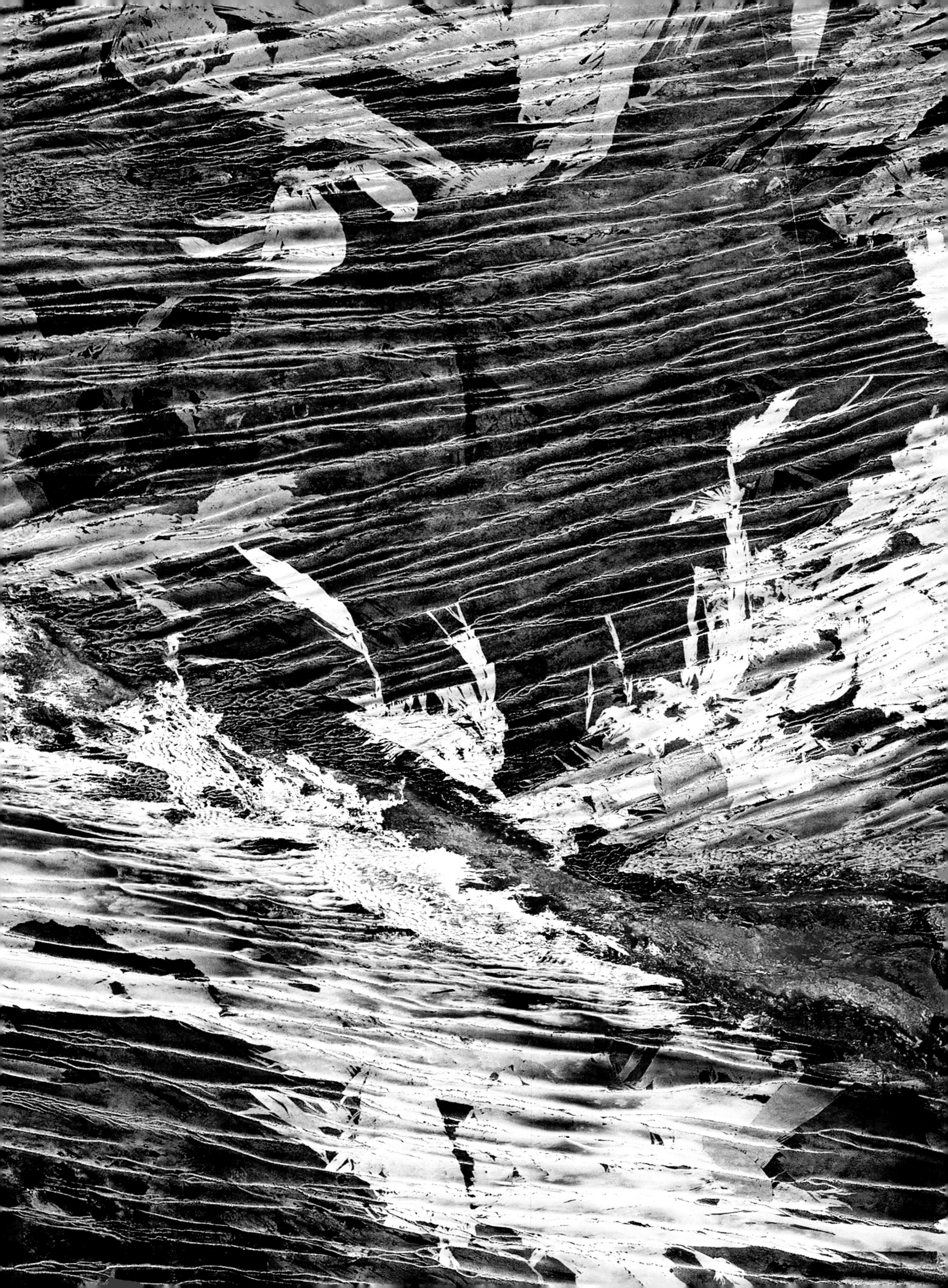

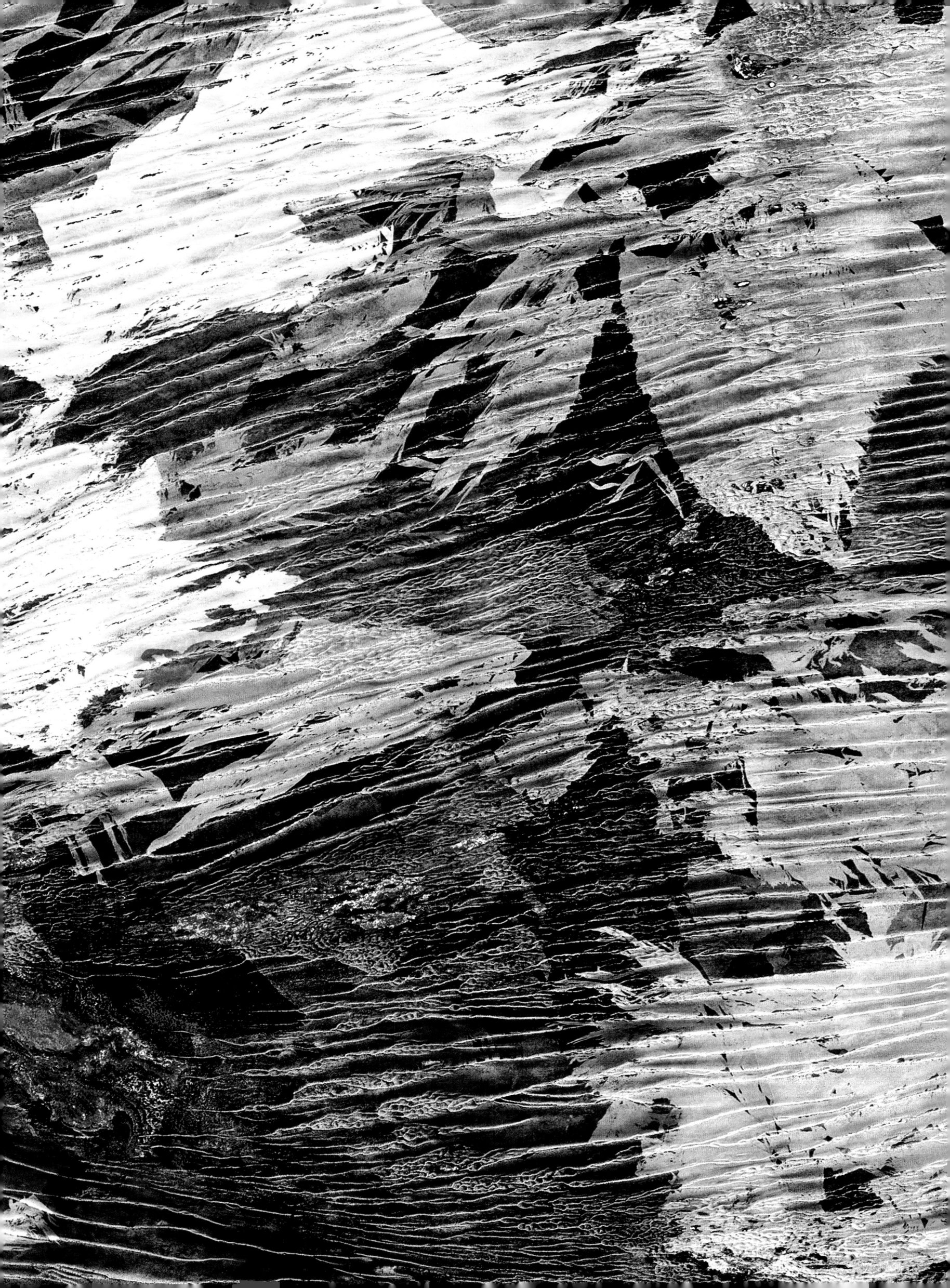

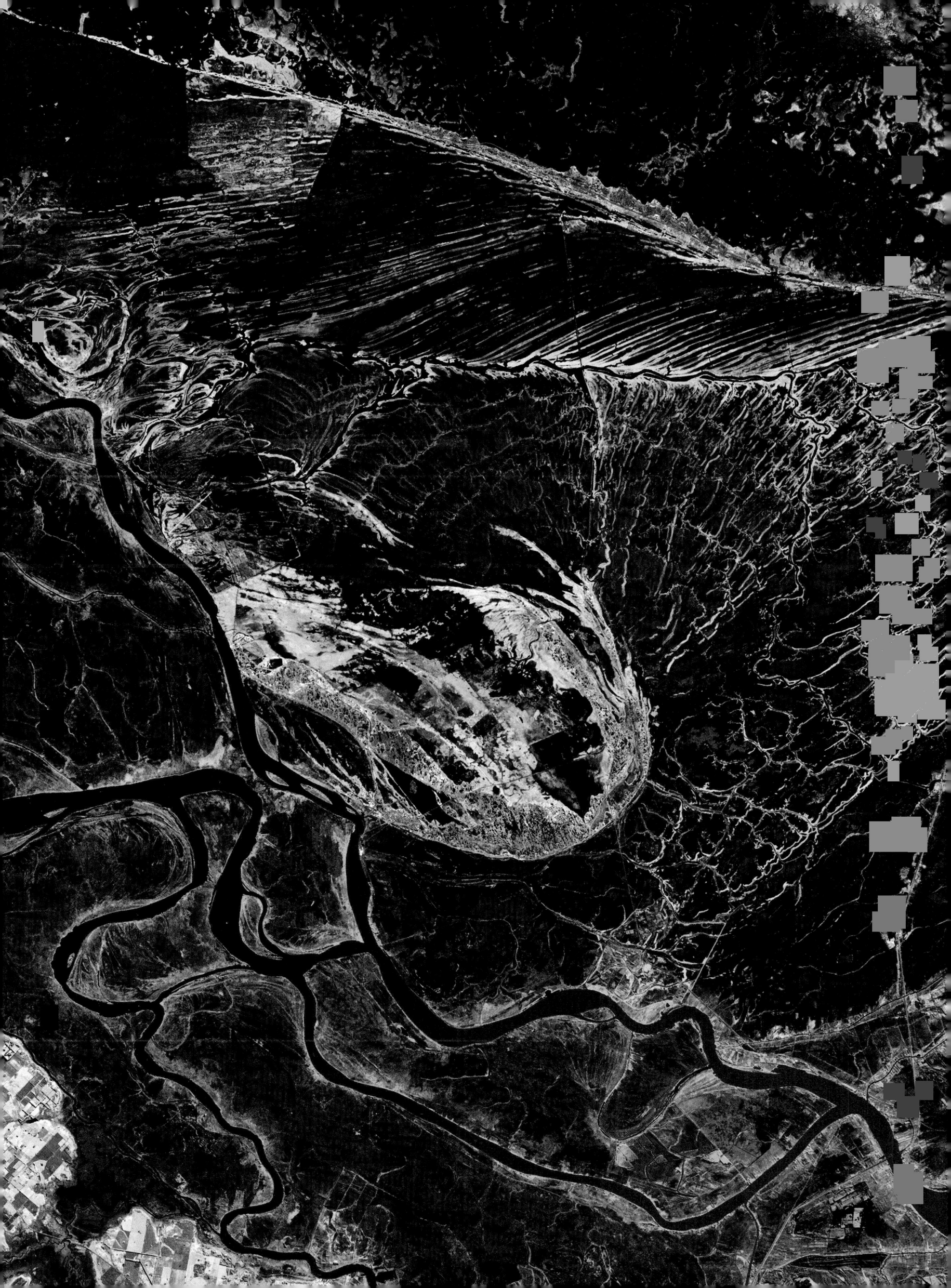

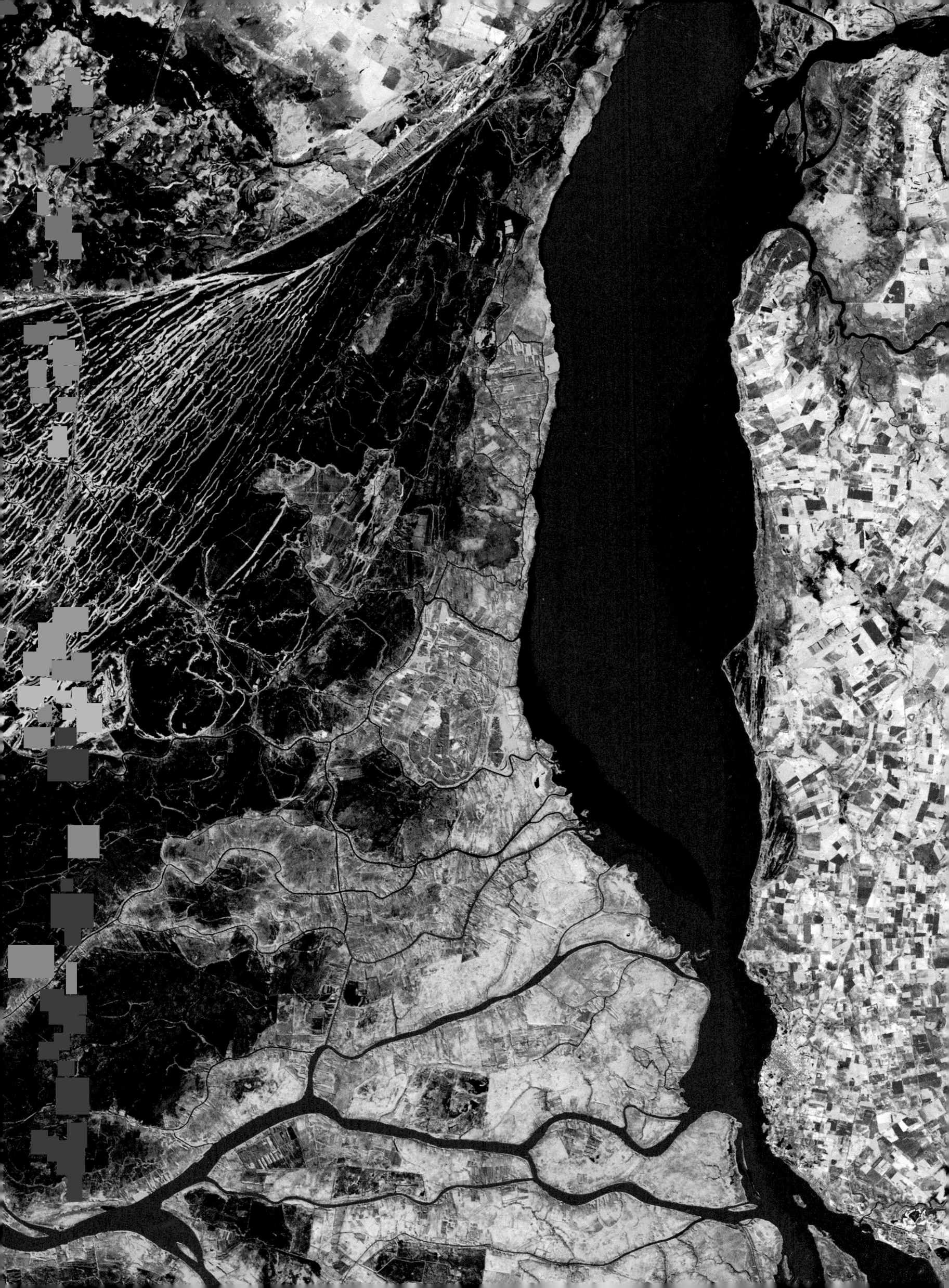

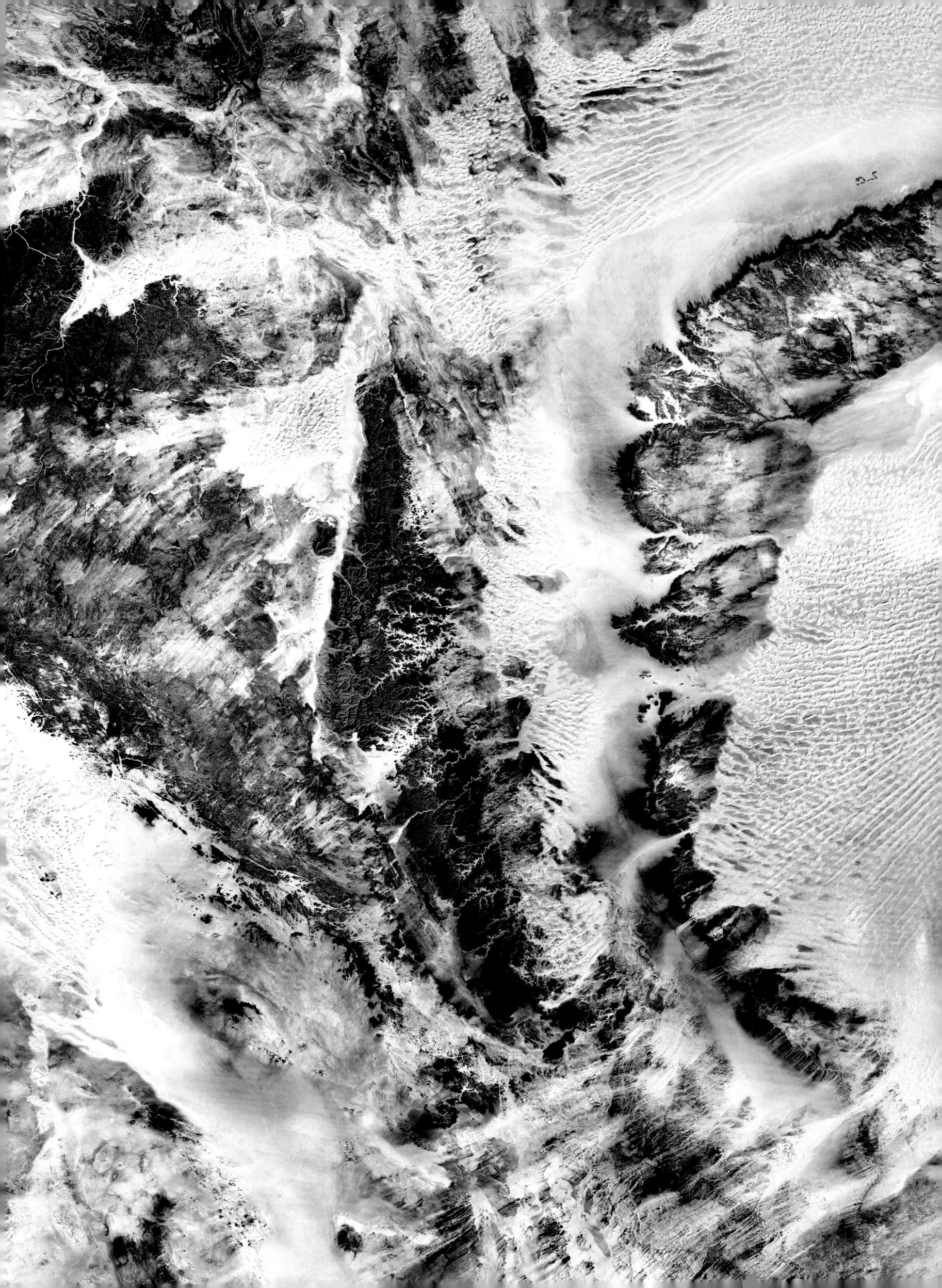

11/13 H

11/13 H